Cellular Nanoscale Sensory Wave Computing

Chagaan Baatar · Wolfgang Porod · Tamás Roska
Editors

Cellular Nanoscale Sensory Wave Computing

Springer

Editors
Chagaan Baatar
Office of Naval Research
Sensors, Electronics
& Networks Research Division
875 N. Randolph Street
Arlington VA 22203
USA
chagaan.baatar@navy.mil

Tamás Roska
MTA Budapest
Computer & Automation Research Institute
Kende ut. 13-17
Budapest 1111
Hungary
roska@sztaki.hu

Wolfgang Porod
University of Notre Dame
Center for Nano Science & Technology
Notre Dame IN 46556
USA
porod@nd.edu

ISBN 978-1-4419-1010-3 e-ISBN 978-1-4419-1011-0
DOI 10.1007/978-1-4419-1011-0
Springer New York Dordrecht Heidelberg London

Library of Congress Control Number: 2009930639

Printed on acid-free paper

Springer is part of Springer Science+Business Media (www.springer.com)

Preface

This book is loosely based on a Multidisciplinary University Research Initiative (MURI) project and a few supplemental projects sponsored by the Office of Naval Research (ONR) during the time frame of 2004–2009. The initial technical scope and vision of the MURI project was formulated by Drs. Larry Cooper and Joel Davis, both program officers at ONR at the time. The unifying theme of this MURI project and its companion efforts is the concept of cellular nonlinear/neural network (CNN) technology and its various extensions and chip implementations, including nanoscale sensors and the broadening field of cellular wave computing.

In recent years, CNN-based vision system drew much attention from vision scientists to device technologists and computer architects. Due to its early implementation in a two-dimensional (2D) topography, it found success in early vision technology applications, such as focal-plane arrays, locally adaptable sensor/processor integration, resulting in extremely high frame rates of 10,000 frames per second. More recently it drew increasing attention from computer architects, due to its intrinsic local interconnect architecture and parallel processing paradigm. As a result, a few spin-off companies have already been successful in bringing cellular wave computing and CNN technology to the market. This book aims to capture some of the recent advances in the field of CNN research and a few select areas of applications.

The book starts with a historical introduction by Larry Cooper and Joel Davis in Chap. 1, who recognized the potential of CNN technology early on and, over the years, encouraged research in various aspects of CNN technology. Chapter 2 by Tamás Roska is an up-to-date review, by one of the pioneers of CNN technology, on the evolution and future outlook of CNN-based computing architecture, including the emerging virtual cellular machine concept. The next chapter, by the principal investigator of the MURI project Wolfgang Porod and his collaborators at the University of Notre Dame, describes the current state of the art in integrating nanoantenna-based sensors in the visible and infrared spectral regions with CNN vision systems to achieve multi-spectral imaging, sensing, and processing capabilities. Chapter 4, by Leon Chua – the inventor of the CNN concept and the driving force behind CNN research for more than 20 years, describes a serendipitous marriage between two of his most influential inventions – CNN and the memristor. This chapter contains an in-depth description of the memristor models. The next chapter,

Chap. 5, describes some of the novel circuit models of nanoscale devices, including equivalent-circuit models for nanoantenna-based infrared detectors.

In Chap. 6, Angel Rodríguez-Vázquez and collaborators, who have been instrumental in turning CNN concepts into VLSI hardware, describe a single mixed-mode CMOS chip implementation of a multi-core vision system on chip, realizing an array of cellular visual microprocessors integrating optical sensing, preprocessing, and final processing on a chip with 25k cores, and providing up to 10,000 frames per second input image flow. The authors of Chap. 7, describes a chip carrier design aimed at integrating nanoantenna infrared sensors on a CNN processor chip with digital processing cells. In Chap. 8, retinal pioneer Frank Werblin, who joined the CNN research early on, explores the circuit-level functional similarities between CNN vision systems on the one hand and the mammalian retina on the other. This is a highly fertile ground for research and focus of much current work, based on the pioneering result of the Berkeley Vision Research Lab published in 2001 showing that the mammalian retina consists of a dozen parallel and interconnected processing layers. This model inspired many CNN algorithms for visual processing. In this context, we wish to mention that another key contributor to the original MURI project, Dr. Botond Roska, currently with the Friedrich Miescher Institute in Basel, made pioneering contribution by elucidating neural circuit pathways, including those connected to individual ganglion cells, by using genetic, viral, and nanotechnology-based tools (*Nature Methods*, 2009;6(2):127–30). The last two chapters discuss some of the algorithmic innovations in solving spatial–temporal tasks via cellular processor arrays for real-world applications, such as multi-target tracking and UAV (Unmanned Aerial Vehicle) surveillance, and ends with some technical consideration and empirical guidelines on the architectural selection choices.

We should emphasize that this book does not discuss the fundamental aspects of CNN concepts and their theoretical underpinning, for which we refer the reader to the numerous textbooks, monographs, as well as comprehensive reviews in the literature.

Finally, we wish to thank Katie Chin of Springer for her patience and constructive suggestions. We would also like to acknowledge our families for their dedication and sacrifices during the preparation of this book.

Arlington, VA *Chagaan Baatar*
Notre Dame, IN *Wolfgang Porod*
Budapest, Hungary *Tamás Roska*

Contents

Chapter 1
A Brief History of CNN and ONR

Larry Cooper and Joel Davis

Cellular Nonlinear Networks and the MURI projects really trace their genesis to the mid-1970s. The ONR Nanoelectronics program was formulated in 1974 to focus basic research on those scientific areas that would influence the development of future electron devices. High speed, high frequency, and radiation hard devices with critical dimensions of less than 1 μm were expected to dominate the Navy's future. Various materials issues were to be considered, including those associated with both silicon and compound semiconductors. Another important component was the development of computer methods that could simulate and evaluate device concepts without huge investments in experiments. It has been said that science research has three legs, theory, experiment, and numerical simulations. Device and circuit simulations were critical components in the early stages of the plans, and the studies in nonlinear circuits was part of that. Leon Chua was a key figure in the ONR programs.

A critical event was the request in 1977 by the R&D office in the Pentagon to prepare a broad plan of research leading to electronic technologies with device dimensions of 20 Å, or 2 nm. A plan was prepared and it was given the name, Ultra Submicron Electronics Research (USER). In 1980, ONR created a special Accelerated Research Initiative program to bring focus on topics that were of high relevance to the Navy. USER became the first program to be funded in the ARI. USER was guaranteed a significant amount of funds over 5 years to focus on technology-changing research, in this case 2 nm electronics. This program was the largest research initiative program ever supported by ONR and it set the stage for the evolution of nanoelectronics in the Navy. It could be shown that elements of the Navy program were involved later, in the creation of another DOD (Department of Defense) program, the ULTRA project of DARPA, a program which ran from 1991 to 1998.

While all of this physics and engineering research was going on, ONR neuroscience was supporting Carver Mead's research on resistive grid networks for retina-like visual information processing. Although relatively simplistic from a

L. Cooper (✉)
Research Assistant, Arizona Institute for Nano Electronics, Arizona State University, Arizona, USA
e-mail: lcooper15@cox.net

J. Davis
Senior Neuroscientist Strategic Analysis, Inc

C. Baatar et al. (eds.), *Cellular Nanoscale Sensory Wave Computing*,
DOI 10.1007/978-1-4419-1011-0_1,

biological point of view, these analog devices began a slow approach to electronic simulation of neural activity. We see the first efforts to bring biology and electronics together.

One of the main components in the ONR program was to explore the development of new and novel approaches for computing architectures based on nanoscale devices. Could anyone conceive of such a computer scheme, having to take into account the issues of interconnect complexity, power dissipation, clock signal distribution, and variability of individual device operation where critical dimension of the devices was only 2 nm? *This question went unanswered at ONR for 15 years.*

The "big event" occurred in 1994. Leon Chua, Tamas Roska, and Frank Werblin visited ONR to arouse support for the CNN-UM development. Chua and Yang had published the first paper on CNN in 1988 and then Chua and Roska elevated the concept to the CNN-Universal Machine in 1993. Chua had been supported by ONR for many years in studies of nonlinear circuits, but this new idea, with an enormous potential for image processing applications, was truly revolutionary. It immediately became clear that CNN could be the answer to the question that had plagued the nanoelectronics program for 15 years, "how can nanoscale electron devices be useful in a computing application?" With only nearest neighbor cellular connections, the complexity of circuit layouts would be minimized. Nanoelectronic devices would dissipate minimal power and could be integrated in large-scale arrays. A quick survey of the various applications, which had been described in conferences and publications, produced an immediate response that CNN could provide the basis for a wide range of image-processing applications of importance to the Navy. The CNN solutions were compared with the conventional PC-based approaches and results were staggering. Improvements by factors of up to 1,000 were projected for speed, power dissipation, and circuit area.

Immediately following this meeting, planning began to utilize the Navy International Cooperative Research Program (NICOP) to provide support for Tamas Roska in Budapest and Angel Rodriguez-Vasquez in Seville. This was one of the first NICOP programs supported by ONR. The London office of ONR would provide part of the funds, and ONR headquarters office would provide the rest. It would strengthen the cooperation and coordination of all of the activities in the CNN program. This program was critical to the design and manufacturing of the first operational CNN-UM processor, the ACE-4k, and the later version ACE-16k.

Separate funding was provided to Chua and Werblin at the University of California at Berkeley. The task of Frank Werblin and Botond Roska was to use patch-clamp microelectrode recording techniques to measure response of living cells using genetic and immunological tracers to illuminate retinal circuitry. Leon Chua would continue to explore various properties of CNN circuits and their relationship to the retinal functions being discovered by Werblin and Roska, in particular, to the complex signal processing in the six layers of retinal neurons.

The trans-Atlantic cooperation between Budapest, Seville, and Berkeley has been the most important feature in the evolution of the CNN technology. From this creature emerged the first realistic processing chips that convinced the Missile Defense

Agency to make their contribution. The Small Business Innovative Research program led to the creation of a new company, Eutecus, which has recently led to new commercial activities. This cooperative environment provided the background that led to the formation of three different MURI projects at ONR.

The first MURI was awarded in 1998 to Arizona State University, with the project title, "Nanoelectronics: Low Power, High Performance Components and Circuits." The support was for a visionary project to incorporate the single-electron transistor (SET) into a CNN cell design. The SET is probably the ultimate device for a charge-sensitive transistor, namely logic functions are determined with the sensing of a single charge. The second MURI, awarded to Princeton University in 2000, was directed toward research on new techniques for nanolithography of three-dimensional integrated circuits. NanoImprint Lithography (NIL) and self-assembled growth of device materials were the focus, and a CNN cell was the test structure. The concept was identified as the "NanoCube" where all the components for an image-processing computer were integrated in a three-dimensional chip. This is exactly the concept selected as one of the four ONR Grand Challenges announced in 1998, "Multifunctional Electronics for Intelligent Naval Sensors". It should also be noted that NIL has become a critical process in many technology developments around the world in the twenty-first century.

The third of the MURI awards went to the University of Notre Dame in 2003. The title of the award contains nearly all of the ideas and visions that had driven the nanoelectronic and neurobiology program for two decades – "Bio-Inspired CNN Image Processors with Dynamically Integrated Multispectral Nanoscale Sensors." Here the goal was to integrate an infrared detector into each cell of a CNN array. The detector is made up of a nanoscale antenna array for tunable infrared radiation detection.

The title of the MURI at Notre Dame makes reference to a major research component embedded in these projects which needs some further comment, namely, bioinspired nanoelectronics. As it was described earlier, and before the MURI projects came into being, Frank Werblin at University of California at Berkeley (UCB) described his studies of the retina in living animals using patch-clamp techniques to monitor the neuronal response following a visual stimulation. These ONR sponsored neuroscience programs were defined by a computational approach leading toward understanding neural function. A further motivator was the computational approach leading to useful devices based on designs and algorithms derived from biology. It was the collaboration of neuroscience with Leon Chua and CNN electronics that provided a "proof of principle" for neuronal function, unavailable in any other way. Leon Chua recognized that the retinal operations that were observed by Werblin could be replicated by CNN functions; this fact led to the cooperation between the Electronics and Biology Divisions at ONR. Botond Roska, a postdoctoral fellow at UCB, carried out the definitive experiments that described how visual information is processed by the mammalian retina. It was a great achievement, and Botond became a prestigious Harvard fellow at Harvard University.

As part of the MURI, centered at the University of Notre Dame, Werblin at UCB and Roska at Harvard continued with studies of the retina, producing a conceptual paradigm shift in understanding the role the eye plays in visual information processing. These ideas are currently influencing disciplines as disparate as retinal prostheses for the blind and the latest generation of low light vision systems, such as represented in the MURI project at Notre Dame. In addition to CNN providing a framework for realistic biological modeling and simulation, the neurally oriented nanoelectronic research supported in these ONR programs has begum to appear in other contexts (e.g., DARPA SyNAPSE). Werblin and Roska have made a dramatic contribution to our understanding of the multimodal processing on images by the different layers of neurons in the retina.

This is not the end of the story. The full capabilities of nanoelectronic devices have not matured, nor have they been exploited in designs for future technologies. The Nanoelectronics program and the MURIs have established some milestones, but at present, the technology is limited to 100 nm electronics. But there is great potential for future progress, as CNN-based sensors provide one of the concepts in which nanoscale devices can have enormous impact. Very early it was shown that CNN-based image processing had great advantages over CMOS-based digital processing, such as 100 times higher frame rates and 1,000 times less power dissipation. CNN-UM-based products have been announced recently for the civilian markets. Technologies for military application are being developed. But, given the huge spectrum of applications for processing of images and patterns, where the advantages of speed, power dissipation, and physical size are relevant, the future should be very exciting, and there has been hardly any use of nanoelectronic device technology.

Just to mention a few of the areas for which the uses of CNN could be advantageous, consider facial recognition, autonomous robots, traffic control, area surveillance, target identification and tracking, collision avoidance, prostheses for the blind, quality control in manufacturing, epileptic seizure control, tactile control in robotics, sound detection and source localization, and many others. All of these have been studied in a CNN context.

Hopefully, this brief history has illuminated some very important principles of scientific research endeavors. The MURI programs provide visionary and creative scientists and engineers from different disciplines opportunity to explore new ideas in a free and cooperative environment.

The pioneers in this expedition of discovery and invention have all received numerous awards and recognitions for their efforts. So what are Leon, Tamas, Frank, and Angel going to do next? Local Activity principles applied in all areas of science? Wave computers? The Artificial Eye prosthesis for the blind? An image processor on a pin?

Chapter 2
Cellular Wave Computing in Nanoscale via Million Processor Chips

Tamás Roska, Laszlo Belady, and Maria Ercsey-Ravasz

Abstract A bifurcation is emerging in computer science and engineering due to the sudden emergence of many-core or even kilo-processor chips on the market. Due to the physical limitations, in CMOS technologies below 65 nm, a drastic power dissipation limit, a major signal propagation speed and distance limit, and a distributed character of the circuit elements are forcing new architectures. As a result, locality, the local connectedness becomes a prevailing property, the cellular, i.e., mainly locally connected processor arrays are becoming the norm, and the cellular wave dynamics can produce unique and practical effects.

In this new world, new principles are needed and new design methodologies. Luckily, the 15 years of research and development in cellular wave computing and CNN technology, we have aquired skills that help establishing some principles and techniques that might lead toward a new computer science and technology in designing mega-processor systems from kilo-processor chips.

In this chapter, we review the architectural development from standard CNN dynamics to the Cellular Wave Computer, showing several practical implementations, introduce the basic concepts of the Virtual Cellular Machine, present a new kind of implementation combining spatial-temporal algorithms with physics, give some architectural principles for non-CMOS implementations, and comment on biological relevance.

2.1 Introduction

When we proposed our MURI project in 2004, cellular computer architectures with thousands of processors (cells, cores) were more or less exceptions, a pioneering direction of research. The study and design of Cellular Wave Computers, we also

T. Roska (✉)
Computer and Automation Institute of the Hungarian Academy of Sciences and the Faculty of Information Technology of the Pázmány University, Budapest
e-mail: roska@sztaki.hu

L. Belady
Eutecus Inc., Berkeley, California, U.S.A.

M. Ercsey-Ravasz
University of Notre Dame, Notre Dame, Indiana, U.S.A.

C. Baatar et al. (eds.), *Cellular Nanoscale Sensory Wave Computing*,
DOI 10.1007/978-1-4419-1011-0_2,

called CNN technology, via mixed mode CMOS (cellular visual microprocessors with 25 k sensing cell processors), digital CMOS, or optical implementations, lead to impressive mission critical applications, including event detection with 30,000 frames per second. Today, however, mainstream products with kilo-processor chips or quarter million processor supercomputers converge to the cellular architectures as well. Indeed, physics is forcing the use of mainly locally connected cell-processor arrays when entering to the kilo-processor chip or mega-processor system arena. Moreover, nano-device arrays have no other choice as well. This trend is manifested in the emerging research architectures (ITRS 2007), as well as in various new products (CELL multiprocessors in games, FPGAs, GPUs, and supercomputers).

Considering the recent trends in computing we might ask:

- Why not place 1 million 8-bit microprocessors on a single 5 billion transistor chip via the new 45-nm CMOS technology?
- Why the most recent supercomputers have cellular, mainly locally connected (toroidal) architecture?
- Why the CELL multiprocessor chip for games, and the latest FPGAs, all have cellular partially locally connected architectures?
- Why the first visual microprocessor with 25 k cell processors has a cellular wave computing architecture?
- Will we have any prototype architecture for multimillion processor nanoscale systems?

The major physical constraints are dissipation and wire delay. This leads to many-processor/core/cell designs and lower clock speed. Hence, the architectural consequence is shown in Fig. 2.1. An essential property is the sparse wiring, mainly local and sparse global (e.g., crossbar).

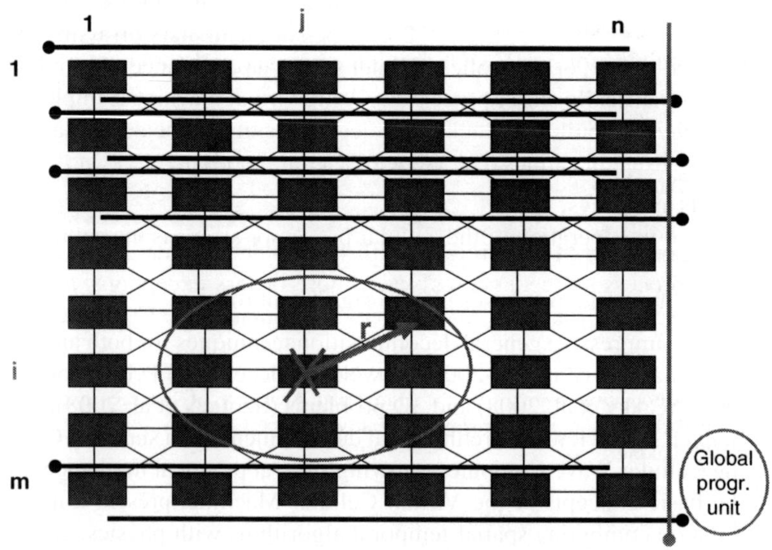

Fig. 2.1 The cellular many-core 2D wave computer architecture

At 60-nm CMOS technology a signal can only traverse 1.5 mm, that is only a small region is reachable within a clock cycle (Matzke 1997). Hence the cellular many-core architecture is a must as a consequence of physical limits. This means that the spatial address of a processor plays a new and very important role, and the dissipation limit is controlling the clock frequency. It seems that there is no adequate computer science for this case.

Historically, well before the seminal paper on *cellular neural/nonlinear networks* known as CNN (Chua and Yang 1988), the two pioneers in computing, already proposed spatially distributed, locally coupled dynamics for computing, during the beginning of the 1950s. In *A. Turing's Morphogenesis* paper (Turing 1952) the locally interacting cells were described by analog second order dynamics and *J. Von Neumann's Cellular Automaton* used discrete state cells, also locally connected. Interestingly, however, they both had only static input patterns, that is initial states. The original *standard CNN* dynamics had *two static inputs*: the initial state pattern and the input pattern.

The introduction of the *CNN Universal Machine* (Roska and Chua 1993) and its generalization, the *Universal Machine on Flows,* UMF, (Roska 2003) represented a major departure from the study of a single spatial temporal dynamics, discrete or continuous valued (digital or analog). Namely, a new stored programmable array computer was constructed with a protagonist spatial-temporal elementary instruction (a spatial-temporal wave) and a new kind of algorithm, the α-recursive function. In this machine, the data are topographic dynamic flows. This is the reason it is called sometimes as *Cellular Wave Computer*.

Our Virtual Cellular Machine architecture is composed of five building blocks, including 1D and 2D Cellular Wave Computers (or processor arrays) as single building blocks (Belady and Roska 2009). A single virtual processor array may be implemented by three different types of physical processor arrays: algorithmic, real-valued dynamical system array, and arrays of a physical dynamic entity defined by a geometric layout. They implement three types of elementary array instruction models: logic, arithmetic, or symbolic. Cellular means a precedence of communication between geometrically closer processors. A big virtual machine is implemented by the smaller physical building blocks: this decomposition is the design task.

Unlike other virtual machines (e.g., Azanovic et al. 2008) here

1. The elements are arrays,
2. The essence of the operation of the cellular processor arrays is the cellular wave dynamics, and
3. The spatial address of a processor plays a significant role.

There are already impressive general decomposition techniques for both analog and digital implementations (Zarándy 2008) as well as successful FPGA implementations for 2D (Rekeczky et al. 2008) and 3D problems (Szolgay et al. 2008).

In this chapter, we review the architectural development from standard CNN dynamics to the Cellular Wave Computer, showing several practical implementations, introduce the basic concepts of the Virtual Cellular Machine, present a new kind of implementation combining spatial-temporal algorithms with physics, give some architectural principles for non-CMOS implementations, and comment on physical relevance.

2.2 From Standard CNN Dynamics to the Cellular Wave Computer

We will introduce first the Cellular Wave Computer, embedding the standard CNN dynamics in it as a special case of an elementary instruction. This introduction is more abstract and rigorous to show next how many different physical implementations can be handled with this machine architecture.

In the Cellular Wave Computer

- *Data are topographic flows* (*cell array* signals on a 1-, 2- or 3-dimensional grid, e.g., a visual, a tactile, or an auditory flow, or the states of atoms in a molecular dynamics calculation).

Data type: *topographic flow*, $\Phi(t)$, in R^n (1D), $R^{n\times m}$ (2D), or $R^{n\times m\times p}$ (3D) as a function in time (continuous or, as a special case, discrete). In a 2D image flow

$$\Phi(t)\colon \varphi\{_{ij}(t)\}, t\in[0,T] \text{ in } R^1, \quad 1\le i\le m, \quad 1\le j\le n$$

$\varphi\{_{ij}(t)\}$ are the *cell signals*.

For example, a 2D image flow could represent the input or the output image flow of a retina.

An $n{\times}m$ map (e.g., an image or picture) $P\colon t = t^*,\ P = \Phi(t^*)$
if P is binary it is a *Mask* M
if $t = t0, t0+\Delta t, t0+2\Delta t, \ldots, t0+k\Delta t$
then we say that it is a *map sequence* (e.g. a video stream).

- *Instructions* are defined in space and time, typically as a spatial-temporal wave acting on the image flow data, the cell signals are *continuous* (real) valued (analog or digitally coded) *and binary*; locally (cell by cell) stored in the cells.
- This local storage, providing for stored programmability in von Neumann sense, may be static or dynamic.
- The *protagonist elementary instruction*, $\Psi(\Phi)$, also called *wave instruction*, is defined as

$$\Phi_{\text{output}}(t) := \Psi\left(\Phi_{\text{input}}(t),\ P, \partial\right);\ t\in[0,\ T]$$

where

Ψ: *a function on image flows or image sequences*
P: a map (image) defining initial state and or threshold (bias) map
∂: boundary conditions, $\partial(t)$ is a boundary input (might be connected to all cells in a boundary row)
T is the finite time interval

A scalar or logic valued functional γ on an image flow is used for branching instructions:

$$q := \gamma\left(\Phi_{\text{input}}(t), P, \partial\right)$$

For example, the so-called global white functional on a binary mask M is logic 1, if the picture is full white, if at least one pixel is black it is logic 0. Another example, the maximum functional on a flow is defined by the highest scalar value at any cell signal value at any time. If the spatial-temporal instruction is of non-equilibrium type, this global output state can also be detected (global fluctuation, GF) or the different types of the nonequilibrium attractors can also be considered as global output parameters. Hence, GF could take values from a multivalued logic set.

We emphasize that the signal and instruction representation in this architecture may have various physical realizations, it might be analog, mixed mode, digital using CMOS, or optical, etc.

- *As a simple special case* for the spatial temporal dynamics, defining the topographic wave instruction, *is the standard* Cellular Nonlinear Network (CNN) dynamics (Chua and Yang 1988), defined for one layer of first order cells.

The output image flow, $Y(t)$ will be calculated from the input image flow $U(t)$ as the solution of the following discrete space, continuous time, nonlinear dynamics (PDDE: partial differential difference equation):

$$\begin{aligned} \mathrm{d}x_{ij}/\mathrm{d}t &= -ax_{ij} + i_t = -ax_{ij} + \sum A\,(ij,kl)\; y_{kl} + \sum B\,(ij,kl)u_{kl} + z_{ij}, \\ y_{ij} &= f(x_{ij}) \text{ for all } i \in [1, M] \text{ and } j \in [1, N] \end{aligned} \tag{2.1}$$

where the spatial summation Σ is made within the r neighborhood of the cell ij, and $u_{ij}(t)$, $x_{ij}(t)$, and $y_{ij}(t)$, are the input, state, and output signals, respectively (elements of U, X, and Y array signal flows).

The standard CNN dynamics, representing the simplest *cellular wave instruction*, is defined by a first-order *cell dynamics*, $r = 1$ neighborhood radius (3×3), feedback (A), and feed forward (B) linear *local interaction patterns* (templates), with threshold (bias) map $z_{ij} = z$ and $x_{ij}(t)$ as the state array, $y_{ij}(t) = \sigma\left(x_{ij}(t)\right)$ being the nonlinear output function, and input image flow defined by $u_{ij}(t)$.

A standard CNN instruction, a template $\{A, B, z\}$, is defined by the 19 $(9+9+1)$ numbers.

The *global control of the computation in a Cellular Wave Computer, in general,* is performed via well-defined *wave algorithms*, as an algorithmic sequence of wave instructions as well as by local and global binary logic instructions. The rigorous definition is given as the *α-recursive function* (Roska 2003).

- The algorithm on a Cellular Wave Computer defined on topographic flows, or rigorously defined as an *α-recursive function* on a (UMF) is as follows

Now we are in a position to define the new recursive function, the α-recursive function. Algorithms of digital computers are defined mathematically via the μ-recursive functions.

The α-recursive function is defined by the

- Initial settings of image flows, pictures, masks, and boundary values: $\Phi(0)$, P, M, ∂;
- Equilibrium and nonequilibrium solutions of PDDE defined via cellular locally connected cell dynamics (a special case is the standard CNN equations on $\Phi(t)$);

- Global (and local) minimization on the above;
- Memoryless arithmetic and logic combinations of the results of the above operations;
- Analog comparisons (thresholding) and logic conditions in branching instructions (via the scalar and logic valued functionals); and
- Recursions on the above operations.
- The CNN Universal Machine is a minimal architecture for α-recursive functions

The Turing Machine is a minimal architecture for the μ-recursive functions. Some additional components make it practically more efficient. Likewise, the CNN Universal Machine (Roska and Chua 1993) without the local logic unit (LLU) and the local analog (arithmetic) output unit (LAOU or LAU) is a minimal architecture for the α-recursive functions. Through the years many different forms of the CNN Universal Machine were implemented in mixed mode and digital CMOS as well as in an optical way. To prove the minimality of the CNN-UM, we can implement all the elements of the $u\alpha$-recursive function step by step on the CNN Universal Machine. The control in the global analogic control unit (GACU) uses global variables, real or logic values, for the entire array (the cell variables are called local). Hence, the GACU is also containing the global detection unit (GDU) determining the global functionals defined above (e.g., the global white or global fluctuation, etc.), as well as the comparison and logic conditions for the branching instructions. The CNN dynamics is the main spatial temporal elementary operation in this abstract CNN-UM. The other side of the minimality can be proved by taking away any component and show some missing element in the α-recursive function.

The extended cell is shown in Fig. 2.2 and the framework of the CNN Universal Machine architecture in Fig. 2.3.

The universality of the CNN-UM can be proved in two senses. Turing Machine universality had been proved via the implementation of the game of life on the CNN-UM. The universality as a nonlinear operator with fading memory for each cell has been proved for feedforward delayed interactions.

In Table 2.1, the summary of properties of the three major classes of Universal Machines operating on integers (UMZ), on reals (UMR), and on flows (UMF) are

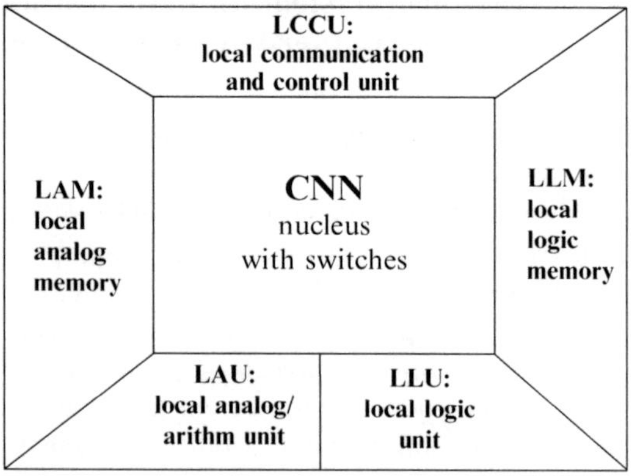

Fig. 2.2 An extended CNN cell

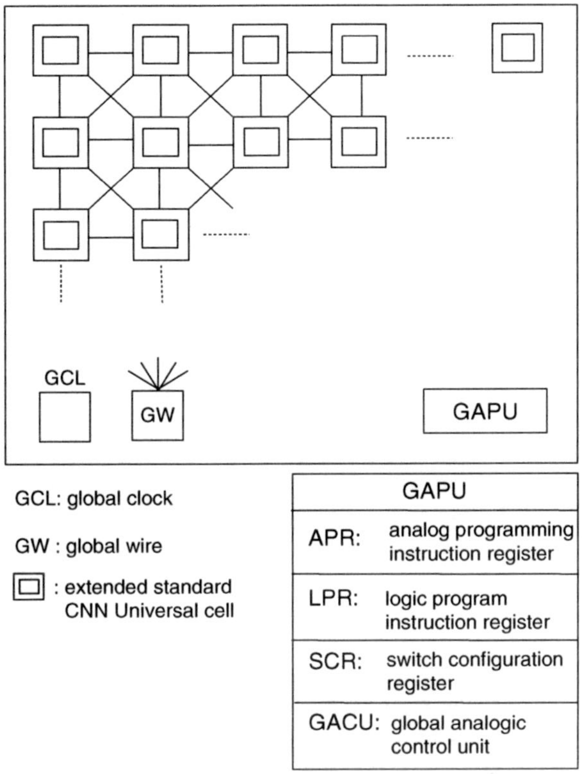

Fig. 2.3 The framework of the CNN Universal Machine Architecture. The global analog/arithmetic-and logic control unit (GACU) hosts also the global control processor and related global memory

shown. These are mathematical machines, on the other hand their significance is that they approximate quite well the real, physical computers used for many years and the ones now emerging.

2.3 Various physical implementations of the Cellular Wave Computer

The first implementations were based strictly *via the CNN Universal Machine.*The range started with a 20 × 22 chip and expanded toward a 128 × 128 processor chip (ACE 16 k), the first full-fledged cellular visual microprocessor hosting optical sensors in each cell processor, operating up to 30,000 frames per second input image flow. It was placed into the Bi-i camera computer, the fastest one in the world in 2003.

The evolution of this technology led to RISC architectures, consisting of those template instructions in different physical forms (step by step, diffusion, digital, etc.) that are optimum in flexibility vs. robustness.

Table 2.1 The three main Universal Machines

Architecture	Universal iterative Machine over Z	Universal, iterative Machine over R	Universal, semi-iterative Machine over image flows
	UMZ	UMR	UMF
Data	Z	R	F (flow $\Phi(t)$ on $R^{n\times n}$)
Elementary operators	Logic maps	Semi algebraic maps	Differential algebraic maps
Mode of operation	Iterative	Iterative	Semi iterative
Sphere of influence of elementary operators (instructions)	Local	Local	Global
Typical machine	Turing Machine	Newton Machine Basin of attraction Machine	CNN Universal Machine
Computing models	Grammar		2D–3D PDDE: artial differential difference and functional equations
	Partial (μ) recursive functions on Z	Register equations on R	α-recursive functions on F

The advent of the kilo processor FPGAs and the several hundred core GPUs (graphic processing units) led to the implementation of the CNN Universal Machine type architectures on these chips.

The optical implementation via a Programmable Optical Analogic CNN (POAC) computer implements the local correlation with a speed of light and the programming of the B template in a 31×31 size is achieved by an acusto-optical modulator. It is interesting to note that the Blue Gene and Cyclops 64 type IBM supercomputers are using also a 3D cellular architecture, as a result of coping with physical constraints.

2.4 Virtual Cellular Machine

2.4.1 Notations and Definitions

2.4.1.1 Core=Cell

Core or cell will be used as synonyms, it is defined as a unit implementing a well defined operator (with input, output, state) on binary, real, or string variables (also defined as logic, arithmetic/analog, or symbolic variables, respectively). Cores/cells are used typically in arrays, mostly with well-defined interaction patterns with their neighbor core/cells, although sparse longer wires/communications/interactions are also allowed.

Core is used if we emphasize the digital implementation, cell is used if it is more general.

2.4.1.2 Elementary Array Instructions

A logic (L), arithmetic/analog (A), or symbolic (S) elementary array instruction is defined via r input ($u(t)$), m output ($y(t)$) and n state ($x(t)$) variables (t is the time instant), operating on binary, real, or symbol variables, respectively. Each dynamic cell is connected mainly locally, in the simplest case, to their neighbor cells.

- L: A typical *logic elementary array instruction* might be a binary logic function on n or n×n (2D) binary variables, (special cases: a disjunctive normal form, a memory look-up table array, a binary state machine, an integer machine),
- A: a typical *arithmetic/analog elementary array* instruction is a multiply and accumulate (add) term (MAC) core/cell array or a dynamic cell array generating a spatial-temporal wave, and
- S: a typical *symbolic elementary array instruction* might be a string manipulation core/cell array, mainly locally connected

Mainly local connectedness means that the local connection has a speed preference compared to a global connection via a crossbar path.

A classical 8-, 16-, or 32-bit microprocessor could be considered as well as an elementary array instruction with iterative or multi-thread implementation on the three types of data. However, the main issue is that we have elementary array instructions, as the protagonist instructions.

2.4.2 *Physical Implementation Types of Elementary Core/Cell Array Instructions (A, B, C)*

We have three *elementary cell processor* (cell core) array implementation types:

D: A digital algorithm with input, state, and output vectors of real/arithmetic (finite precision analog), binary/digital logic, and symbolic variables (typically implemented via digital circuits).

R: A real-valued dynamical system cell with analog/continuous or arithmetic variables (typically implemented via mixed mode/analog-and-logic circuits and digital control processors), placed in a mainly locally connected array.

G: A physical dynamic entity with well-defined Geometric *Layout* and I/O ports (function in layout) – (typical implementations are CMOS and/or nanoscale designs, or optical architectures with programmable control), placed in a mainly locally connected array.

2.4.3 *Physical Parameters of Array Processor Units (Typically a Chip or a Part of a Chip) and Interconnections*

Each of these array units is characterized by its

- g, geometric area,
- e, energy,
- f, operating frequency,
- w = e f local power dissipation, and
- The signals are traveling on a wire with length l, width q, and with speed v_q introducing a delay of $D = l\, v_q$.

Ω *cores/cells* can be placed on a single Chip, typically in a square grid, with input and output physical connectors typically at the corners (sometimes at the bottom and top "corners" in a 3D packaging) of the Chip, altogether there are K input/output connectors. The maximal value of dissipation of the Chip is W. The physics is represented by the maximal values of Ω, K, and W (as well as the operating frequency). The operating frequency might be global for the whole Chip, Fo, or could be local within the Chip, f_i (some parts might be switched off, $f_i = 0$), may be a *partially local frequency* fo > Fo.

The *interconnection pathways* between the arrays and other major building blocks are characterized by the delay and the bandwidth (B).

2.4.4 *Virtual and Physical Cellular Machine Architectures and Their Building Blocks*

A Virtual Cellular Machine is composed of five types of building blocks:

1. *Cellular processor arrays/layers* with simple (L, or A, or S type) or complex cells and their local memories, these are the protagonist building blocks,
2. *Classical* digital stored program *computers* (microprocessors),
3. Multimodal topographic or nontopographic *inputs* (e.g., scalar, vector, and matrix signals),
4. *Memories* of different data types, organizations, and qualitatively different sizes and access times (e.g., in clock cycles), and
5. *Interconnection pathways* (buses).

The tasks, the algorithms to be implemented, are defined on the Data of the Virtual Cellular Machines.

We consider two types of Virtual Cellular Machines: single- or multi-cellular array/layer machines, also called homogeneous and heterogeneous cellular machines.

In the *homogeneous Virtual Cellular Machine*, the basic problem is to execute a task, for example a Cellular Wave Computer algorithm, on a bigger topographic

Virtual Cellular Array using a smaller size of physical cellular array. Four different types of algorithms have already been developed (Zarándy 2008).

Among the many different, sometimes problem oriented *heterogeneous Virtual Cellular Machine* architectures we define two typical ones. Their *five building blocks* are as follows.

1. Cellular processor arrays of one dimensional, CP1, and two dimensional, CP2, ones;
2. P – classical digital computer with memory and I/O, for example a classical microprocessor;
3. T – topographic fully parallel 2D (or 1D) input;
4. M – memory with high speed I/O, single port or dual port (L1, L2, L3 parts as cache and/or local memories with different access times);
5. B – data bus with different speed ranges (B1, B2, ...).

The CP1 and CP2 types of cellular arrays may be composed of cell/core arrays of simple and complex cells. In the CNN Universal Machine, each complex cell contains logic and analog/arithmetic components, as well as local memories, plus local communication and control units. Each array has its own controlling processor; we called it in the CNN Universal Machine as Global Analog/arithmetic-and-logic Programming Unit (GAPU).

The size of the arrays in the Virtual Cellular Machines is typically large enough to handle all the practical problems that might encounter in the minds of the designers. In the physical implementation, however, we confront the finite, reasonable, cost effective sizes, and other physical parameters.

The *Physical Cellular Machine* architecture is defined by the same kind of five building blocks, however, with well-defined physical parameters, either in a similar architecture like that of the Virtual Cellular Machine or a different one.

A building block could be physically implemented as a separate chip or as a part of a chip. The *geometry of the architecture* is reflecting the physical layout within a chip and the chips within the Machine (multi-chip machine).

This architectural geometry defines also the communication (interacting) speed ranges, as well. Hence physical closeness means higher speed ranges and smaller delays.

The *spatial location or topographic address* of each elementary cell or core, within a building block, as well as that of each building block within a chip, and each chip, within the Virtual Cellular Machine (Machine) architecture, plays a crucial role. *This is one of the most dramatic differences compared to classical computer science.*

In the Physical Cellular Machine models we can use exact, typical, or qualitative values for size, speed, delay, power, and other physical parameters. The simulators can use these values for performance evaluation.

We are not considering here the problems and design issues within the building blocks, it was fairly well studied in the Cellular Wave Computing or CNN Technology literature, as well as implementing a virtual 1D or 2D Cellular Wave Computer on a smaller physical machine. The decomposition of bigger memories on smaller physical memories is the subject of the extensively used virtual memory concept.

We mention that sometimes a heterogeneous machine can be implemented on a single chip by using the different areas for different building blocks (Rekeczky et al. 2008).

The architecture of the Virtual Cellular Machine and the Physical Cellular Machine might be the same, though the latter might have completely different physical parameters. On the other hand they might have completely different architectures.

The *internal functional operations* of the cellular building blocks are not considered here. On one hand, they are well studied in the recent Cellular Wave Computer literature, as well as in the recent implementations (ACE 16 k, ACE 25 k = Q−Eye, XENON), etc.), on the other hand, they can be modeled based on the Graphics Processing Units (GPU) and FPGA literature. Their functional models are described elsewhere (see also the Open CL language description).

The *two basic types of multi-cellular heterogeneous Virtual Machine architectures* are defined next.

1. *Global system control and memory* architecture is defined in Fig. 2.4.
2. *Distributed system control and memory* architecture is shown in Fig. 2.5.

The thick buses are "equi-speed" with much higher speed than the connecting thin buses.

2.4.5 The Design Scenario

There are three domains in the design scenario:

- The *Virtual Cellular Machine* architecture based on the data/object and operator relationship architecture of the problem (topographic or nontopographic),

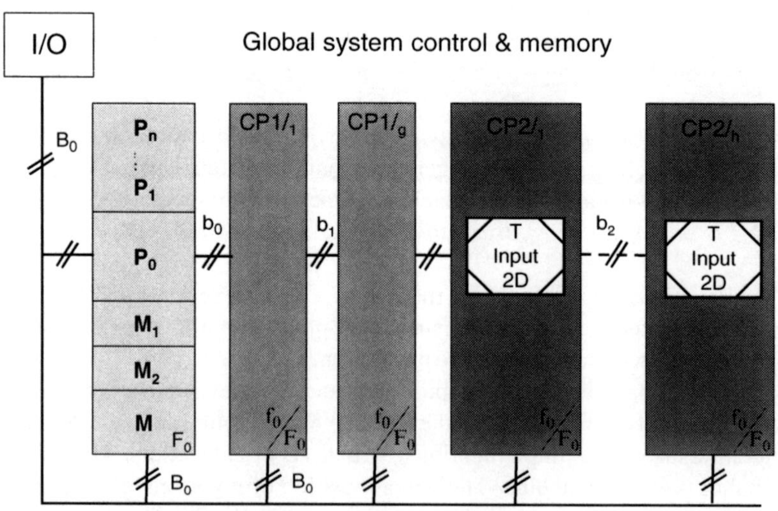

Fig. 2.4 Global system control and memory architecture

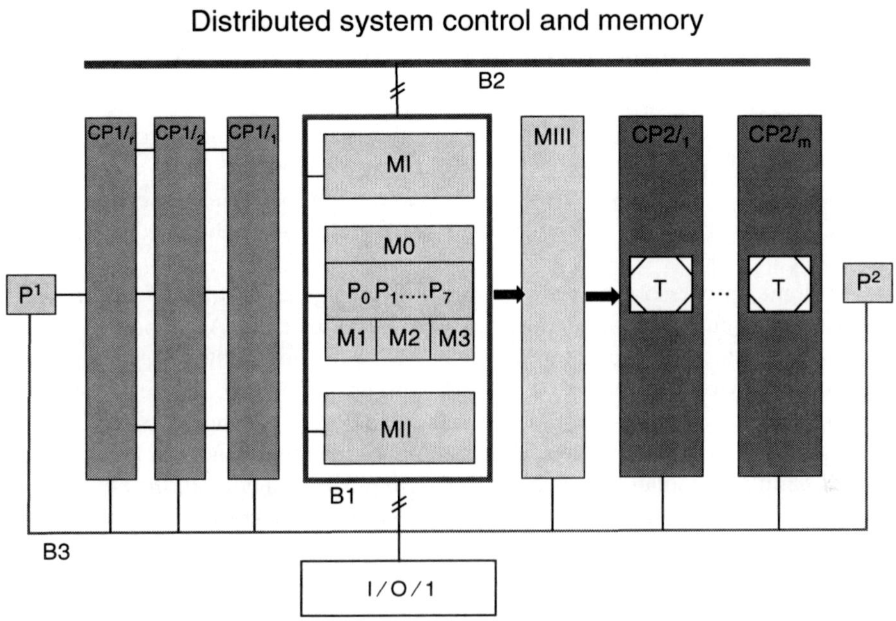

Fig. 2.5 Distributed system control and memory architecture

- The physical processor/memory topography of the *Physical Cellular Machine*, and the
- *Algorithmic domain* connecting the preceding two domains.

The design task is to map the algorithm defined on the Virtual Cellular Machine into the Physical Cellular Machine, e.g., the decomposition of bigger virtual machine architectures into smaller physical ones, as well as to transform nontopographic data architectures into topographic processor and memory architectures.

2.4.6 The Dynamic Operational Graph and its Use for Acyclic UMF Diagrams

Extending the UMF diagrams (Roska 2003) describing Virtual Cellular Machines leads to digraphs, with processor array and memory nodes, and signal array pathways as branches with bandwidth weights. These graphs with the dissipation side-constraint define optimization problems representing the design task, under well-defined equivalent transformations.

In some well-defined cases, especially within a 1D or 2D homogeneous array, the recently introduced method via Genetic Programming with Indexed Memory (GP-IM) using UMF diagrams with Directed Acyclic Graphs (DAG) seems a promising tool showing good results in simpler cases (Pazienza 2008).

2.5 Recent, Non-Standard Architecture Combining Spatial-Temporal Algorithms with Physical Effects

A strikingly new direction in designing Cellular Wave Computer algorithms is the combination of spatial-temporal CNN algorithms on a mixed-mode visual microprocessor with on-chip physical effects, such as random noise. In this section we will present the generation of true random binary patterns using this method (Ercsey-Ravasz et al. 2006).

On digital processors there is no possibility to generate quickly real random events, only pseudo-random number generators can be used. An important advantage of the analog architecture of the CNN-UM is the possibility to use the natural noise of the device to generate true random numbers. The natural noise of the CNN-UM chip is usually highly correlated in space and time, so it cannot be used directly to obtain random binary images. This true random number generator is based on a chaotic cellular automaton (CA) perturbed with the natural noise of the chip in each time step. Due to the used chaotic cellular automation the correlations in the noise will not induce correlations in the generated random patterns. Meanwhile the real randomness of the noise will kill the deterministic properties of the chaotic CA.

There were several studies developing random number generators on the CNN-UM, but all of them were generating pseudo-random binary images with $1/2$ probability of the black and white pixels (logical 1 and 0 were generated with the same probability). As starting point we used one of these relatively simple but efficient chaotic CA (Crounse et al. 1996; Yalcin et al. 2004) called PNP2D. This chaotic CA is based on the following update rule:

$$x_{t+1}(i,j) = (x_t(i+1,j) \vee x_t(i,j+1)) \oplus x_t(i-1,j) \oplus x_t(i,j-1) \oplus x_t(i,j)$$

where i, j are the coordinates of the pixels, the index t denotes the time-step, and x is a logic value 0 or 1 representing white and black pixels, respectively. Symbols $\vee$ and $\oplus$ represent the logical operations or and exclusive-or (XOR). This chaotic CA is relatively simple and fast, it passed all important RNG tests and shows very small correlations so it is a good candidate for a pseudo-random number generator. It generates binary values 0 and 1 with the same $1/2$ probability, independently of the starting condition.

The way we transform this into a true random number generator is relatively simple. After each time step the $P(t)$ result of the chaotic CA is perturbed with a noisy $N(t)$ binary picture (array) so that the final output is given as:

$$P'(t) = P(t) \oplus N(t)$$

The symbol $\oplus$ stands again for the logical operation XOR, i.e. pixels which are different on the two pictures will become black (logic value 1). This operation assures that no matter how $N(t)$ looks like, the density of black pixels remains the same $1/2$. Because the used noisy images contain only very few black pixels (logic values 1) we just slightly sidetrack the chaotic CA from the original deterministic path and all the good properties of the pseudo-random number generator will be preserved.

The $N(t)$ noisy picture is obtained by the following simple algorithm. All pixels of a gray-scale image are filled up with a constant value a and a cut is realized at a threshold $a + z$, where z is a relatively small value. In this manner all pixels which have smaller value than $a + z$ will become white (logic value 0) and the others black (logic value 1). Like all the logic operations this operation can be also easily realized on the CNN-UM. Due to the fact that the used CNN-UM chip is an analog device, there always will be natural noise on the gray-scale image. Choosing thus a proper z value one can generate a random binary picture with few black pixels. Since the noise is time dependent and generally correlated in time and space, the $N(t)$ pictures might be strongly correlated but will fluctuate in time. These time-like fluctuations cannot be controlled, these are caused by real stochastic processes in the circuits of the chip and are the source of a convenient random perturbation for our RNG based on a chaotic CA.

We performed our experiments on the ACE16K chip (128×128 cells) included in a Bi-i v2 (Zarandy and Rekeczky 2005). No significant correlations appeared in the generated patterns and the density of black and white pixels remained the same. A random image with $1/2$ density generated by this method is shown in Fig. 2.6a.

Perturbing the CA with this noise also assures that our true RNG started each time from the same initial state will always yield different results $P_1'(t)$, $P_2'(t)$, $P_3'(t)$ etc. Starting from the same initial condition (initial random binary picture) the patterns generated after several time-steps are shown on Fig. 2.7 On this figure two different sequences ($P_1'(t)$ and $P_2'(t)$) are compared. The third column represents the image resulting from an XOR operation performed on $P_1'(t)$ and $P_2'(t)$. For a simple deterministic CA this operation would yield a completely white image for any time step t. In our case however, the picture is white in the beginning because the two sequences started from the same initial condition, but as time passes the small $N(t)$ perturbation propagates over the whole array and generates completely different binary patterns. For $t > 70$ time-steps the two results are already totally different.

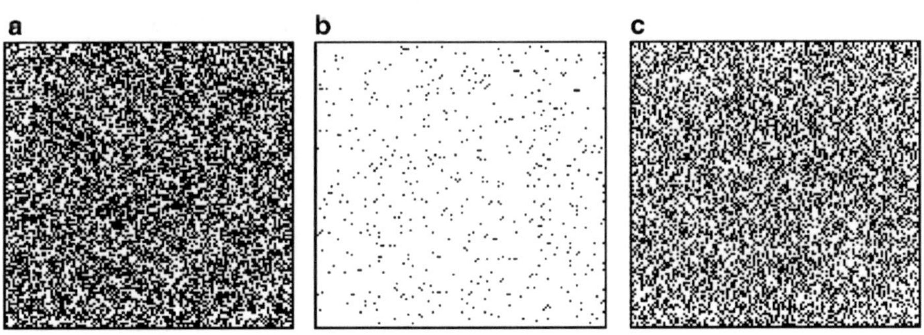

Fig. 2.6 Three random binary images with (**a**) $p = 1/2$, (**b**) $p = 0.03125$, (**c**) $p = 0.375$ probability of the black pixels, generated on the ACE16K chip

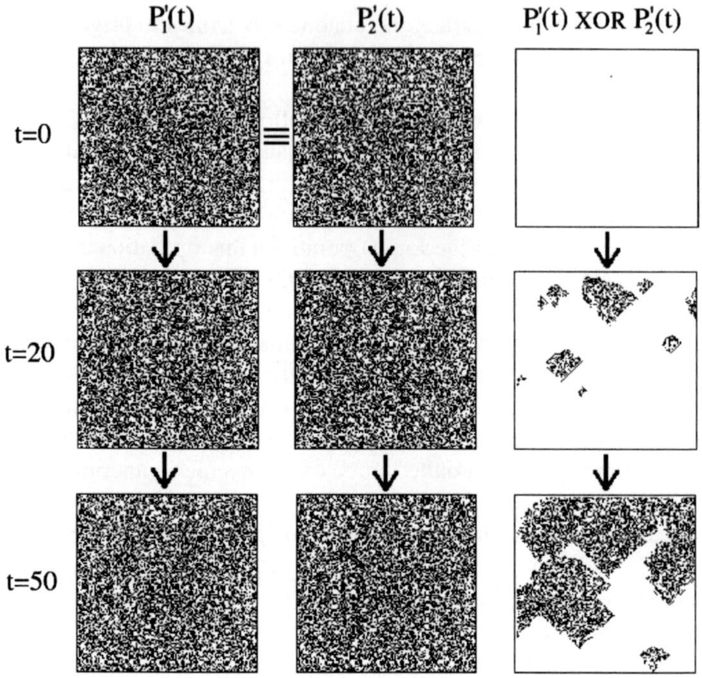

Fig. 2.7 Illustration of the nondeterministic nature of the generator. The figure compares two different sequences $P_1'(t)$, $P_2'(t)$ with the same initial condition in the $t = 0, 20, 50$ iteration steps, respectively

Due to the parallel nature of the CNN-UM the speed of this RNG also shows important advantages compared to other pseudo-random RNGs used on digital computers (Ercsey-Ravasz et al. 2006).

Up to now the method presented generates black and white pixels (1 and 0) with equal $1/2$ probabilities. In many applications however one needs to generate binary values with any probability p. On digital computers this is done by generating a real value in the interval [0,1] with a uniform distribution and making a cut at p. Theoretically it is possible to implement similar methods on CNN-UM by generating a random gray-scale image and making a cut-off at a given value. However, on the actual chip it is extremely hard to achieve a gray-scale image with a uniform distribution of the pixel values between 0 and 1 (or -1 and 1).

Our solution for generating a random binary image with p probability of the black pixels is by using many independent binary images with $p = 1/2$ probability of the black pixels. Let p be a number between 0 and 1

$$p = \sum_{i=1}^{8} x_i \frac{1}{2^i}$$

represented here on 8 bits by the x_i binary values. One can approximate a random binary image with any fixed p probability of the black pixels by using 8 images I_i, with probabilities $p_i = 1/2^i, i \in \{1, \ldots, 8\}$ of the black pixels, with the condition that they do not overlap: comparing any of the two images there are no black pixels occupying the same position. Once these 8 images are generated one just have to unify (perform OR operation) all I_i images for which $x_i = 1$ in the expression of p. These 8 basic I_i images can be obtained with a simple algorithm using 8 independent images (P_i) with $p = 1/2$ probabilities of the black pixels (for details see Ercsey-Ravasz et al. 2006).

This algorithm was also implemented on the ACE16K chip and reproduced the expected probabilities nicely. The differences between the average density of black pixels (measured on 1,000 images) and the expected p probability were between 0.01% and 0.4%. Normalized correlations in space between the first neighbors were measured between 0.05% and 0.4%, correlations in time between 0.7% and 0.8%.

Two random images with different probabilities of black pixels ($p = 1/2^5 = 0.03125$ and $p = 1/2^2 + 1/2^3 = 0.375$) are shown in Fig. 2.6b, c. Since the presented method is based on our previous true RNG, the images and binary random numbers generated here are also non-deterministic.

2.6 Hints for Architectural Principles for Non-CMOS Nano-Scale Implementations

In view of the multi billion CMOS chips, one can ask what kind of non-CMOS nano-device architectures might lead to competitive chips. Probably, in the digital processing domain, those elementary logic functions that have been developed for the CMOS circuits during the last 40 years are physically optimal for the CMOS implementation.

Therefore, in searching for non-CMOS architectures the elementary instructions should be nano-friendly, that is the starting point is to consider those nano-device arrays that can easily be implemented. Considering the implementation-friendly patterns of the device arrays and their nano-friendly interactions, these arrays, by nature, are of typical Cellular mainly locally coupled types with possible crossbar connections. In this cellular architecture, the main question is: what are the competitive elementary array instructions and the decomposition techniques to implement more complex instructions in space and time. These nano-friendly elementary array-input-array-output instructions might drastically differ from the ones we are accustomed in CMOS architectures. Even the 2-input logic might differ (instead of the AND, NAND, OR, NOR, it might be an XOR or other elementary functions). More interestingly, many-input–many-output functions, even dynamic functions might be more competitive with much less dissipation and more complex tasks. Then, unlike the decomposition techniques with disjunctive normal forms will not be useful anymore. Hence, new functional decomposition techniques are needed, adjusted to the physical capabilities of the component arrays of the nano-devices. The first results in different fields are already emerging.

In some cases the embedding of the non-CMOS nano-device arrays into CMOS chips is a way to success. Hence, the design of an interaction between the nano-scale and deep submicron CMOS scale arrays are of crucial importance. A promising way is the CMOL array concept (Likharev et al. 2002).

2.7 Biological Relevance

The several layer cellular architectures with different sizes of receptive fields as well as with global bus-like interconnections are seemingly reflecting several uncovered Neuromorphic architectures. The retina is one of them, may be the most prominent one. The surprising discovery of the "multi-screen" parallel channel operation of the mammalian retina (B Roska and Werblin 2001) ignited a new way of thinking in visual signal processing architectures. Its approximate implementation via CNN Universal Machine architectures and cellular visual microprocessor based Bi-i camera computers (Bálya et al. 2002) signaled the first step in multi-channel dynamic visual computing.

Acknowledgments The supports of the Office of Naval Research, the Future and Emerging Technology program of the EU, the Computer and Automation Research Institute of the Hungarian Academy of Sciences, the Hungarian National Research Fund (OTKA), the Pázmány P. Catholic University, Budapest, the University of California at Berkeley, and the University of Notre Dame are gratefully acknowledged.

Appendix

The UMF (Universal Machine on Flow) diagrams – a representation of Virtual Cellular Machines

A single cellular array/layer:

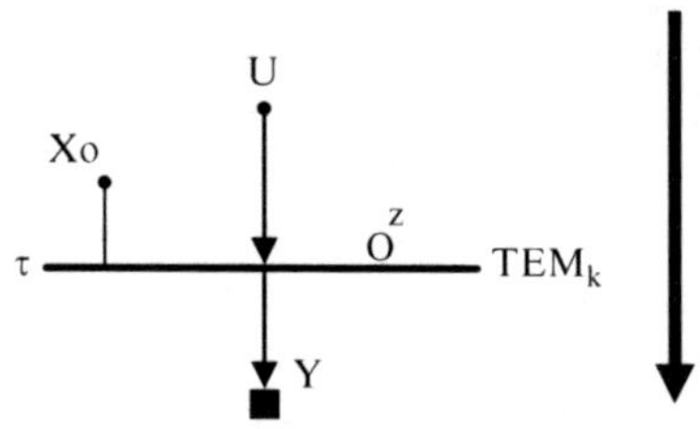

U: input array, Xo Initial state array, z: threshold or mask array, Y: output array, τ: time constant or clock time, TEM_k: local interaction pattern between cells or cores

Array Signals, variables

- logic/symbolic array
- logic/symbolic value
- arithmetic/analog array
- arithmetic/analog value

Boundary conditions

Left side : Input boundary condition

Right side : Output boundary condition

Constant : 

Zero Flux :

Period/toroidal: 

Boundary conditions are optional, if not given, it means “don’t care”

Decisions/Branching

On global analog parameter

Is the value of functional *q* less then 0.5?

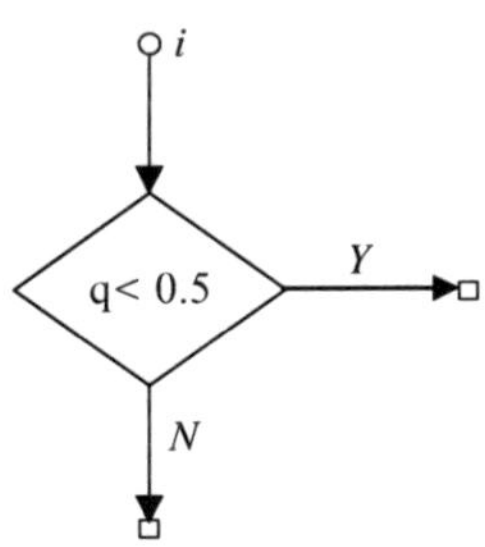

On global Logic parameter set, including global Fluctuation

Does the logic value of functional q refers to white?

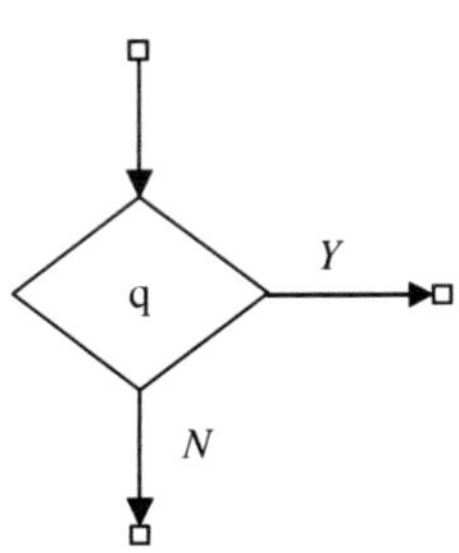

Algorithmic structures in terms of arrays/layers

Cascade

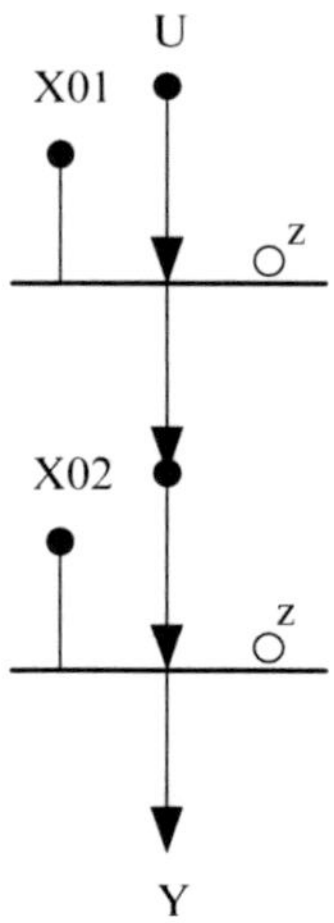

Parallel

A typical parallel structure with two parallel flows is shown below, by combining them in the final layer

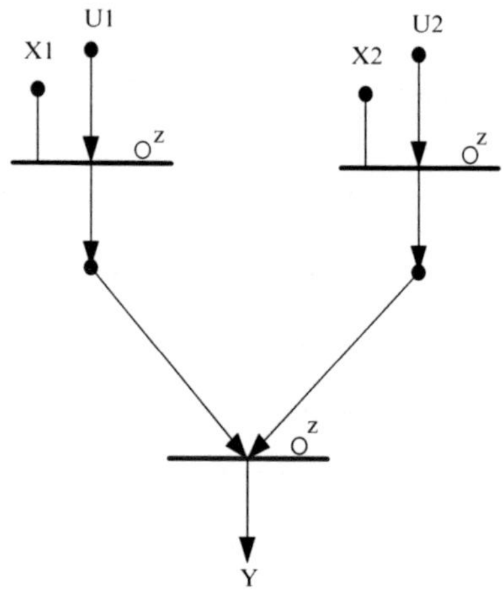

References

Bálya, D, Petrás I, Roska T, Carmona R, Rodríguez-Vázquez Á (2004) Implementing the multilayer retinal model on the complex-cell CNN-UM chip prototype. Int J Bifurcation Chaos 14:427–451

Bálya D, Roska B, Roska T, Werblin FS (2002) A CNN framework for modeling parallel processing in the mammalian retina. Int J Circuit Theor Appl 30:363–393

Chua LO (1999) A paradigm for complexity. World Scientific, New York, Singapore

Chua LO, Roska T (2002) Cellular neural networks and visual computing. Cambridge University Press, Cambridge, UK

de-Souza SX, Suykens JAK, Vandewalle J (2006) Learning of spatiotemporal behavior in cellular neural networks. Int J Circuit Theor Appl 34:127–140

Ercsey-Ravasz M, Roska T, Néda Z (2006) Stochastic simulations on the cellular wave computers. Eur Phys J B 51:407–412

Fodróczi Z, Radványi A (2006) Computational auditory scene analysis in cellular wave computing framework. Int J Circuit Theor Appl 34:489–515

Halfhill TR (2007) Faster than a blink. Microprocessor, www.MPRonline.com, 2/12/07, 2007

ITRS (2007) International technology roadmap for semiconductors 2003, 2005, 2007

Kék L, Karacs K, Zarándy Á, Roska T (2007) CNN template and subroutine library for cellular wave computing. Report DNS -1 – 2007, Computer and Automation Research Institute of the Hungarian Academy of Sciences, Budapest

Kunz R, Tetzlaff R, Wolf D (2000) Brain electrical activity in epilepsy characterization of the spatio-temporal dynamics with cellular neural networks based on a correlation dimension analysis. IEEE Int Symp Circuits Syst (ISCAS 00)

Mozsáry A, et al (2007) Function-in-layout: a demonstration with bio-inspired hyperacuity chip. Int J Circuit Theor Appl 35(3):149–164

Porod W, et al (2004) Bioinspired nano-sensor enhanced CNN visual computer. In Roco MC, Montemagno C (eds) The coevolution of human potential and converging technologies. Ann NY Acad Sci 1013:92–109

Rekeczky CS, Szatmári I, Bálya D, Timár G, Zarándy Á (2004) Cellular multiadaptive analogic architecture: a computational framework for UAV applications. IEEE Transact Circuits Syst I 51:864–884

Rodriguez-Vázquez A, Linan Cembrano G, et al (2004) ACE 16 k: The third generation of mixed signal SIMD CNN ACE chips toward VsoCs. IEEE Transact Circuits Syst I 51:851–863

Roska B, Werblin FS (2001) Vertical interactions across ten parallel, stacked representations in the mammalian retina. Nature 410:583–587 (see also in Scientific American, April, 2007)

Roska T (2003) Computational and computer complexity of analogic cellular wave computers. J Circuits Syst Comput 5(2):539–562

Roska T (2005) Cellular wave computers for brain-like spatial-temporal sensory computing. IEEE Circuits Syst Magazine 19(2): 5–19

Roska T (2007a) Cellular wave computers for nano-tera-scale technology – beyond boolean, spatial-temporal logic in million processor devices. Electron Lett 43:427–429 (Insight Letter)

Roska T (2007b) Circuits, computers, and beyond boolean logic. Int J Circuit Theor Appl 35: 427–429

Roska T, Chua LO (1993) The CNN Universal Machine – an analogic array computer. IEEE Transact Circuits Syst II 40:163–173

Szatmári I (2006) Object comparison using PDE-based wave metric on cellular neural networks. ibid, vol. 34, pp. 359–382, 2006.

Tetzlaff R, Niederhöfer Ch, Fischer Ph (2006) Automated detection of a preseizure state: non-linear EEG analysis in epilepsy by cellular nonlinear networks and volterra systems. Int J Circuit Theor Appl 34: 89–108

Turing A (1952) The chemical basis of morphogenesis. Phil Trans R Soc Lond 237B:37–72

Von Neumann J (1987) Papers of John von Neumann on computing and computer theory. In Aspray W, Burks A (eds) Section IV: Theory of natural and artificial automata. The MIT Press and Tomash Publications, Los Angeles/San Francisco

Zarándy Á, Dominguez-Castro R, Espejo S (2002) Ultra-high frame rate focal plane image sensor and processor. IEEE Sensors J 2:559–565

Zarándy Á, Rekeczky CS (2005) Bi-i: A standalone ultra high speed cellular vision system. IEEE Circuits Syst Magazine 5(2):36–45

Chapter 3
Nanoantenna Infrared Detectors

Jeffrey Bean, Badri Tiwari, Gergo Szakmány, Gary H. Bernstein, P. Fay, and Wolfgang Porod

Abstract This project focuses on devices that can be used for detection of thermal or long-wave infrared radiation, which is a frequency range for which developing detectors is of special interest. Objects near 300 K, such as humans and animals, emit radiation most strongly in this range, and absorption is relatively low in the LWIR atmospheric window between 8 and 14 μm. These facts provide motivation to develop detectors for use in this frequency range that could be used for target detection, tracking, and navigation in autonomous vehicles. The devices discussed in this chapter, referred to as dipole antenna-coupled metal-oxide-metal diodes (ACMOMDs), feature a half-wavelength antenna that couples electromagnetic radiation to a metal-oxide-metal (MOM) diode, which acts as a nonlinear junction to rectify the signal. These detectors are patterned using electron beam lithography and fabricated with shadow evaporation metal deposition. Along with offering CMOS compatible fabrication, these detectors provide high-speed and frequency-selective detection without biasing, a small pixel footprint, and full functionality at room temperature without cooling. The detection characteristics can be tailored to provide for multi-spectral imaging in specific applications by modifying device geometries. This chapter gives a brief introduction to currently available infrared detectors, thereby providing a motivation for why ACMOMDs were chosen for this project. An overview of the metal-oxide metal diode is provided, detailing principles of operation and detection. The fabrication of ACMOMDs is described in detail, from bonding pad through device processes. Direct-current current–voltage characteristics of symmetrical and asymmetrical antenna diodes are presented. An experimental infrared test bench used for determining the detection characteristics of these detectors is detailed, along with the figures of merit which have been measured and calculated. The measured performance of fabricated ACMOMDs is presented, including responsivity, noise performance, signal-to-noise ratio, noise-equivalent power, and normalized detectivity. The response as a function of infrared input power, polarization dependence, and antenna-length dependence of these devices is also presented.

J. Bean (✉), B. Tiwari, G. Szakmány, G.H. Bernstein, P. Fay, and W. Porod
Department of Electrical Engineering, University of Notre Dame, Notre Dame, IN 46556
e-mail: Jeffrey.A.Bean.5@nd.edu

C. Baatar et al. (eds.), *Cellular Nanoscale Sensory Wave Computing*,
DOI 10.1007/978-1-4419-1011-0_3,

3.1 Introduction

The purpose of this research project is to develop prototype CMOS compatible devices capable of high-speed detection in the thermal, or long-wave infrared (LWIR), band between 8 and 14 μm at room temperature without cooling. Developing detectors capable of functioning in the LWIR is of special interest for two reasons: the peak radiation of an object with a temperature around 300 K, such as a human or animal, is centered in this range and atmospheric absorption is relatively low between 8 and 14 μm (Lord 1992).

Figure 3.1 shows a comparison of two images, one taken in the visible band and one taken in the thermal infrared. In the visible image, a man can be seen wearing a shirt with his hand inside a black plastic bag. His facial features and the surroundings in the room are visible. However, in the thermal infrared band, the man's shirt and the plastic bag are transparent, allowing the thermal radiation from his body to be imaged.

While some details apparent in the visible image are lost, it is clear that the thermal IR image contains other valuable information not available in the visible image. This type of imaging is useful for detecting humans, animals, or any heat source (e.g., engines, machinery, etc.) in a scene where recognition may be difficult in the visible band. Employing multispectral imaging would clearly provide for powerful and robust information gathering.

The transmission of the earth's atmosphere for wavelengths between 200 nm and 28 μm (Lord 1992) is shown in Fig. 3.2. The absorbing gas species are noted on the plot above valleys in transmission. There is a very large "window" between 8 and 14 μm, where there is little absorption, noted by the shaded area. The fact that peak thermal radiation of objects near 300 K is centered in the LWIR and also that this radiation corresponds to a band where there is little absorption not only make the LWIR an interesting frequency range but also one that can be efficiently utilized. Possible applications for this type of detector include target detection and tracking, navigation in autonomous vehicles, and on-chip radio frequency (RF) interconnects.

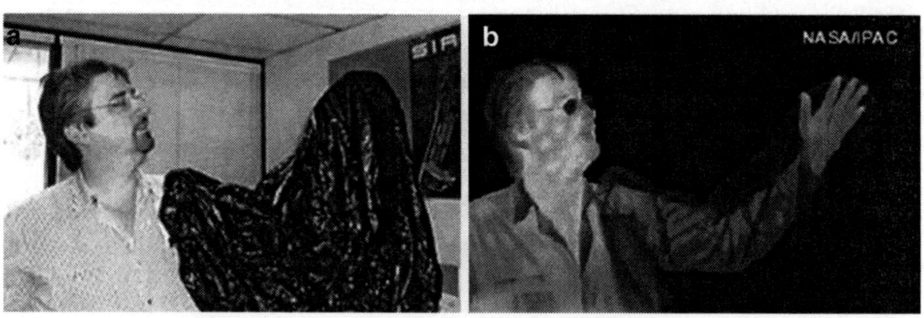

Fig. 3.1 Comparison of a human imaged in the (**a**) visible band and (**b**) LWIR (Courtesy NASA/IPAC). In (**a**), facial features, shirt, plastic bag, and the surroundings are visible, whereas in (**b**), the man's thermal signature, including his arm, is visible

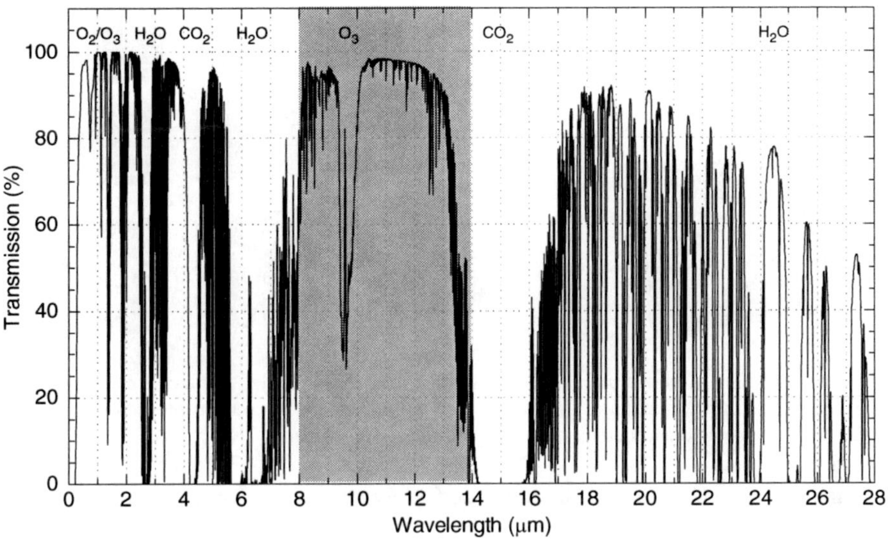

Fig. 3.2 Transmission of the earth's atmosphere for wavelengths between 200 nm and 28 μm. The species of absorbing gases at various wavelengths is noted. The wavelength range of interest for this work is between 8 and 14 μm, where there is little absorption, denoted by the shaded area

The ultimate goal of this research is to develop, fabricate, and characterize detectors that could be integrated with prefabricated CMOS imaging or Cellular Nonlinear/Neural Network (CNN) chips. In this chapter, the guidelines for this project will be outlined along with a short introduction to the CNN paradigm. In addition, infrared detectors from technologies that are currently available will be discussed, including the type that will be utilized for this project and the motivation behind this selection.

3.1.1 Project Overview

The primary goal of this project involves the development of high-speed infrared detection devices capable of functioning in the LWIR at room temperature without cooling. These devices could then be integrated with CMOS imaging chips. One such example of an imaging chip would be the CNN variety, which are inherently parallel computing devices due to their architecture and offer high-speed image processing (Chua and Yang 1988; Chua and Yang 1988; Chua and Roska 2002). A CNN array consists of $M \times N$ identical cells that each individually contains processing and sensing elements (Chua and Yang 1988). Each cell is a multiple-input, single-output processor, meaning that multiple sensors could be connected to each processor. Each cell is connected to neighboring cells (Chua and Yang 1988), which provides an interface between cells so the images may be captured and processed

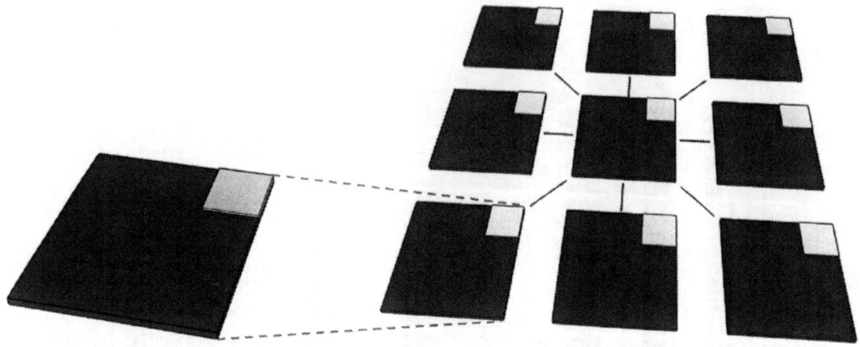

Fig. 3.3 CNN Cell Architecture and Neighboring Connections. In the single cell, the dark region represents the computational area and the light region represents the sensor integration area. This single cell is replicated and connected to neighboring cells to form a CNN array

in various ways. Figure 3.3 illustrates the CNN architecture for a single cell and the connection scheme of neighboring cells utilized in the CNN chip to be used for the project.

The dark area of each cell denotes where processing, memory, and control elements, whereas the light areas indicate locations for integrated sensors and detectors.

Because of the parallel processing architecture of the CNN paradigm, these chips are known for their high-speed image processing capabilities (Nossek et al. 1992). Commonly implemented detection arrays, such as charge-coupled devices, are composed of an array of sensors whose outputs are read and processed serially by a single computing element. However, with a CNN chip, each sensor is integrated into a cell containing its own processing architecture, so all pixels are read and processed in parallel (Nossek et al. 1992). This parallel processing allows image processing capability of 10,000 frames per second or more.

The requirement to integrate these detectors leads to certain device constraints, related to both fabrication and operation, which must all be met for a successful integration. In order to fully utilize the image processing capability of a CNN chip, sensing devices capable of detection at 10,000 frames per second or greater must be exploited in the design. Since the detectors developed in this project will be integrated onto a prefabricated CNN chip, the processes used to fabricate them must be compatible with standard complementary metal-oxide-semiconductor (CMOS) fabrication procedures. In addition, the chip area available for these detectors within each cell dictates that the detectors fit within a $10\,\mu m \times 10\,\mu m$ pixel area. Finally, the detectors must offer full functionality at room temperature without cooling.

3.1.2 Infrared Detectors

There are numerous types of devices available that can be employed to detect infrared radiation. These devices can be divided into three broad categories: thermal detectors, photon (quantum) detectors, and radiation-field detectors (Rogalski

2000). Each type is capable of detecting incident infrared radiation and converting it into some measurable signal. However, depending on the way in which the detector functions, each type has characteristics that suit it for use in specific applications.

The three infrared detector types are grouped depending on the physical mechanisms that give rise to their operation. When subjected to infrared radiation, the response of a thermal detector is based on its material properties, which are dependent on temperature. Photon detectors respond to infrared radiation by creating free carriers from the interactions between incident photons and electrons bound within the sensing material. Radiation-field detectors feature an antenna element that detects incident electromagnetic waves at a designed frequency.

3.1.2.1 Thermal Infrared Detectors

Thermopiles, bolometers, microcantilevers, ferroelectric, and pyroelectric detectors are types of thermal detectors, meaning that some material property changes in response to a temperature change, in this case thermal infrared radiation. Thermal devices are generally operational over a wide range of wavelengths and can offer uncooled functionality. However, these detectors have low detectivity relative to photon detectors. The sensitivity of thermal detectors can be increased by thermally insulating them from their surroundings. However, the trade-off for this increased sensitivity is an increased response time.

A thermopile is a series combination of multiple thermocouples. A thermocouple is composed of a junction of two dissimilar thermoelectric materials, commonly metals or semiconductors (Yamashita 2004). A temperature difference present between the dissimilar materials produces a voltage potential, known as the Seebeck Effect (Bramley and Clark 2003; Yamashita 2004). For a thermocouple used as a detector, one side of the junction is generally connected to a heat sink or cooling source. The other side of the junction, the "sensing" side, is subjected to the incident radiation. The comprising materials of a thermocouple determine the voltage derived from a temperature difference between the two sides of the junction. The output of a thermopile detector is proportional to the incident radiation energy and can simply be monitored by reading the potential across the junction. Responsivity of a thermocouple can be increased by connecting more thermocouples in series and/or by thermally insulating the junction pairs from their surroundings (Lahiji and Wise 1982). However, there is a trade-off between sensitivity and response time; the more sensitive the device, the slower it will respond to incident radiation.

Bolometers and microbolometers are detectors that utilize materials whose resistance varies as a function of temperature. The material chosen for the active element determines the magnitude and sign of the resistance change in response to a temperature change. When the detector is subjected to infrared radiation, the detector's temperature changes and, consequently, so does the resistance of the active element (Allen et al. 1969; Codreanu 2003). Detection of incident infrared radiation can be determined by using a constant voltage supply and monitoring current through the bolometer, or by using a constant current supply and monitoring the voltage developed across the bolometer's sensing element. The sensitivity of a bolometer can be increased by thermally insulating the device from detector substrate. Sensitivity is also controlled by the material chosen for the resistive element in the detector (Summers and Zwerdling 1974). Metals have low-temperature coefficients of resistivity, but exhibit low noise figures (Block and Gaddy 1973). On the contrary, semiconductors have a much higher temperature coefficient of resistivity, but have higher associated device noise (Noda et al. 2002). The main drawback to this type of device is the trade-off between response time and detector sensitivity.

Bolometers can also be coupled with an antenna to provide added responsivity and frequency selectivity (Schwarz and Ulrich 1977). These detectors operate by utilizing a planar antenna, commonly the bow tie variety, to couple electromagnetic radiation to the bolometer. The induced antenna current heats the bolometer and causes a change in the resistance of the detector element, just as in the case of the conventional bolometer.

Microcantilever detectors are microelectromechanical systems devices that feature a cantilever structure composed of layers of two different materials of dissimilar thermal expansion coefficients. As the temperature of the detector changes due to incident infrared radiation, the lengths of the layers within the structure change by different amounts, causing a deflection or bending of the cantilever (Corbeil et al. 2002). This deflection due to the resulting stress is known as the bimaterial effect (Datskos 2004). The deflection can be measured by numerous techniques including optical, capacitive, piezoresistive, and electron tunneling, each with extremely high precision. One drawback to this type of detector is that physical vibrations of the detector also cause cantilever deflections and sensor excitation unrelated to incident radiation. Therefore, this type of device cannot be used in remote or portable sensing applications where vibration isolation is not possible, such as an autonomous vehicle.

Ferroelectric and pyroelectric detectors comprise a category of detectors that contain an element composed of a material that changes polarization when subjected to temperature changes (Beerman 1969; Glass 1969). Pyroelectric detectors are composed of a material that generates an electric potential or surface charge when exposed to infrared radiation. When the intensity of irradiation changes, so does the surface charge. Ferroelectric detectors function in a similar manner: when subjected to infrared radiation, the active material exhibits a spontaneous electrical polarization. This polarization is dependent on the intensity of the infrared radiation. Because of the sensing nature of these detectors, they must operate in a chopped system to facilitate spontaneous polarization changes (Lang et al. 1969). A chopped system employs a mechanical wheel that spins similar to a fan blade. The chopper is placed between the illumination source and the detector, and alternatively either blocks the irradiation or allows it through to the detector. When radiation is incident on the detector, the periodic modulation due to the chopper creates an alternating signal that can be monitored with external circuitry.

3.1.2.2 Quantum Infrared Detectors

Quantum, or photon, long-wavelength infrared detectors consist of photovoltaic (PV), photoconductive (PC), and quantum well detectors; each of these technologies exploits semiconductors for sensing infrared radiation. When subjected to infrared radiation, photons interact with electrons within the semiconductor to create mobile charge carriers. Responsivity of each type of detector is wavelength dependent, and is determined by the energy band structure of the detector. The energy band gap can be varied in a ternary alloy by adjusting the compositions of the comprising elements that gives rise to the ability to tune the wavelength of desired peak responsivity within the range of the binary materials, and the use of quantum wells with well-defined intersubband transitions allows additional degrees of freedom for detection of long-wavelength infrared radiation. Although quantum detectors have fast response time, they generally must be cooled to cryogenic temperatures to minimize background noise, or dark currents, when detecting 3 μm or longer wavelengths (Rogalski 2000). In the context of this project, cryogenic temperatures cannot be supported since the detectors will be integrated with a CNN chip. In addition, cryogenic cooling imposes severe constraints on functionality in remote or portable applications, which would likely be the conditions for use in autonomous vehicles for the purposes described above.

PV are semiconductor-based devices, composed of a nonlinear junction, where photoinduced currents are created when subjected to infrared radiation (Cohen-Solal and Riant 1971; Long 1977). This occurs when incident photons create an electron–hole pair either near to or within a potential barrier (Long 1977; Tidrow et al. 1999). Two barrier types commonly chosen are reverse-biased p-n junctions or Schottky barrier types. The built-in field created by the potential barrier separates the photo-generated electron–hole pair to create the photoinduced current. An intrinsic PV detector must have incident photon energies of at least the band gap of the semiconductor, or the Schottky barrier height, respective of the junction. For high-speed operation, a bias is applied to a PV detector and the photocurrent is measured. The photocurrent of a PV detector is proportional to the absorption rate of incident photons, not by the incident photon energy, given that the incident photon energy is greater than the potential barrier height.

PC detectors are similar to PV detectors and function by the photo-generation of charge carriers in the semiconductor due to incident electromagnetic radiation. When incident on the structure, electromagnetic radiation is absorbed and the conductivity of the detecting material changes (Long 1977). This change in conductivity, or resistivity, can be monitored similar to that in the case of the thermal bolometric detector.

Quantum well infrared photodetectors (QWIPs) are composed of superlattice structures, typically grown by molecular beam epitaxy or metal organic chemical vapor deposition (Tidrow et al. 1999; Fastenau et al. 2001; Matsukura et al. 2001). Alternating layers of doped or undoped compound semiconductors create quantum wells in which infrared radiation is absorbed (Richards et al. 2003). When incident photons are absorbed, intersubband transitions within the valence or conduction band take place and the excited carriers induce a current. QWIPs are generally cryogenically cooled since thermionic emission from one quantum well to the next produces large dark currents. However, room-temperature operation is possible with the sacrifice of response time and sensitivity (Richards et al. 2003).

3.1.2.3 Radiation-Field Infrared Detectors

The least-developed and smallest class of infrared detectors studied to date and the subject of this work are those of the radiation-field variety, which directly detect a radiation field similar to radio or television receivers (Capper and Elliott 2000). These devices feature an element that couples an incident electromagnetic wave at a specific frequency to sensing circuitry. Responsivity of these types of devices is generally frequency dependent, with the characteristics dependent on the element that couples radiation to the sensing element. Depending on the frequency of the detected wave, a nonlinear junction, such as a diode, may be used as the sensing element to provide rectification of the AC signal.

One type of a rectifying sensor that can be used to detect electromagnetic radiation is an antenna-coupled diode (Esfandiari et al. 2005). Antennas are commonly used to collect radio and television signals, but can be tailored to detect infrared radiation by scaling the antenna dimensions. Radiation from an electromagnetic wave is coupled to a nonlinear rectifying junction by the antenna. Various antenna types have been coupled to diodes, including dipole antennas (Fumeaux et al. 2000), bowtie antennas (Chong and Ahmed 1997), log-periodic antennas (Chong and Ahmed 1997), spiral antennas (Boreman et al. 1998), microstrip patch (Codreanu et al. 1999), and microstrip dipole (Codreanu and Boreman 2001; Codreanu et al. 2003) antennas. Various diodes are available such as semiconductor p-n, Schottky, and MOM varieties. Which of these diode types is most appropriate for a given detection application depends on the desired operating characteristics. These diodes provide for rectification of the coupled signal. Semiconductor-based diodes are generally suitable for rectifying signals of frequency up to approximately 1 THz, whereas MOM types must be used for signals with frequencies of greater than 1 THz.

Antenna-coupled diodes are frequency selective, have a small pixel "footprint," and operate with full functionality without cooling. Depending on the type of diode chosen, antenna-coupled diodes can also have fast response times. Therefore, based on these characteristics, these detectors are an excellent candidate for infrared radiation detection.

3.1.3 Detector Characterization

There are four main figures of merit that are used to characterize infrared detectors: responsivity, signal-to-noise ratio (S/N or SNR), noise-equivalent power (NEP), and normalized detectivity. These characteristics will be used to compare the detectors fabricated in this research to infrared detectors currently available on the market. This section will describe the definitions of the figures of merit, an explanation of device noise used in calculating the figures of merit, and a comparison of detectors that are currently available. The types of noise that may impact the performance of ACMOMDS will must be determined so that figures of merit can accurately calculated.

3.1.3.1 Figures of Merit

Responsivity relates the output of the infrared detector, as a current or voltage, to the intensity of the incident radiation. Responsivity can be defined as either spectral responsivity or blackbody responsivity, depending on the type of illumination. Spectral responsivity is defined as the detector output per watt of monochromatic radiation (Dereniak and Boreman 1996). Blackbody responsivity is defined as the detector output per watt of broadband incident radiation (Dereniak and Boreman 1996). For blackbody responsivity, the radiant power on the detector contains all wavelengths of the radiation, independent of the spectral response characteristics of the detector (Dereniak and Boreman 1996). Responsivity, both monochromatic and broadband, can be defined as:

$$\Re_{\mathrm{V}} = \frac{V_{\mathrm{S}}}{E_{\mathrm{e}} A_{\mathrm{d}}} = \frac{V_{\mathrm{S}}}{\varphi_{\mathrm{e}}} \text{ or } \Re_{\mathrm{I}} = \frac{I_{\mathrm{S}}}{E_{\mathrm{e}} A_{\mathrm{d}}} = \frac{I_{\mathrm{S}}}{\varphi_{\mathrm{e}}}$$

where R_{V} and R_{I} are the voltage and current responsivities, respectively, V_{s} and I_{s} are the signal voltage and current, respectively, E_{e} is the irradiance in watts/cm^2, A_{d} is the detector area in cm^2, and φ_{e} is the radiant flux in watts. Although responsivity is a figure of merit that can be used to describe the sensitivity of a detector in terms of its output for a given input, it is often difficult to compare detectors using this figure of merit alone. For example, depending on measurement conditions and device technologies, larger detectors may have higher responsivities than those of a smaller variety. In addition, responsivity is a function of frequency, or wavelength, making it more difficult to compare detectors with this measure.

The minimum level of radiant power that a detector can detect is dependent on the noise level. The signal output must be above the noise level to be easily detected (Dereniak and Boreman 1996). SNR relates the output of a detector to the internal detector noise. This is expressed as:

$$\frac{S}{N} = \frac{V_{\mathrm{S}}}{V_{\mathrm{N}}} = \frac{I_{\mathrm{S}}}{I_{\mathrm{N}}}$$

where S/N is a dimensionless ratio and V_{N} and I_{N} are the noise voltage and current, respectively. Like responsivity, it is difficult to compare detectors on SNR alone. In many cases, the SNR of a detector can be increased by increasing the infrared input power. As such, reported SNR values should always be accompanied by the irradiance.

NEP relates the radiant flux incident on a detector to its SNR (Dereniak and Boreman 1996). NEP is defined as the radiant power incident on a detector, not the absorbed power, which yields a SNR = 1. This can be expressed as:

$$\mathrm{NEP} = \frac{\phi_{\mathrm{e}}}{V_{\mathrm{S}}/V_{\mathrm{N}}} = \frac{V_{\mathrm{N}}}{\Re_{\mathrm{V}}} = \frac{I_{\mathrm{N}}}{\Re_{\mathrm{I}}}$$

where NEP is in watts. Just as in the case of responsivity, NEP can be either blackbody or spectral, depending on if the incident radiation is monochromatic or broadband. Unlike responsivity, however, a small NEP is desired. (Dereniak and Boreman 1996). Like responsivity, it is difficult to compare detectors based solely on NEP, since it is typically dependent on the square root of the detector area and the square root of the electrical bandwidth of the measurement. Other factors, such as chopping frequency, biasing conditions, and operating temperature can affect NEP. However, general comparisons of detector NEPs can be made if the value is accompanied by detector area and the bandwidth of the measurement (Dereniak and Boreman 1996).

Since it is difficult to compare detectors based on responsivity, SNR, and NEP, Jones defined a new term called normalized detectivity in 1953 (Dereniak and Boreman 1996). Normalized detectivity, or D^*, normalizes NEP to a 1 cm^2 detector are and 1 Hz bandwidth. D^* is expressed as:

$$D^* = \frac{\sqrt{A_{\mathrm{d}}\Delta_{\mathrm{f}}}}{\mathrm{NEP}}$$

where D^* is in $\mathrm{cm\,Hz^{1/2}watt^{-1}}$, or Jones. The interpretation of D^* is that it yields SNR, given a detector with 1 W of radiant power incident on it, and are of 1 cm^2, and a noise-equivalent bandwidth of 1 Hz. Depending on whether the NEP was calculated using monochromatic or broadband radiation, the associated D^* will also observe that form. Assumptions in this calculation are that the noise has a flat, white-noise spectrum, that noise is proportional to the square root of the bandwidth and detector area, that the detector is operating at optimum biasing and operating temperature, and that the spectral response is flat (Dereniak and Boreman 1996).

3.1.3.2 Electrical Noise Considerations

Electrical noise is defined as the random current or voltage fluctuations in electrical circuits (Dereniak and Boreman 1996). Since the electrical noise of devices impacts various metrics used to compare different detector technologies, the associated noise of a device must be examined. There are various entities that can impact measurements of an electrical device and can be classified as either interferences or noise sources. Interference can be of the man-made variety, which can be generated by transformers, motors, radio signals, and other electrical equipment, or the natural variety, which from phenomena of nature, such as lightning, earthquakes, or sunspots.

However, the noise sources associated with the operation of the infrared detector impact the calculation of figures of merit and must be accurately determined.

The intrinsic noise associated with infrared detection can classified as external or internal. External sources of noise include noise in the interface or operation of the measurement electronics, such as that from the preamplifier and lock-in amplifier. The low-noise current preamplifier and lock-in amplifier that are used for the measurements are specifically designed for low-noise detection of signals, but there is still noise that impacts the measurements. A detailed analysis of the preamplifier noise will be presented in Section *Figures of Merit.*

There are several types of internal noise that may impact infrared detectors. However, the sources that may impact the performance of the antenna-coupled MOM diodes fabricated for this research that will be discussed include Johnson, shot, and 1/*f* noise.

Johnson, or Nyquist, noise is the fluctuation caused by the thermal motion of charge carriers in a resistive element (Dereniak and Boreman 1996). Even though charge neutrality is maintained for the overall device or structure, local random thermal motion of carriers can cause charge gradients. The rms Johnson noise voltage across the resistor, R, at an absolute temperature, T, can be shown to be:

$$v_{\mathrm{J}} = \sqrt{4kTR_{\mathrm{D}}\Delta f}$$

where k is Boltzmann's constant, T is the absolute temperature of the detector in Kelvin, R_{D} is the diode resistance, and Δf is the electrical bandwidth of the measurement in Hz. The Norton equivalent rms Johnson noise current can be expressed as:

$$i_{\mathrm{J}} = \sqrt{\frac{4kT\Delta f}{R_{\mathrm{D}}}}.$$

Shot noise is associated with a dc current i flowing across a potential barrier (Dereniak and Boreman 1996). This is related to thermionic emission of electrons over the barrier in a MOM diode. Since the charge carriers, in this case electrons in the metal, cross a potential barrier when the incident radiation heats the structure and increases the energy of the electrons, the current through the diode possesses this type of noise. The current fluctuations associated with shot noise are given by:

$$i_{\mathrm{S}} = \sqrt{2qi\Delta f}$$

where i_{S} is the shot current-noise, q is the charge of an electron, and i is the average device current.

1/*f* noise is a noise source that is a strong function of the frequency of operation. The current noise is inversely proportional to the square root of the frequency, expressed by the relation:

$$i_{\mathrm{f}} \propto \sqrt{\frac{i_{\mathrm{dc}}^{\sim 2}\mathrm{d}f}{f^{\sim 1}}}$$

where i_{dc} is the dc bias current, and f is the frequency of the operation of the device. Although the causes for this type of noise are not fully understood, hence the expression in terms of a proportionality, potential causes include carrier fluctuations, trap occupancy variations in the oxide barrier, and the associated variations in trapping time constants (Dereniak and Boreman 1996). Assuming that the antenna-coupled MOM diode devices have ohmic contacts to the electrical leads, which in turn have ohmic contacts to the bonding wires and the LCC and socket, $1/f$ noise is always zero for unbiased detection of a device in equilibrium. However, if a dc bias is applied to the device, $1/f$ noise will be encountered and must to be taken into account.

These three types of noise will be experimentally addressed in Section *Figures of Merit*. The noise study will include those internal to the device as well as external, such as the preamplifier and lock-in amplifier.

3.1.3.3 Detector Comparison

Table 3.1 compares infrared detectors currently available in the market along with those currently being researched (Dereniak and Boreman 1996; Fumeaux et al. 2000). The highest D^* is desired, but in most application contexts D^* is compared in light of the following parameters: detector area, operating temperature, and response time. Although some detector types may be functional in spectral ranges outside of the LWIR, each detector's D^* is reported for LWIR detection to provide comparison in the spectrum of interest.

HgCdTe, or MCT, detectors are the best thermal IR detectors available on the market today. The detectivity value is an order of magnitude greater than any other technology and the response time is very short. However, there are several factors that make them impractical for use in some applications. The largest issue is that MCT detectors require cryogenic cooling, which would be not only impractical for portable applications, but does not allow for direct integration with commercial integrated circuits without major modifications. In addition, the detector area is quite

Table 3.1 D^* comparison for various infrared detectors types as a function of device characteristics

Detector Type	Area (mm^2)	Operating Temperature (K)	Response Time (ms)	D^* (cm Hz$^{1/2}$W^{-1})
HgCdTe	~0.25	77	10^{-3}	1×10^{10}
Pyroelectric	0.78–63.6	300	10^{-4}	1×10^{9}
Thermistor	0.25–25	300	5	3×10^{8}
Thermopile	~5	208–343	5	7×10^{8}
Bolometer (Ge)	~5	0.3–2.0	0.5	3×10^{8}
Antenna-coupled bolometer (Nb)	0.0001	300	10^{-4}	6×10^{5}
Antenna-coupled MOM diode	0.0001	300	10^{-10}	1×10^{6}

large, also presenting a road-block to integration in high pixel-count imager applications. Bolometers suffer the same disadvantages; while offering a reasonable D^* with fast response times, large detector areas and the requirement of cryogenic cooling prevent bolometers from being usable for many applications.

Pyroelectric detectors offer a high D^*, comparable to MCT detectors, offer full functionality at room temperature without cooling, and have fast response times. The barrier issues lie in the fact that the detector area is large and that most pyroelectric materials are not CMOS compatible. Thermistors and thermopiles offer reasonable D^* values and full functionality at room temperature. The response times are relatively slow, however, and again detector area prevents this type of detector from being integrated into large-format imagers.

Antenna-coupled bolometers offer a small detector area, room-temperature operation, and a fast response time. D^* is relatively low, though. Higher D^* values of 2.89×10^7 and $1.08 \times 10^8\,\text{cm}\,\text{Hz}^{1/2}\,\text{W}^{-1}$ have been reported (Middlebrook et al. 2006), but these values were obtained by thermally isolating the devices on a membrane and measuring the values in air and under vacuum, respectively. The need for a fabrication process that includes membrane formation is problematic for integration with CMOS, and placing the detector in a vacuum to improve performance is not viable for many potential applications.

Antenna-coupled MOM diodes, like antenna-couple bolometers, offer small detector areas, room temperature operation, and fast response. However, previous research has reported lower than desired D^* values at $1 \times 10^6\text{cm} \cdot \text{Hz}^{1/2} \cdot \text{W}^{-1}$ (Abdel-Rahman et al. 2004). In order for a detector to be commercially viable, D^* values in the 10^8 range or higher are required. This research focuses on antenna-coupled MOM diode fabrication, their detection characteristics, and viability for commercial applications.

3.2 Antenna-Coupled MOM Diodes

For high-performance imaging applications, a high-speed, frequency-selective detector is desired that offers full functionality at room temperature without cooling, fits within approximately a $10 \times 10\,\mu\text{m}$ area (in order to supply high pixel counts in a practical imager size), and offers CMOS compatible fabrication. Of the technologies discussed in Chap. 2, the only category of infrared detector capable of meeting all of these stipulations is the radiation-field type.

The infrared radiation-field detectors fabricated in this research are composed of two parts: an antenna and a nonlinear junction. The antenna is a half-wavelength dipole formed by two quarter wavelength metal lines, separated by an oxide barrier that forms a MOM tunnel diode. This diode rectifies radiation-induced terahertz antenna currents.

In this chapter, an introduction to dipole antennas is given, along with design considerations and parameters for this project. An overview of MOM diodes is presented next, including a discussion of the various types, such as the point-contact and thin-film varieties, and energy band diagrams of these structures. Then, the device fabricated for this project, the dipole antenna-coupled MOM diode, is discussed. Principles of operation for these devices in response to incident radiation are provided, including an explanation of the functionality behind their operation.

3.2.1 Dipole Antenna

A dipole antenna is a common type of antenna with a center-fed element that can either transmit or receive electromagnetic radiation. The usable frequency range of dipole antennas is vast; NASA has studied the dynamics of the magnetosphere using a 1,647-ft-long dipole antenna with a center frequency of 300 kHz, and capable of operating between 3 kHz and 3MHz (Gallagher and Adrian 2007). This project focuses on frequencies eight orders of magnitude higher, using dipole antennas with a length of 3.1 μm for detecting 28.3 THz radiation. Although the former case utilizes a dipole antenna in space and the latter antenna functions on a silicon substrate, the operating principle behind each remains the same.

When electromagnetic radiation is incident on a thin antenna, electromagnetic waves propagate along the length of the antenna. Waves traveling in opposite directions form standing waves on the antenna. In the case of a half-wavelength dipole, a voltage node forms at the center and antinodes form at each end of the antenna; conversely, a current antinode forms at the center and nodes form at each end (Balanis 2005). A half-wavelength dipole is shown in Fig. 3.4. Below the antenna, an approximation of the voltage distribution along the antenna is shown as a function of time. In addition, the corresponding approximation of the current distribution is also shown.

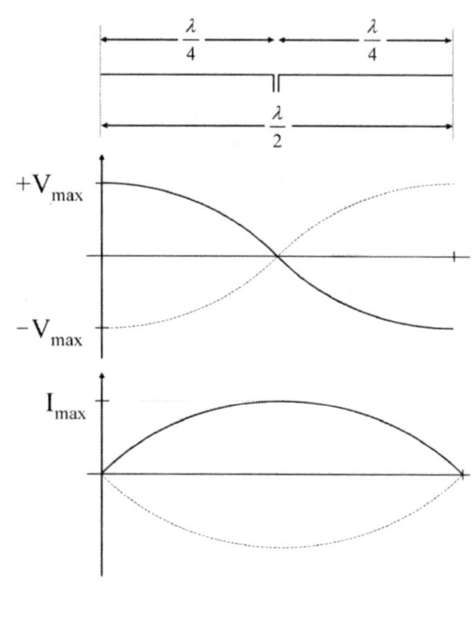

Fig. 3.4 Dipole antenna with corresponding voltage and current distributions. Each line indicating the voltage distribution along the antenna corresponds to the current distribution with the same line type, showing one half of a wave cycle in total

In this research, the dipole antenna serves as a means of coupling 28.3 THz (10.6 μm wavelength) infrared radiation to a rectifying junction. A rectifying junction, in this case a MOM diode, is used instead of the more conventional transistor-based received architectures common in radio-frequency applications is that no conventional circuit technologies currently operate fast enough. However, a MOM diode is capable of rectifying the terahertz antenna currents so that a DC current may be measured across the junction.

The dipole full-length L is 3.1 μm, which corresponds to the equivalent substrate half-wavelength in silicon dioxide of the incident radiation. Since the detector is positioned at the interface of two different materials, the effective dielectric constant ε_{eff} (Hasnain et al. 1983) is approximately:

$$\varepsilon_{\text{eff}} = \frac{\varepsilon_{\text{SiO}_2} + \varepsilon_{\text{air}}}{2}$$

The dielectric constant of SiO_2 $\varepsilon_{\text{SiO}_2}$ at 28.3 THz is 4.84 (Gonzalez and Boreman 2005), which gives an effective dielectric constant of 2.92. The effective wavelength of an electromagnetic wave in a substrate λ_{eff} (Balanis 2005) is given by:

$$\lambda_{\text{eff}} = \frac{\lambda_o}{\sqrt{\varepsilon_{\text{eff}}}}$$

where λ_o is the wavelength of interest in air. For an electromagnetic wave with wavelength 10.6 μm propagating in SiO_2, λ_{eff} equals 6.2 μm. In addition, Rakos simulated the design to calculate the radiation resistance and reactance of the antenna (Rakos 2006). For an incident wavelength of 10.6 μm, the reactance of the antenna goes to zero at a dipole length of 3.1 μm, which denotes the first resonance of the antenna. The radiation resistance of the 3.1 μm antenna at 28.3 THz is approximately 80 Ω from simulations (Rakos 2006; Sun 2006).

The width of the dipole antenna must also be considered in the design process. Transverse currents, those perpendicular to the length of the dipole, cannot be neglected if the width of the antenna is too large compared with the free-space wavelength of radiation λ_o (Rutledge et al. 1978). Simulation results at 10 and 46 GHz show that the antenna width must be less than $\lambda_o/35$ can suppress transverse currents so that longitudinal currents, those along the antenna axis, can be dominant as desired (Rutledge et al. 1978). Therefore, for radiation at 10.6 μm, the dipole antenna width must be less than 303 nm. The width of the dipole antennas fabricated in this research is 50 nm, and therefore longitudinal currents are able to propagate along the antenna length.

3.2.2 *MOM Diodes*

A MOM diode diode, also commonly referred to as metal-insulator-metal and metal-barrier-metal diodes, is schematically illustrated in Fig. 3.5. It consists of a thin

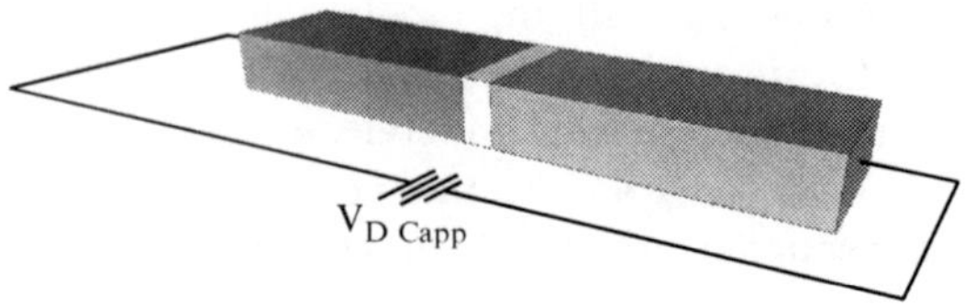

Fig. 3.5 Simplified illustration of a MOM diode, which is a thin insulator of tunneling thickness sandwiched by two parallel metal electrodes

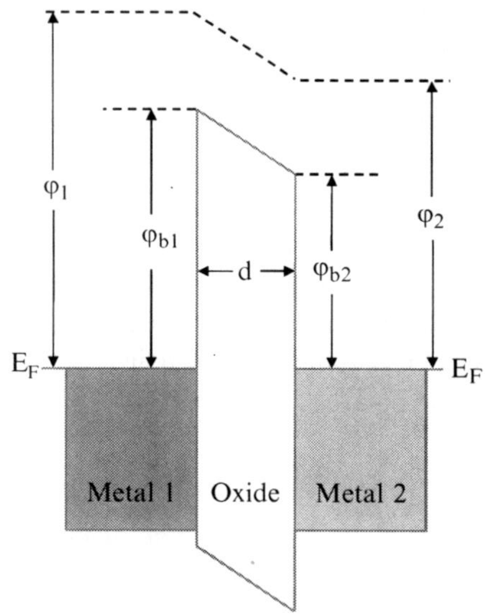

Fig. 3.6 Energy band diagram of an unbiased MOM diode. In an unbiased device, the Fermi levels align to reach equilibrium. Any difference in work function of the metals causes bending in the energy bands within the barrier

($\sim$20 Å) insulator, denoted by the white area, sandwiched between two parallel conductors. There have been numerous types of MOM diodes studied over the past half century. These consist of both point-contact and thin-film MOM diodes.

An energy band diagram of a MOM diode under no applied bias is shown in Fig. 3.6. The structure in this figure features dissimilar metals of work functions φ_1 and φ_2, interface barrier heights φ_{b1} and φ_{b2}, and an oxide barrier of thickness d. The work function of a metal is defined as the minimum energy required to move an electron from the Fermi level into vacuum (Michaelson 1977). The barrier height of a metal–insulator interface is defined as the potential difference between the Fermi level of the metal and the band edge of the insulator. Although the figure shows a diode formed with dissimilar metal electrodes, referred to as an asymmetric diode, the same metal can be used for each electrode to form a symmetric diode. In an unbiased structure, such as Fig. 3.6, the Fermi levels of the metals in the unbiased structure align to reach equilibrium and the energy bands bend within the oxide layer, resulting in a built-in field. This built-in field is equal to the difference in the work functions of the metals divided by the thickness of the barrier.

In theory, there is no field present in the oxide for symmetrical MOM diodes, that is with the same metal used in each electrode, since the work functions are the same. However, in practice, even MOM diodes fabricated with the same metals can

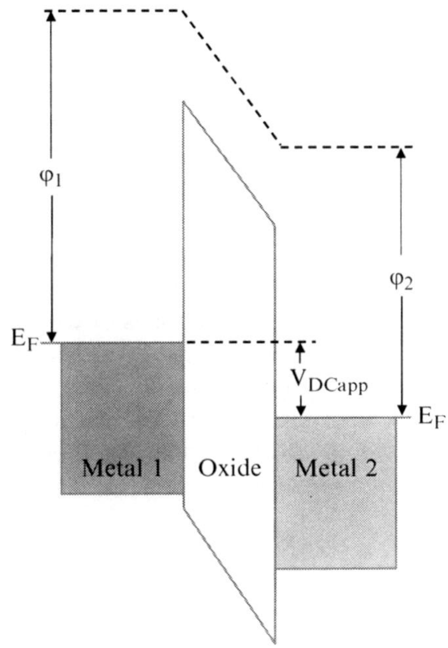

Fig. 3.7 Energy band diagram of MOM diode under small applied bias. The difference between the two metal's Fermi levels is equal to the applied bias

exhibit behavior that gives rise to an asymmetrical barrier. Any impurities or charges trapped at the metal-oxide interface or within the oxide can cause asymmetries in the potential barrier shape (Simmons 1963). However, for the purposes of illustration and explanation, a simplified trapezoidal barrier will be assumed throughout this analysis.

The band diagram of a MOM diode when a small DC voltage bias VDCapp is applied is shown in Fig. 3.7.

The positively biased electrode moves down in energy, causing a difference in the Fermi levels equal to the applied bias VDC_{app}. The electric field across the oxide layer of the structure changes by an amount equal to the applied bias divided by the thickness of the layer and adds to any field already present in the barrier.

For an MOM diode with an oxide barrier thickness on the order of 20 Å, even small biases of approximately one volt can cause extremely high fields (5 MV/cm), and exceed the breakdown field of the oxide. Depending on the growth or deposition method, this can be vary from 1.4 MV/cm for deposited films (Vanbesien et al. 2006) to over 500 MV/cm for ultrathin (5 Å) native oxide barriers grown in oxygen (Gloos et al. 2003). Gloos et al. (2003) found that the breakdown field increases for decreasing barrier thickness. If the breakdown field of the insulator in the MOM diode is exceeded, the device may be destroyed.

When a small bias, much less than the barrier height φ_b, is applied between the two conductors, electrons can traverse a sufficiently thin barrier by quantum mechanical tunneling (Small et al. 1974; Heiblum et al. 1978; Sanchez et al. 1978). For a finite potential barrier, as in the case for MOM diodes, the wave function of an electron on one side of the barrier penetrates the barrier and yields a finite probability of the electron being on the opposite side of the barrier (Heiblum et al. 1978; Codreanu et al. 2003). The probability for electrons to tunnel through a potential barrier decreases exponentially with increasing barrier thickness (Fisher and Giaever 1961). For tunneling to take place in an MOM structure, the oxide thickness must less than approximately 50 Å.

For the case of a MOM diode with dissimilar metal electrodes, as shown in Fig. 3.7, electrons on the left side of the barrier have a higher probability of tunneling through the potential barrier to the right side than in the reverse direction, because electrons on the left have available states of which to tunnel, whereas those on the right have no available states of which to tunnel on the left. The nonlinear current voltage (I-V) characteristic of a MOM diode arises from this quantum mechanical process. The nonlinearity depends on the work functions of each metal as well as the oxide type and thickness (Mead 1962; Simmons 1963; Nelson and Anderson 1966; Kwok et al. 1971). Further detail on the antenna-coupled MOM diodes fabricated for this project can be found in Sect. 3.3.

3.2.2.1 MOM Diode Design

There are many varieties of MOM diodes that can be utilized as nonlinear junctions. However, the desired diode characteristics must first be analyzed with respect to the desired frequency of operation. Since MOM diodes are tunneling devices, if the tunneling transit time is slower than one wave cycle of the 28.3 THz incoming wave, then the device will fail to respond and will not rectify the incoming signal efficiently. The overlap area of the MOM diode must be small enough so that the RC time constant is less than one infrared wave cycle (Fumeaux et al. 2000).

An equivalent circuit of an antenna-coupled MOM diode under incident infrared radiation can be modeled as an antenna and diode connected in series (Sanchez et al. 1978; Yngvesson 1991), as shown in Fig. 3.8.

The MOM diode can be described by a junction capacitance C_D in parallel with a nonlinear voltage-dependent resistance R_D (V). This parallel combination is in series with the resistance r, which represents metal lead and/or spreading resistance (Yngvesson 1991).

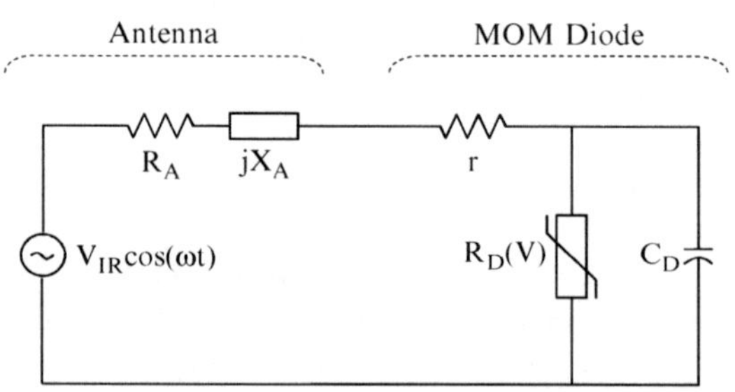

Fig. 3.8 Equivalent-circuit model of an antenna-coupled MOM diode. As a receiver, the antenna is represented by a voltage source with series impedance. The diode is represented by a parallel combination of a capacitor and voltage controlled current source in series with the lead impedance

An antenna functioning as a receiver can be represented by an alternating current source $V_{IR}\cos(\omega t)$ with induced voltage V_{IR} at angular frequency ω, operating at the frequency of the incident electromagnetic radiation. This is connected in series with impedance $R_A + jX_A$ where R_A is the real impedance of the source. For this circuit, the RC time constant is the product of the diode capacitance and the equivalent resistance, which is R_D in parallel with the series combination of R_A and r. This leads to a cutoff frequency f_c of:

$$f_c = \frac{R_A + r + R_D(V)}{2\pi (R_A + r) R_D(V) C_D}.$$

While rectification and mixing are still observed above this frequency, it is with diminished efficiency. Although there is some disagreement on this issue, the signal amplitude has been theoretically calculated to decrease as ω^{-1} (Bradley et al. 1972; Small et al. 1974), $\omega^{-3/2}$ (Sanchez et al. 1978), and ω^{-2} (Green 1971).

To minimize the response time of the diode and attain a high cutoff frequency, the diode capacitance must be small. If the capacitor is considered a small parallel plate capacitor, the diode capacitance C_D is:

$$C_D = \varepsilon_{ox}\,\varepsilon_o \frac{A}{D}$$

where ε_{ox} is the relative permittivity of the oxide in the MOM diode, ε_o is the permittivity of free space, A is the junction area, and D is the thickness of the dielectric. The dielectric constant of Al_2O_3 at 28.3 THz is approximately 1 (Momida et al. 2007). For a diode with a 25 Å barrier composed of Al_2O_3, a 50 × 50 nm overlap area, and an antenna resistance of 1 kΩ, the cutoff frequency is 180 THz. The exact composition of the oxide barrier is not known, so it will be written as AlO_x, where x is not necessarily an integer. The cutoff frequency will change depending on the dielectric constant of the AlO_x.

3.2.2.2 Point Contact MOM Diodes

The first MOM diode structures fabricated in the laboratory were of the point-contact, or cat-whisker, variety and were introduced in the early 1960s (Green 1971; Yasuoka et al. 1979). These diodes consisted of a thin, sharp tipped tungsten wire with a thin (~35 Å) oxide covering that mechanically contacted a polished metal, commonly nickel, post. An illustration of a point-contact MOM diode can be seen in Fig. 3.9, where the dark center denotes the tungsten wire and metal post, whereas the light area surrounding the wire indicates the native oxide coating on the tungsten wire.

When a small DC bias was applied between the thin wire and the post, electrons tunneled from one metal electrode to the other through the insulating oxide barrier (Green 1971). The contact area was controlled by both the shape of the whisker and

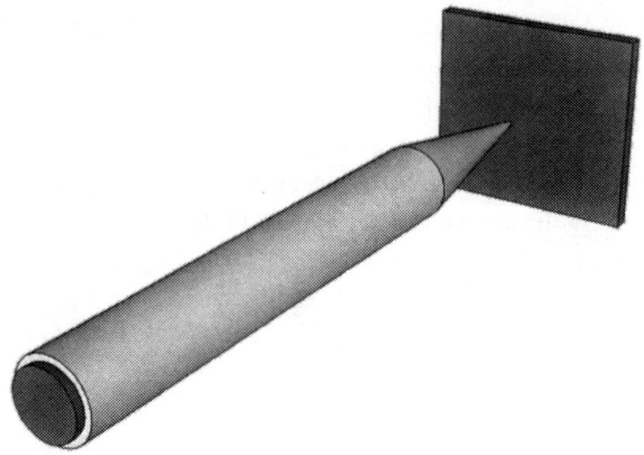

Fig. 3.9 Point-contact MOM diode. The whisker is composed of tungsten and has a native oxide coating. The MOM diode is formed when this is brought into contact with the metal base. The contact pressure controls diode properties

its contact pressure on the polished post (Yasuoka 1979). This contact area also has an effect on diode resistance, so this parameter can be modulated if desired with a point-contact MOM diode. A nonlinear current–voltage characteristic arises from electrons tunneling through this oxide barrier.

The metallic wire also functioned as a receiving antenna when subjected to incident optical or infrared radiation. It has been shown that these structures are capable of coupling incident radiation into an optically induced voltage across the diode (Heiblum et al. 1978; Yngvesson 1991). The drawback to this type of device lies in its size and mechanical instability. Physical vibrations can cause the contact pressure and area of the wire to change. In extreme cases, the wire may lose contact with the post and produce instable current–voltage and detection characteristics. While very small contact areas can be achieved and the whisker can act as an in-air receiving antenna, this type of MOM diode and antenna structure would not be suitable for many imaging applications due to its mechanical instabilities and non-CMOS compatible processing.

3.2.2.3 Thin-Film MOM Diodes

With the aid of higher resolution lithography and new fabrication techniques, MOM diodes were able to take a thin-film form in the late 1970s as "edge" MOM diodes (Wang et al. 1975; Heiblum et al. 1977). Thin-film diodes are fabricated on the surface of a semiconductor wafer. Since they do not rely on contact pressure between separate components, as in the case of point-contact MOM diodes, they do not suffer from mechanical instabilities.

Edge MOM diode structures were fabricated using an optical lithographically defined sandwich structure of a metal layer between two layers of insulating material. The edge of the thin sandwich structure was then coated with an oxide layer thin enough to allow tunneling behavior. Then, an overlying metal was laid over the original structure. An edge MOM diode is illustrated in Fig. 3.10. The medium gray metal is sandwiched between the light gray insulators. An oxide of tunneling

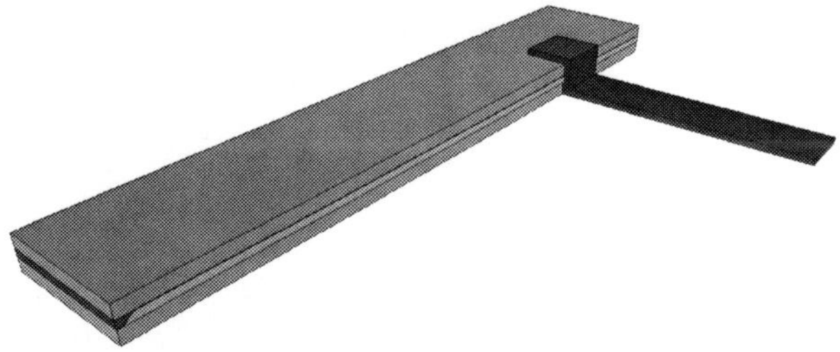

Fig. 3.10 Illustration of an edge MOM diode structure. In this structure, a metal layer is sandwiched by two insulators. A thin oxide is grown or deposited on the sandwiched metal layer (which is cut away at the *left* to indicate its presence) and the overlapping metal is patterned to complete the MOM structure

thickness covers this sandwiched metal layer and is shown as the light gray material. The MOM diode is formed between the sandwiched metal layer, the thin oxide coating layer (which is cut away at the left end to indicate the presence of the oxide), and the overlying dark gray metal structure, which provide a small contact area for tunneling.

These structures can also function as a rectifying antenna; the leads of the overlying metal can be structured to serve as an antenna and the edge MOM forms the nonlinear junction. The overlap areas of edge-MOM diodes are commonly too large for optical-frequency detection, however, since they are defined using optical lithography. However, another similar type of thin-film MOM diodes can be made with even smaller overlap areas. Overlap diodes fabricated using electron beam lithography (EBL) simply consist of a metal line with an oxide covering that is overlapped by another metal line. These can function as rectifying antennas as well, with the metal leads forming the antenna and the diode serving as the nonlinear junction.

3.2.3 Conduction Mechanisms

There are many possible mechanisms that can describe the flow of electrons in an antenna-coupled MOM diode. In every case, current flows from one metal electrode to the other. However, this electrical conduction can arise from various mechanisms that can be either classical or quantum mechanical. The specific properties and times for each mechanism can be used to determine which mechanism is predominant, but several conduction mechanisms often occur simultaneously in a device. Each operation will be described in this section, and the detection mechanisms that give rise to the signals in the structures fabricated for this project are indicated.

When photons of energy $h\nu$ are incident on a MOM diode structure, they can interact with electrons within the metal. Electrons resident at the Fermi level near the barrier can gain energy and tunnel across the barrier to the other metal with increased probability (Heiblum et al. 1978; Sanchez et al. 1978; Tucker and Millea

1978). The energy associated with electromagnetic radiation comes in indivisible packets called quanta, each associated with a single photon. This can be defined by:

$$E = h\nu = \frac{hc}{\lambda}$$

where E is the energy associated with electromagnetic radiation in joules, h is Planck's constant, ν is the photon's frequency in hertz, c is the speed of light in vacuum in meters per second, and λ is the wavelength in meters.

Figure 3.11 illustrates an incident photon of energy $h\nu$ interacting with electrons near the barrier in a MOM structure. Excited electrons gain energy from the incident photon that increases the probability of tunneling through the oxide to the other metal electrode. Those electrons that do tunnel due to energy gained from incident photons give rise to photon-assisted tunneling current in the MOM diode (Tucker and Millea 1978).

The electrons that tunnel through the oxide without photon excitation lead to an associated dark current for the device (Heiblum et al. 1978). However, since the incident photon energy of infrared photons is relatively low, ~0.1 eV for 10 μm wavelength, this is not the detection mechanism for the devices presented in this work.

If the incident photon energy $h\nu$ is greater than that of the barrier height, electrons can move to the other metal electrode without tunneling by surmounting the barrier. Incident photons excite electrons into states above the barrier and cross the barrier with a probability close to unity (Diesing et al. 2004). The interaction of a photon of energy $h\nu$ with an electron near the barrier is shown in Fig. 3.12.

The associated dark current of the structure is still present, shown by the electrons tunneling through the barrier without excitation. Again, since infrared photon

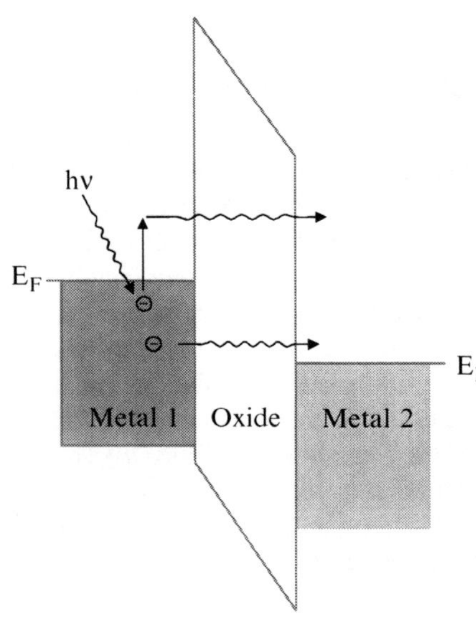

Fig. 3.11 Photon-assisted tunneling in MOM diode structure. Electrons near the barrier in Metal 1 can gain energy from incident photons, which increases their probability of tunneling through the barrier to Metal 2

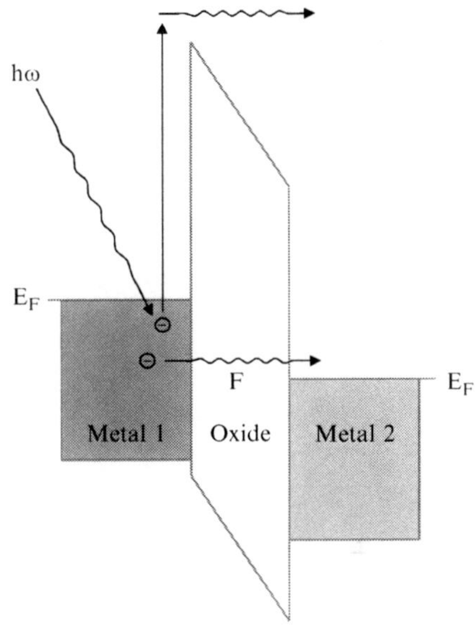

Fig. 3.12 Electrons surmounting barrier in MOM diode structure. Electrons in Metal 1 can gain enough energy from incident photons to surmount the barrier in order to get to Metal 2

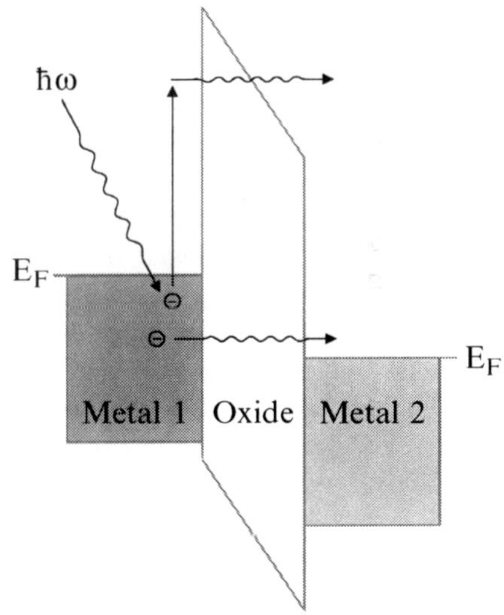

Fig. 3.13 Excited electrons tunnel across the barrier in MOM diode structure. Electrons in Metal 1 can gain energy from incident photons and tunnel to Metal 2 near the top of the tunnel barrier

energies are much less than the metal-oxide barrier height in the MOM diode, this is not typically observed for infrared detection.

There exists another conduction mechanism that is a combination of the cases of photon-assisted tunneling and electrons surmounting the barrier. In this case, photon-excited electrons with energy well above the Fermi level but still below the top of the tunnel barrier can tunnel from one metal to the other (Diesing et al. 2004), as shown in Fig. 3.13. This can substantially contribute to the tunneling current because the effective barrier height is reduced for these electrons (Thon et al. 2004).

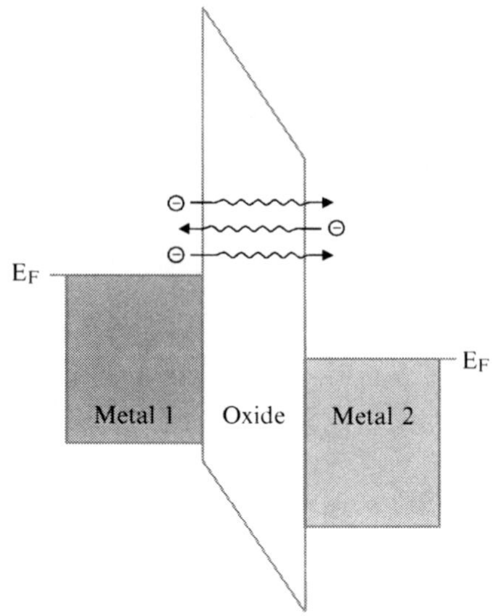

Fig. 3.14 Thermally assisted tunneling in MOM diode structure. Electrons with increased energy from heating of the structure have an increased probability of tunneling from one metal to the other through the barrier

When optical or infrared radiation is incident on a structure, photons can interact with the material, causing lattice vibrations also known as phonons. These vibrations can cause heating within the metal in a MOM diode. Electrons close to the barrier gain energy as $k\Delta T$ (Codreanu et al. 2003) where k is Boltzman's constant. According to the density of states of electrons in a metal and the temperature-dependent Fermi distribution function, a carrier distribution can be determined. As temperature increases, the Fermi function becomes less sharp, and an increasing portion of the electron distribution have probabilities of energies above the Fermi level, hence increasing the tunneling current (Pierret 2002). When electrons tunnel through the barrier due to the extra energy gained by heating of the structure, it is known as thermally assisted tunneling, shown in Fig. 3.14. If a temperature difference between the metals is present, an even greater contribution of thermally assisted tunneling can be expected. However, because of the poor absorptivity of metals in the LWIR (Twu and Schwarz 1974) and the fact that the entire structure is illuminated, it is not believed that a temperature difference is developed. Thermally assisted tunneling increases monotonically with heating due to photon absorption in the substrate and also with the applied DC bias (Green 1971). Thermally assisted tunneling does not contribute to the rectification at infrared frequencies because it is too slow (Heiblum et al. 1978; Wilke et al. 1994). While thermally assisted tunneling may give rise to electrons tunneling through the barrier in the devices, it is not a detection mechanism that would produce a polarization-dependent response in response to infrared photons.

Wilke et al. determined that the heating of the structure due to incident radiation can cause spreading resistance nonlinearity and thermal currents that are as strong as electron tunneling currents, even at room temperature (Simmons 1963; Wilke et al. 1994). As heat is generated in the vicinity of the MOM structure on a Si/SiO_2 substrate, thermionic emission of electrons over the barrier occurs. Joule heating

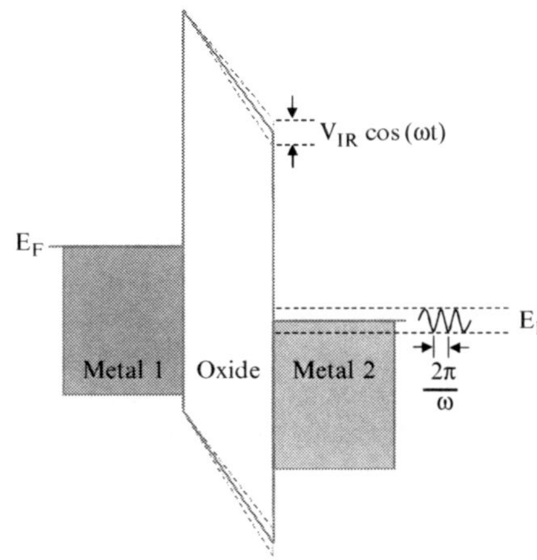

Fig. 3.15 Fermi level modulation in MOM diode structure. Incident radiation on the structure induces a time-dependent bias that leads to rectification of radiation induced antenna currents in an ACMOMD

due to the dissipation of laser-induced ac current in the antenna structure can also contribute to this effect (Wilke et al. 1994).

The last principle of operation for antenna-coupled MOM diodes is Fermi level modulation or field-assisted tunneling. When infrared radiation is incident on the structure, an optical or infrared voltage is induced (Heiblum et al. 1978; Fumeaux et al. 1998). This oscillating perturbation of the barrier can lead to multiple types of conductions mechanisms (Thon et al. 2004). This is shown in Fig. 3.15, where the radiation is incident on the structure, causing an alternating current bias that is in summation with the bias voltage.

This time-dependent bias $V(t)$ can be expressed as:

$$V(t) = V_{DC\mathrm{app}} + V_{\mathrm{IR}} \cos(\omega t)$$

where V_{IR} is the amplitude of the induced voltage and ω is the angular frequency of the incident radiation (Heiblum et al. 1978; Yngvesson 1991). The oscillation of the tunnel barrier leads to a degeneracy of electronic states that are separated by multiples of the photon energy (Thon et al. 2004).

When the induced alternating current infrared voltage has the same polarity as the applied DC bias, the separation between the Fermi levels of the two metal electrodes is increased. The initial state on the left side of the barrier is directly coupled to final states on the right side that are separated by the incident photon energy (Thon et al. 2004). Therefore, the probability of an electron to tunnel through the potential barrier from left to right increases, as does the tunneling current in the structure. On the contrary, when the induced infrared voltage is of opposite polarity to the DC applied bias, the Fermi levels come closer together and the tunneling probability of an electron through the potential barrier decreases. The increased probability of tunneling when the overall bias is largest is greater in magnitude than the decreased probability of tunneling when the overall bias is smaller. This nonlinear tunneling behavior allows the MOM diode to act as a rectifier. Since tunneling is an inherently fast process (Nagae 1973), 10^{-15} s to traverse a barrier on the order of 10 Å, MOM diodes are capable of rectifying high frequency signals. This applies in the LWIR (Daneu et al. 1969; Twu and Schwarz 1972; Gustafson and Bridges 1974), mid-IR (Sokoloff et al. 1970; Sakuma and Evenson 1974), and even up to optical frequencies (Faris et al. 1973; Gustafson et al. 1974).

Any number of the aforementioned conduction mechanisms can occur simultaneously in an MOM diode (Diesing et al. 2004). In addition, trap states in the tunnel barrier can impact the conduction mechanisms as well, since electrons can occupy vacant trap states and tunnel from one metal to the other in multiple steps (Gupta and Van Overstraeten 1975). However, the wavelength of the incident radiation impacts which conduction mechanisms are expected to occur. Photon assisted tunneling is expected to dominate for photon energies that are on the on the order of the barrier height (Diesing et al. 2004). For infrared radiation in the LWIR, tunneling due to Fermi-level modulation is likely to dominate.

3.2.4 Substrate and Antenna Effects

The radiation pattern of a dipole antenna is significantly influenced by the dielectric environment surrounding it. Since the antennas in this research will be placed at an air–SiO_2 interface, the radiation pattern with respect to the antenna axis will be strongly asymmetrical (Wilke et al. 1994). As a result of this asymmetry, the dipole antenna is more sensitive to radiation from the half-space with the higher dielectric constant. The power received by the antenna from each half-space can be approximated by:

$$\eta_{\mathrm{rel}} = \frac{P_1}{P_2} \approx \frac{\varepsilon_2^{3/2}}{\varepsilon_1^{3/2}}$$

where η_{rel} describes the relative efficiency of coupling of an antenna at an interface of materials 1 and 2, P_1 and P_2 is the power coupled to, or received by, the antenna from material 1 and 2, respectively, and ε_1 and ε_2 are the dielectric constants of materials 1 and 2, respectively (Wilke et al. 1994).

Given the dielectric constant of silicon, and the relatively small thickness of the silicon dioxide matching layer in terms of the wavelength of the incident radiation, simulations indicate that the coupling of the radiation to the antenna will be approximately 40 times more efficient for illumination from the substrate side as compared to the air-side (Alda et al. 2000). Although substrate-side illumination is theoretically much more efficient than air-side illumination, it presents some difficulty from a practical point of view. For example, a standard leadless chip carrier (LCC) cannot be used, necessitating the use of a flip-chip arrangement, and in an imaging system this would preclude placing circuitry below the antenna.

However, a strong antenna response that is on the same order as for substrate-side illumination can still be observed when utilizing air-side illumination (Wilke et al. 1994; Alda et al. 2000) if a well-designed SiO_2 surface layer is used. The relevant mechanism of the antenna response is the coupling of the incident energy from within the top SiO_2 layer. Alda et al. (2000) reported that the ratio of antenna responses between the air-side and substrate-side illumination $\Delta V_{air}/\Delta V_{substrate} = 0.84$ for 10.6 μm wavelength illumination. Therefore, it can be concluded that air-side illumination can be made to be nearly as efficient depending on the material properties and thickness of the SiO_2 layer. This verifies that the integration of the antenna-coupled MOM diodes fabricated in this research with a prefabricated CMOS chip is possible.

The material properties and thickness of the SiO_2 play an important role in the antenna signal. The spectral dependence of the complex part of the refractive index combined with the role of reflections within the SiO_2 layer impact the spectral dependence of the antenna (Alda et al. 2000). It should be noted that reflections from the back side of the wafer have been found to be of negligible importance (Alda et al. 2000). For this research project, this is certainly the case since the roughness of the backside of the wafers is on the order of the wavelength of incident radiation.

Surface impedance has an influence on the current distribution I along an antenna. When subjected to a specific wavelength of irradiation, antenna currents will propagate along an antenna. There is an exponential attenuation of these antenna currents due to the surface resistance of metals at infrared frequencies, which is related to the skin effect (Wilke et al. 1994). The surface attenuation constant Γ_{sr} is given by:

$$\Gamma_{sr} = -\frac{1}{2}\ln\frac{I(L)}{I(0)} = \frac{2\pi}{\lambda_o}\kappa$$

where L is the antenna length, λ_o is the illumination wavelength, and κ is the imaginary part of the complex refractive index of the antenna metal.

There is further attenuation of the antenna currents due to the placement of the dipole between two materials of different dielectric constants ε_1 and ε_2. The phase velocity of the electromagnetic waves, in this case antenna currents propagating along the length of the metal dipole antenna, can be described by:

$$v_p = c\left(\sqrt{\frac{\varepsilon_1+\varepsilon_2}{2}}\right)^{-1}$$

where v_p is the phase velocity and c is the speed of light in free space. The phase velocity of the antenna currents is intermediate between the phase velocities of the two surrounding dielectrics (Coleman 1950). Therefore, the antenna currents are exponentially attenuated as a function of the length, with the attenuation constant due to the dielectric difference Γ_C expressed as:

$$\Gamma_C = -\frac{1}{L}\ln\frac{I(L)}{I(0)} = \frac{2\pi}{\lambda_o}\sqrt{\frac{\varepsilon_1+\varepsilon_2}{2}},$$

which is known as the Coleman effect. The effect of this attenuation can be determined by studying its influence on dipole antennas of different lengths L (Wilke et al. 1994). This will be experimentally addressed in Sect. 3.4.2.3.

3.3 Fabrication

The processes utilized in the fabrication of antenna-coupled MOM diodes (ACMOMDs) include thermal oxidation of silicon wafers, optical and EBL, metal deposition, and multiple developing and etching steps. Optical lithography is used to create bonding pads for the devices, whereas EBL is used to pattern the ACMOMD devices. In this chapter, the fabrication processes for both the bonding pads and the ACMOMDs are discussed in detail.

3.3.1 Substrate

The substrate for the ACMOMDs fabricated in this research is a 625 μm thick, single-side polished p-type silicon wafer of 13–16 Ω-cm resistivity. A thermal oxide of 1.5 μm is grown on both sides of the wafer using an oxidation furnace to provide an insulating substrate for the device and to also serve as a quarter wave matching layer for 10.6 μm irradiation.

Surface roughness of the substrate must be minimized since these devices are composed of very thin metal lines on the order of 30 nm. If the surface roughness of the substrate is too great, the antenna, leads, or both could be broken and cause an open circuit. Thermal oxidation of silicon was chosen because of its convenience, ability to provide consistently smooth films, and offer low optical loss.

3.3.2 Bonding Pad Fabrication

In order for the characteristics of completed nanoscale devices to be determined, a means for electrically contacting them must be provided. In the case of both electrical and infrared measurements for this project, bonding pads serve as a way to provide a larger contact area than the nanoscale ACMOMDs. The bonding pads are defined using optical lithography and fabricated by utilizing a lift-off process.

The bonding pads for each device are composed of a two 4 μm × 10 μm metal leads, between which an ACMOMD will be placed. A tapered metal line connects these leads to two 120 μm × 120 μm metal bond pads that can be contacted with a probe station or wire bond.

The lift-off process for fabricating bonding pads begins by spin-coating the wafer with 1.40 μm of negative photoresist. The bonding pads are defined with a UV exposure through a mask on a contact aligner. This is followed by an image reversal procedure. The wafer is then developed and 50 Å of titanium and 200 Å of gold are deposited using an electron beam evaporator. By using a lift-off solvent, in this case acetone, the photoresist and overlying metal is removed and the bonding pads remain on the substrate. The bonding pads can also be fabricated using positive photoresist and etching deposited layers of gold and titanium.

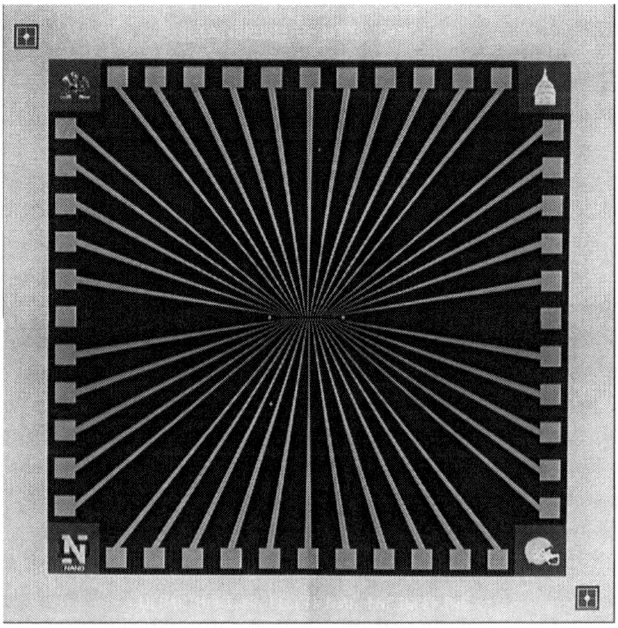

Fig. 3.16 Titanium-gold bonding pads on Si/SiO_2 substrate. These bonding pads provide electrical contacts to the ACMOMDs so that they may be characterized. Note: leprechaun in the *upper left*, Golden Dome in the *upper right*, interlocking ND Nano logo in the *lower left*, and gold helmet of the Fighting Irish in the *lower right*

The bonding pad design was chosen to allow for easy integration with a LCC, which is used in the infrared testing setup. Figure 3.16 is an optical micrograph of one set of titanium-gold bonding pads fabricated on a silicon wafer with 1.5 μm thermally grown silicon dioxide layer.

3.3.3 ACMOMD Fabrication

The ACMOMD fabrication process utilizes EBL to create the antenna-coupled MOM diode pattern in a positive radiation resist. The pattern in the resist serves to define the device when metal is deposited on the sample. Each step performed during the fabrication of the ACMOMD is explained in detail, including EBL, development, metal deposition, and lift-off.

3.3.3.1 Electron Beam Lithography

Electron beam resist is a material that is sensitive to high energy electrons and most commonly used as a high-resolution resist for direct-write EBL. The resist is composed of long molecular chains suspended in a solvent. This liquid solution can be applied to a wafer using spin coating, just as in the case for optical lithography. The spin speed determines the thickness of the resist on the sample.

For this project, two types of EBL resist are utilized: polymethyl methacrylate (PMMA) and copolymer methyl methacrylate (MMA). PMMA is a high-contrast, high-resolution resist, whereas copolymer MMA–methacrylic acid (MMA–MAA) is a lower resolution mixture of methyl methacrylate and 8.5% PMMA.

For the experiments in this research, copolymer is used as the bottom layer of resist that lies on the sample substrate. The copolymer MMA layer is applied using a wafer spinner to yield a 4500 Å layer after a hotplate bake. The sample is then covered by a 700 Å top layer of PMMA, totaling 5200 Å for the resist stack.

The reason the bi-layer resist stack was chosen can be explained with a profile image, shown in Fig. 3.17, of the resist after EBL and development. The PMMA is a high-resolution, high-contrast resist that will be used to define the desired pattern for the device on the substrate. Incident electrons break the long molecular chains that can be then removed using a developing solution. The copolymer MMA layer, which is a more sensitive resist, simply serves the purpose of a spacing layer. This layer of resist lifts the PMMA off of the substrate, while providing an undercut beneath the PMMA opening to facilitate a double-angle metal deposition procedure.

This double-angle metal deposition has been utilized to fabricate nanoscale tunnel junctions (Dolan 1977) and single electron transistors (Fulton and Dolan 1987; Fulton et al. 1989; Orlov et al. 2000). The amount of undercut is directly dependent on the thickness of the resist as well as the spread of the electron beam within the resist layer, a function of the accelerating voltage of the EBL system.

The pattern for the antenna-coupled MOM diode is created using the supplied Elionix software. Figure 3.18 shows a screenshot of the pattern array. For each device, there are two rectangles, one at each end, which contact the bonding pads that are discussed in Sect. 3.3.2 and shown in Fig. 3.16. Leads from these rectangles provide the electrical connection to the two halves of a dipole antenna. There is a small gap between each half of the dipole antenna. The length of the dipole antenna was designed to correspond to the equivalent wavelength of the desired detection wavelength, in this case 10.6μ m radiation, in the silicon dioxide layer, on which the antenna is located.

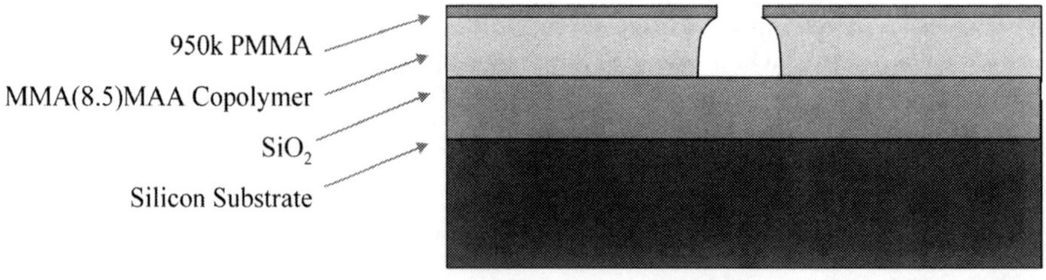

Fig. 3.17 Profile of developed bi-layer resist stack on sample substrate. The copolymer layer is a lower resolution, higher sensitivity resist than the PMMA, which provides an undercut after development and forms the foundation for the fabrication method of the ACMOMDs

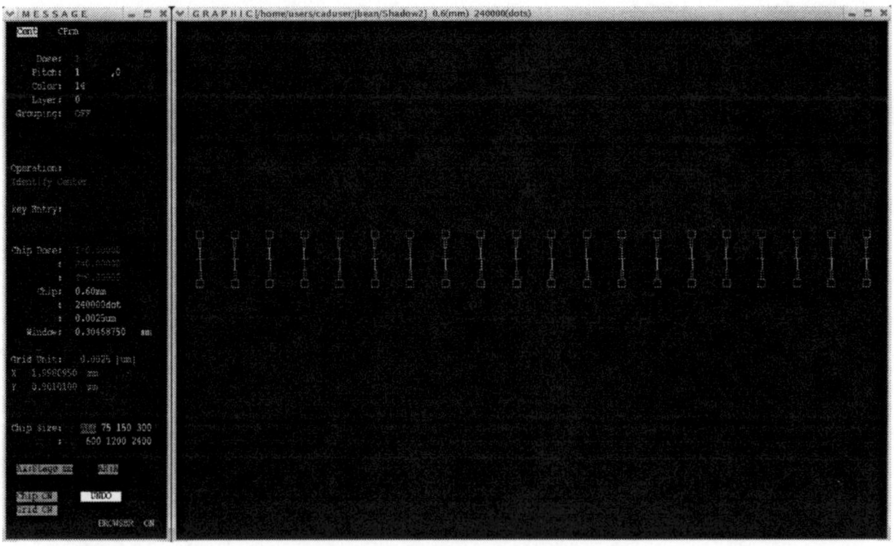

Fig. 3.18 Screenshot of EBL pattern using Elionix software. The devices shown are aligned and exposed into the resist over the bonding pads shown in Fig. 18

With a beam accelerating voltage of 75 kV, the undercut for the aforementioned resist stack shown in Fig. 3.17 is roughly 25 nm in each direction. That is, for a 50 nm line in the PMMA, the line width in the underlying copolymer MMA at the substrate level would be 100 nm. The gap between the dipole halves is 65–85 nm, as shown by the rectangle in the screenshot in Fig. 3.19. Since the undercut of a 75 kV exposure is approximately 25 nm and the PMMA bridge width ranges from 65–85 nm, there is some residual copolymer left beneath the PMMA bridge after development. Therefore, a 50 $\mu C/cm^2$ areal dose is applied to this gap, represented by the rectangle between the two halves of the dipole antenna.

The exposed pattern is developed in a mixture of methyl isobutyl ketone (MIBK), isopropanol (IPA), and methyl ethyl ketone (MEK), of ratio 1:3:1.5% MIBK:IPA:MEK (Bernstein et al. 1992) and then rinsed in IPA to stop the development and clean the sample. The areal dose applied to the gap is high enough to expose the underlying copolymer MMA layer but low enough to keep the PMMA bridge intact after development. Since the PMMA is left intact and the MMA is removed below, a shadow evaporation technique can be utilized that allows these antenna-coupled MOM diodes to be fabricated using a single EBL step. An illustration of the developed structure on the sample is shown in Fig. 3.20.

There is a small PMMA "bridge" between each half of the antenna, created by the gap in the EBL pattern. A cutaway view of the sample and resist profile, showing the PMMA bridge, is shown in Fig. 3.21.

The line widths of the antenna leads and antenna in the PMMA are 50 nm after development. The sample is placed in a reactive ion etcher for a descum procedure.

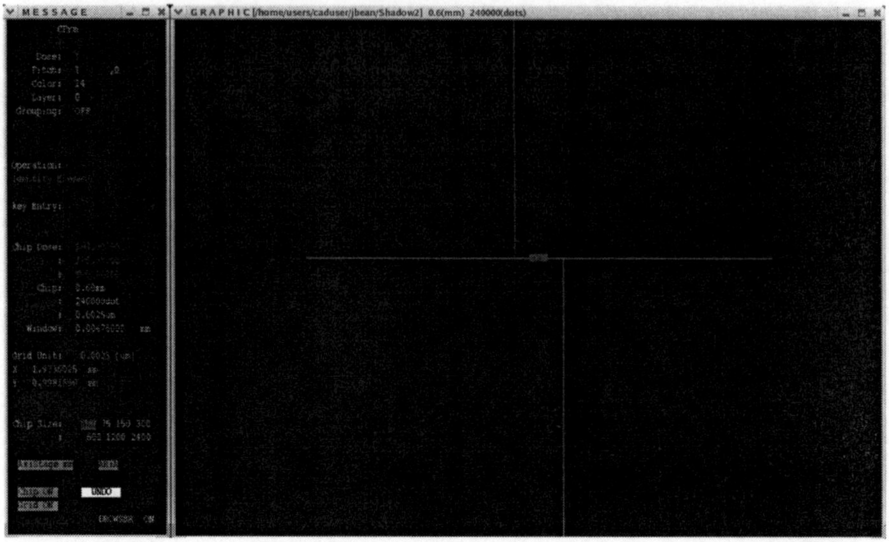

Fig. 3.19 Screenshot of dipole antenna. This magnified view of a single device pattern shows the dipole antenna (*oriented horizontally*) and the lead structures (*oriented vertically*). A rectangle is shown in gap in the pattern of the dipole antenna, where the small areal dose is applied

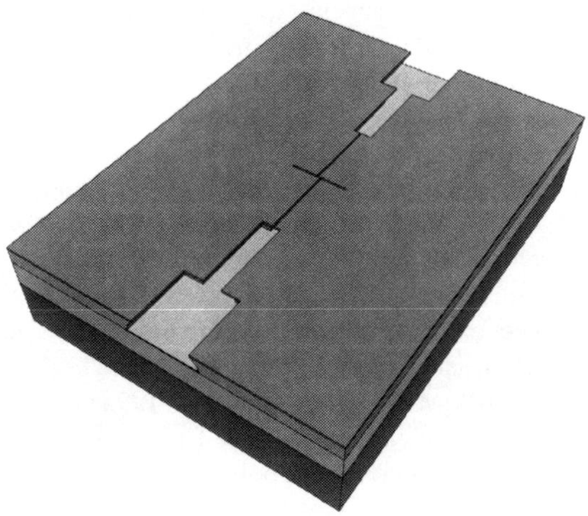

Fig. 3.20 Illustration of the EBL pattern on substrate after development. The largest rectangles on either end of the pattern represent where contact will be made with the bonding pads. The electrical leads lead to each half of the dipole antenna, which are separated by a small PMMA bridge

The purpose of this step is to clean any residual resist from the substrate in the patterned areas to facilitate improved adhesion of the metal during deposition. An optical micrograph of the electrical leads and developed EBL pattern is shown in Fig. 3.22.

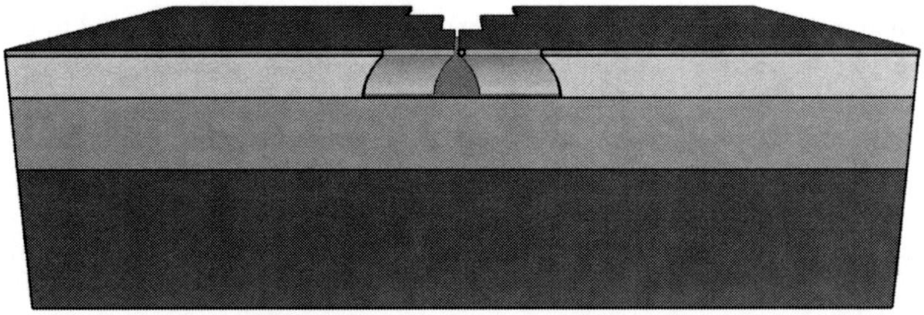

Fig. 3.21 Cutaway view of EBL pattern on sample after development. This cross-section is taken along the antenna shown in Fig. 3.24 and through the PMMA bridge. This bridge and undercut forms the basis for how the ACMOMDs presented in this work are formed

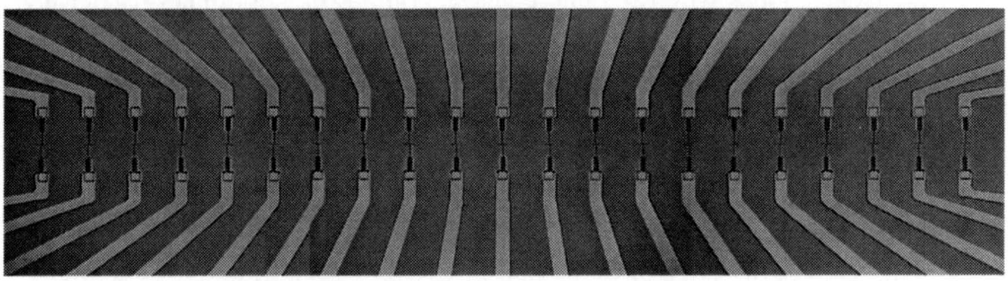

Fig. 3.22 Optical micrograph of developed EBL pattern overlaying electrical leads. Each connected device has a resist profile similar to that shown in Fig. 3.22

3.3.3.2 Metal Deposition

All metal evaporations for devices fabricated in this project are performed with an electron beam evaporator. Evaporated particles travel in a straight path as they leave the source, leading to a directional, nonconformal deposition that is essential to the fabrication of the devices.

Shadow Evaporation ACMOMDs

ACMOMDs can be fabricated using a single EBL step by utilizing a shadow evaporation that involves two metal depositions at opposing 7° angles and an intermediate oxidation that forms the tunnel barrier. For shadow evaporation devices, the oxide tunnel barrier is not deposited, but rather grown on the first deposited metal. As such, the first metal layer must be a metal that readily forms a native oxide in the presence of oxygen.

In this case, the first metal is aluminum that is deposited at −7° to a final thickness of 300 Å. To create the tunnel barrier of the MOM diode, which is formed by oxidizing the aluminum layer, oxygen is allowed into the evaporator chamber. This can be done by venting the chamber to atmosphere, referred to as an air oxidation, or by bleeding oxygen into the chamber, referred to a controlled oxidation. For controlled oxidations, the total oxygen exposure of the sample is equal to the oxygen pressure in the chamber times the total exposure time. When the desired oxygen exposure is reached, the evaporator chamber is evacuated. Before the second metal deposition, the sample angle is switched to the second deposition angle, $+7°$. A cross-section of the sample, along the antenna arms and through the PMMA bridge, can be seen in Fig. 3.23.

The dashed lines represent the two angles of evaporation and the resulting locations of metal deposition for each angle. The darker metal represents the first deposition, whereas the lighter metal represents the second deposition. The deposition portion of the fabrication is complete; however, the resist and overlying metal layers are still present on the sample. The second metal layer is then deposited to a 300 Å thickness.

Given MOM diodes with the same material composition, a thinner barrier layer provides a lower resistance. This can be accomplished by utilizing a controlled oxidation, one where the amount of oxygen introduced into the evaporator chamber to oxidize the first deposited metal can be controlled. This provides the ability to tailor the barrier layer depending on the desired thickness. By controlling the temperature of the sample, the base chamber pressure, and the oxygen introduced into the chamber, a level of fabrication repeatability can be established between fabrication runs.

Once the MOM diode has been formed, electrostatic discharge precautions are made any time the sample is handled. A grounding wrist strap is worn and is connected through a high impedance cord to ground. Static dissipative gloves are also worn to protect the devices. The sample is removed from the sample holding plate and placed in methylene chloride, which serves as the lift-off solvent. The PMMA and MMA dissolve, yielding the resultant structure on the substrate as shown in Fig. 3.24. The circled area shows where the MOM diode is formed. The overlap area of the two metal electrodes is roughly 50×50 nm.

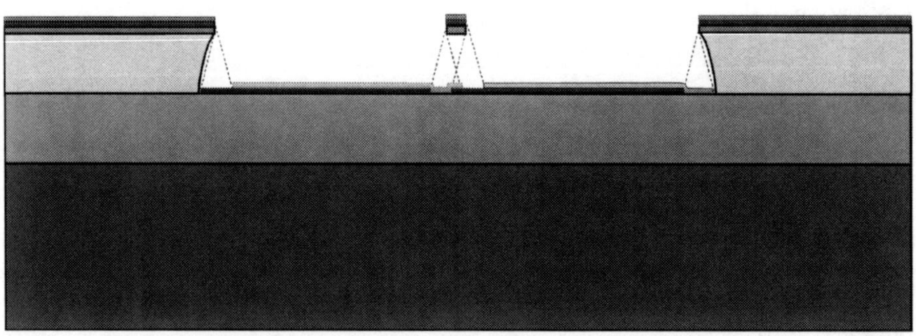

Fig. 3.23 Cross-section of sample after metal deposition. The *dotted lines* represent the angles at which the depositions occur. The first deposition, with aluminum, is performed at an angle of 7° from normal and is then oxidized. The second metal deposition, with platinum, is performed at an opposing 7° angle

Fig. 3.24 Cross-section of device after lift-off. The PMMA bridge shown in Fig. 3.28 causes a break in the aluminum and platinum layers, but the aluminum and platinum layers are electrically connected through the circled MOM diode

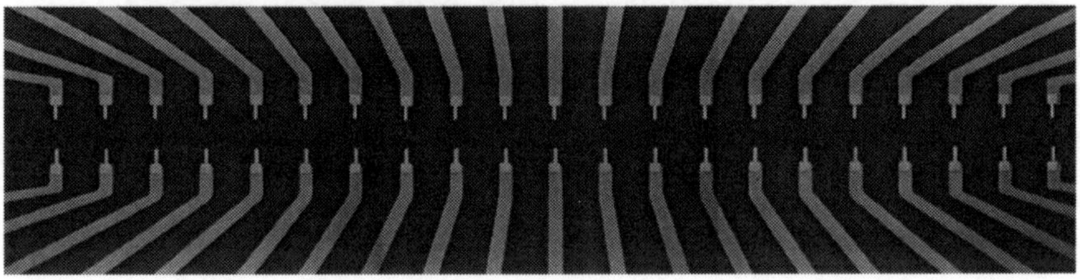

Fig. 3.25 Completed antenna-coupled MOM diodes connected to electrical leads. The titanium-gold leads provide electrical connection for *I*-*V* and IR characterization

The completed antenna-coupled diodes connected to the titanium-gold electrical leads are shown in the optical micrograph in Fig. 3.25.

A completed antenna diode can be seen in the scanning electron micrograph (SEM) in Fig. 3.26. The bonding pads of the device can be seen at the top right and bottom left of the image. The large rectangles ensure proper connection between the bonding pad and the antenna lead. The antenna arms are each 1.55 μm long, yielding an antenna full-length of 3.1 μm. The overlap between the two evaporations, which forms the MOM antenna-coupled diode, can be seen in the inset SEM in Fig. 3.16. The double-angle evaporation is evident in this image, as parallel leads can be seen above and below the antenna structure.

Two-Lithography ACMOMDs

The double-angle evaporation technique detailed in the previous section allows fabrication of the antenna-coupled MOM diodes in this research using just one evaporation step. However, these devices can also be fabricated using two separate lithography and metal deposition steps, one for each antenna arm (Fumeaux et al. 1998, 1999, 2000; Gonzalez and Boreman 2005; Tiwari 2009). Although this requires more processing steps, it provides for more latitude in terms of the comprising materials of the ACMOMDs.

The substrate and resist stack remain unchanged, whereas the metal deposition procedure is simply modified so that deposition is normal to the sample. The first

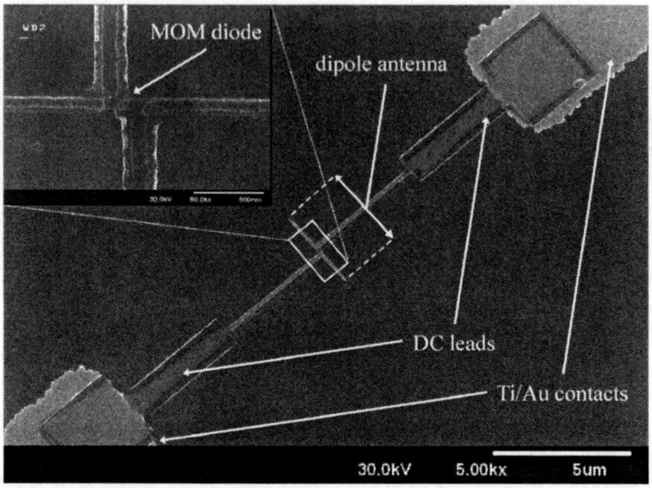

Fig. 3.26 Electron micrograph of completed ACMOMD. The connection of the EBL defined device to the titanium gold contacts can be seen at the *lower left* and *upper right corners*. The DC leads provide the electrical connection to the MOM diode at the center of the dipole antenna. Inset image: MOM diode overlap

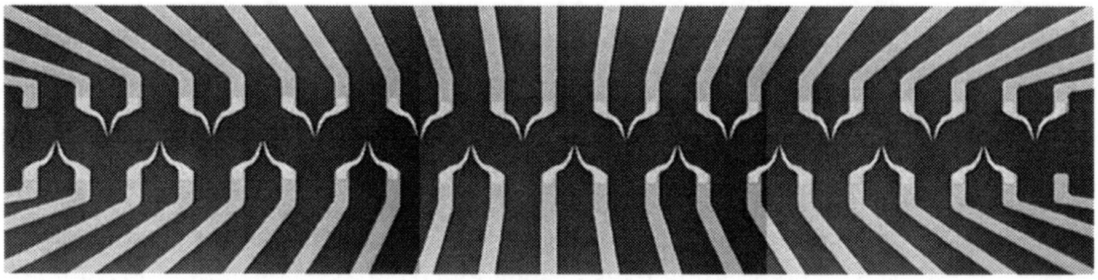

Fig. 3.27 Optical micrograph of two-step lithography ACMOMDs. Each half of the dipole is fabricated with separate lithography and metal evaporation steps. The tunnel barrier can either be grown on the first metal or deposited by various means

antenna arm is defined using EBL on a PMMA/MMA coated bonding pad array sample. The pattern is composed of a lead that connects to the bonding pad along with one half of the dipole. Metal is then deposited on the sample, followed by a lift-off procedure. The sample is then coated again with the same resist stack as in the case of the first layer. The second antenna arm is then patterned with EBL. If the metal chosen for the first layer readily forms a native oxide in air, this will serve as the oxide barrier of the MOM diode. If the first metal layer does not form an oxide, one must be deposited, such as SiO_2, Al_2O_3, or HfO_2, before the second metal layer can be deposited. Once the second metal layer is deposited, a lift-off procedure is performed and the antenna-coupled MOM diode devices are complete. Figure 3.27 shows an optical micrograph of the two-step lithography devices connected to the electrical leads.

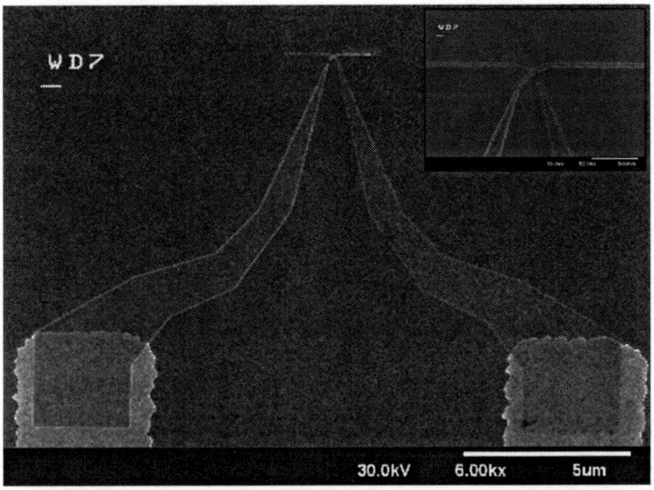

Fig. 3.28 Scanning electron micrograph of two-step lithography $Al/AlO_x/Pt$ ACMOMD. Each half of the dipole antenna is fabricated separately. Inset: MOM overlap area, which is approximately 50 nm × 50 nm

Figure 3.28 is a SEM of the structure that shows the overlap of the two antenna arms.

3.3.3.3 Packaging

Once the fabrication is complete, the ACMOMDs can be characterized. In this project, current–voltage characteristics for the diode are first obtained and then the samples are prepared for infrared testing. Each array of bonding pads is separated into single 44-pin chips. This sample is placed in a 44-pin LCC and held in place using an adhesive. Aluminum wire bonds then connect each 120 × 120 μm bonding pad to the corresponding pin on the 44-pin LCC. A set of ACMOMDs wirebonded within a 44-pin LCC is shown in Fig. 3.29.

3.4 Detector Characterization

In order to compare the ACMOMDs fabricated in this research to other infrared detector technologies, their response to infrared radiation must be characterized. First, the DC current–voltage characteristics are provided. Then, the layout of the experimental infrared testing arrangement is provided and the purpose of each component is explained. The figures of merit detailed in Sect. 3.1.3 will be calculated and compared with various types of infrared detectors available. Measurements that characterize the performance of ACMOMDs will then be presented that determine

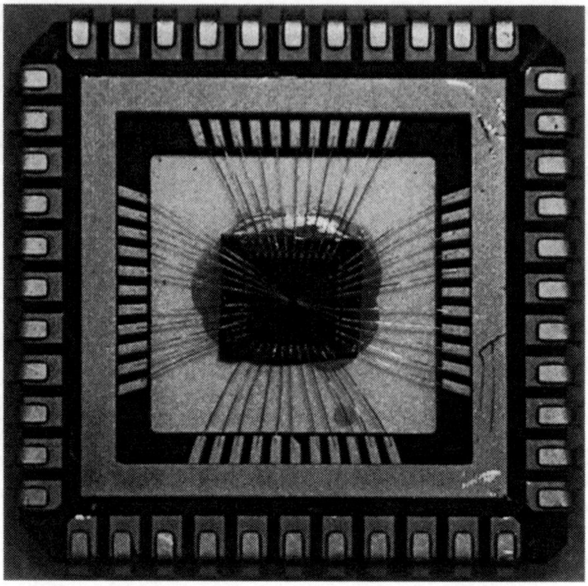

Fig. 3.29 Set of completed ACMOMDs in a 44-pin LCC. This LCC fits into a socket for the infrared characterization of the device. This arrangement allows up to 20 devices to be measured on each sample

polarization dependence, antenna-length dependence, and spatial response. Noise sources will be explained in detail and accompanied with an experimental noise analysis.

3.4.1 Current-Voltage Characteristics

The current–voltage characteristics of the MOM diodes are obtained using a probe station. The nonlinear current–voltage characteristic of the MOM diode depends on metals used, oxide type, and oxide thickness, as described in Sect. 3.2.

Three entities have been examined and compared for each diode type: zero-bias diode resistance, zero-bias curvature, and peak curvature. From the measured device characteristics, the low-pressure oxygen alumina growth characteristics have been inferred, and the relationship between oxygen exposure, and hence oxide thickness, and diode resistance, zero-bias curvature, and peak curvature has been determined.

Zero-bias resistance is $1/(\mathrm{d}I/\mathrm{d}V)$ at zero-bias. Curvature is found by dividing the second derivative of the current by the first derivative (Kale 1985):

$$\gamma = \frac{\partial^2 I}{\partial V^2} \Big/ \frac{\partial I}{\partial V}.$$

Zero-bias curvature is the value of the curvature coefficient at zero-bias. Peak curvature is the value at which the maximum curvature of the diode is reached across the entire biasing voltage range.

Various metals have been utilized to create asymmetrical MOM diodes. The first metal for the diodes must always be one that readily forms a thin oxide in the presence of oxygen, whether at atmospheric partial pressure or lower pressures. The difference in work functions of the metals creates a built-in electric field in the oxide between the electrodes, since the Fermi levels of the materials align to reach an equilibrium state. Table 3.2 outlines work functions of metals used in the fabrication of the MOM diodes, as well as other metals of interest (Michaelson 1977).

The metal combination of choice for the controlled-oxidation MOM fabrication was aluminum and platinum. The aluminum serves as the base metal, AlO_x as the tunnel barrier, and platinum as the top metal. The energy band diagram for this structure is similar to Fig. 3.5, where Metal 1 is the platinum electrode, with $\varphi_{b1} = 5.65\,\text{eV}$, and Metal 2 is the aluminum electrode, with $\varphi_{b2} = 4.28\,\text{eV}$. The reason for this choice of metals is because of the large separation of the work functions and more nonlinearity can be obtained in the I-V characteristic. This is the largest separation of work functions of materials that are commonly used for microelectronics fabrication.

3.4.1.1 Air-Oxidation

Figure 3.30, upper left, shows an I-V characteristic for an Al/AlO_x/Al symmetric ACMOMD, whose fabrication was discussed in Section *Shadow Evaporation ACMOMDs*. The acquired data points are plotted along with a fifth-order polynomial fit. The resistance, upper right, is calculated by taking the inverse of the first derivative of the current. The second derivative of the current represents the diode nonlinearity, lower left, which is useful for detection. The curvature of a diode, lower right, is obtained by dividing the second derivative of the current by the first derivative of the current.

The I-V characteristic of this Al/AlO_x/Al diode is fairly symmetrical about zero-bias, as the metal electrodes are the same. The small asymmetry is due to charges trapped in the oxide near the metal-oxide or oxide-metal interfaces as discussed in

Table 3.2 Work functions of commonly evaporated materials used in microelectronics fabrication

Metal	Work function φ_b (eV)
Al	4.28
Ti	4.33
W	4.50
Cr	4.50
Cu	4.65
Au	5.10
Ni	5.15
Pt	5.65

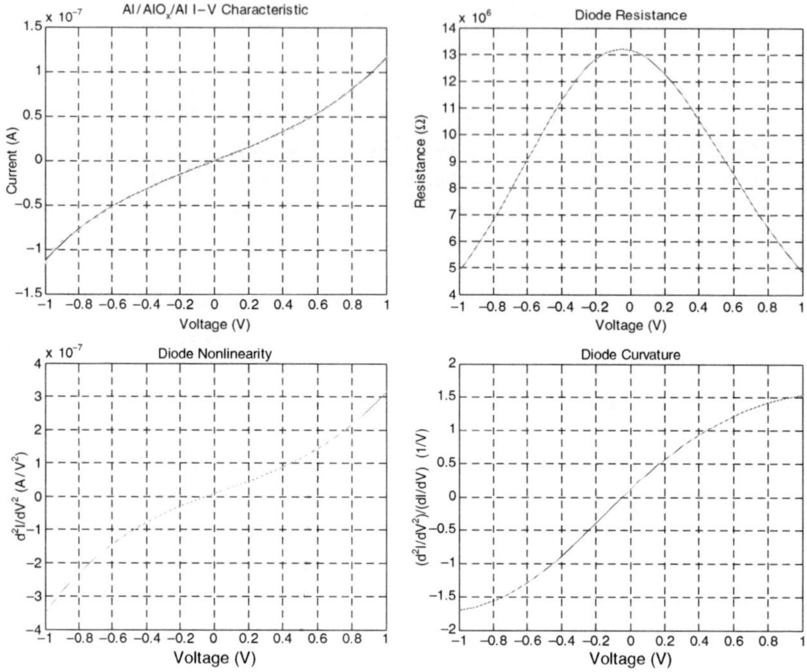

Fig. 3.30 *I-V* characteristic (*upper left*), resistance (*upper right*), diode nonlinearity (*lower left*), and curvature (*lower right*) for an air oxidation $Al/AlO_x/Al$ ACMOMD. This *I-V* characteristic is fairly symmetric about zero bias since both metals in the MOM diode are aluminum

Sect. 3.2.2. In addition, different oxygen content in the electrodes can lead to asymmetries in the I-V characteristic (Kadlec and Gundlach 1975). The zero-bias diode resistance is 13.2 MΩ, the zero-bias curvature is 0.13 V^{-1}, and the peak curvature coefficient is 1.7 V^{-1} at a biasing point of −1 V.

A current–voltage characteristic of an Al/AlO_x/Pt MOM diode fabricated using an air oxidation is shown in Fig. 3.31.

This structure's energy band diagram, under equilibrium and with an applied bias, is shown in Figs. 3.6 and 3.7, respectively.

The I-V characteristic of this ACMOMD is not symmetric about zero-bias, as the metal electrodes are dissimilar. The zero-bias diode resistance is 312 MΩ, the zero-bias curvature is 1.4 V^{-1}, and the peak curvature coefficient is 4.9 V^{-1} at a bias of 0.50 V. While the curvature of this air-oxidation device is promising, for high-performance detectors it is desirable to reduce the resistance while maintaining high curvature. S. Yngvesson derived that the current sensitivity S_I of an antenna-coupled diode (Yngvesson 1991), based on the equivalent circuit shown in Fig. 3.14, is:

$$S_{\rm I} = \frac{\alpha}{2\left(1+\frac{r}{R_{\rm D}}\right)^2\left(1+\frac{\omega^2 C_{\rm D} r R_{\rm D}^2}{r+R_{\rm D}}\right)}$$

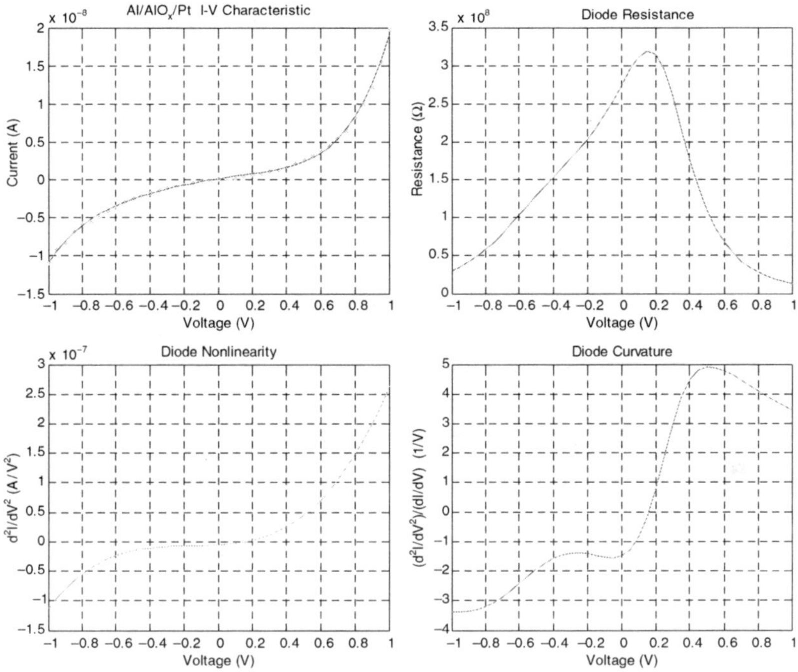

Fig. 3.31 *I-V* characteristic (*upper left*), resistance (*upper right*), diode nonlinearity (*lower left*), and curvature (*lower right*) for an air oxidation Al/AlO$_x$/Pt ACMOMD. The asymmetry about zero-bias is due to the fact that the comprising metals of the MOM diode are aluminum and platinum, which have different work functions

Table 3.3 Material type and curvature coefficient for ACMOMDs

ACMOMD composition	Curvature coefficient (V^{-1})	Standard deviation (V^{-1})
Al/AlO$_x$/Al	0.00	0.02
Al/AlO$_x$/Ti	0.08	0.03
Al/AlO$_x$/Ni	0.45	0.15
Al/AlO$_x$/Pt	0.74	0.17

where α is the inverse of the thermal energy. It is clear that the current sensitivity decreases as R_D increases and as such, it is desirable to reduce R_D.

Reducing diode resistance can be achieved by varying the parameters of the fabrication process, as discussed below. As mentioned in Sect. 3.4.1, numerous material combinations were pursued in fabricating ACMOMDs. The zero-bias curvature will be used to determine the best diode materials to use for reduced resistance devices. Table 3.3 shows those metal combinations that were found to be repeatable, which were the aluminum-based devices, and their average zero-bias curvature. In each case, 20 ACMOMDs fabricated with an air oxidation were used to calculate the values in Table 3.3.

Since Al/AlO_x/Pt devices have the highest zero-bias curvature for air oxidation devices, they were chosen to be pursued for controlled-oxidation reduced resistance devices. Table 3.3 follows the assertion that be using metals with larger work function separations, listed in Table 3.2, would create more asymmetry in the I-V characteristic and hence a greater zero-bias curvature.

3.4.1.2 Controlled Oxidation

The I-V characteristic of a typical Al/AlO_x/Pt controlled-oxidation MOM diode, discussed in Section *Two-Lithography ACMOMDs*, is shown in Fig. 3.32. The oxygen exposure used for this oxidation was 1,200 s at 50 mTorr, yielding a 60 Torr-s exposure. The zero-bias resistance has been reduced to 220 kΩ and the zero-bias curvature and peak curvature have decreased to $0.62\,V^{-1}$ and $0.82V^{-1}$, respectively.

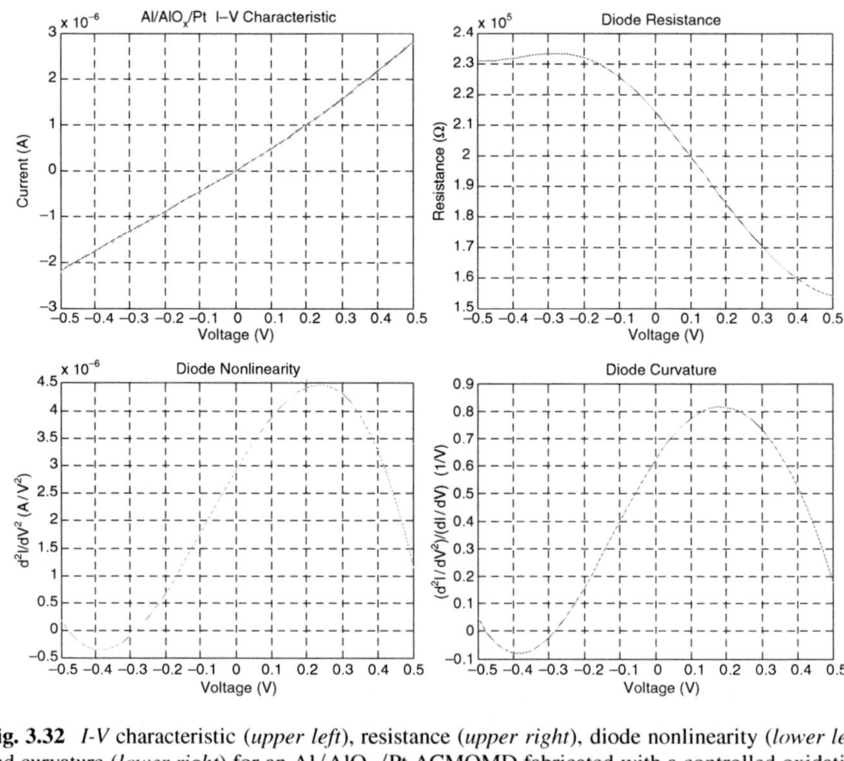

Fig. 3.32 *I-V* characteristic (*upper left*), resistance (*upper right*), diode nonlinearity (*lower left*), and curvature (*lower right*) for an Al/AlO_x/Pt ACMOMD fabricated with a controlled oxidation. There is still some asymmetry in the I-V characteristic about zero-bias, but it is less than that of the air oxidation ACMOMD

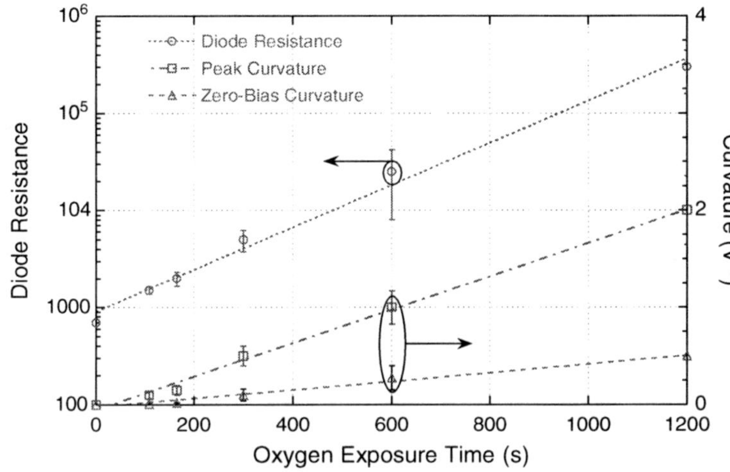

Fig. 3.33 Diode resistance and curvature coefficient as a function of oxygen exposure time. In each case, the oxygen pressure was held at 50m Torr and the time was varied to reach the different oxygen exposures. The thickness of the barrier increases for increasing oxygen exposures. The resistance increases exponentially and curvature increases linearly with increasing oxygen exposure

As can be seen, the controlled oxidation diodes have much lower resistance, but also lower zero-bias and peak curvature. The relationship between oxygen exposure and these three diode parameters has been studied with devices fabricated with various oxygen exposures. In each case, the partial O_2 pressure was 50 mTorr and the time was varied to reach different oxygen exposures. The results of this study are shown in Fig. 3.33.

Diode resistance has an exponential relationship with oxygen exposure time. On the contrary, peak curvature and zero-bias curvature, have a linear relationship with oxygen exposure time. By reducing the oxygen exposure and hence the thickness of the alumina layer, the resistance can be decreased exponentially, whereas curvature reduces in a linear fashion. The results shown in Fig. 3.33 were found to be repeatable, offered a yield of approximately 90%, and provided a high level of precision between devices and fabrication runs, due to the in-situ control of oxygen pressure.

3.4.2 *Infrared Response Characteristics*

The experimental testing arrangement that provides for the infrared spectroscopic measurements of the antenna-coupled MOM diodes fabricated in this research consists of six main parts: a CO_2 laser, a beam splitter and attenuator, a mechanical chopper, a power meter, a micrometer stage, and electrical signal sensing equipment. A block diagram of the testing arrangement is shown in Fig. 3.34.

Using a CO_2 laser with 10.6 μm linearly polarized radiation, the detection characteristics and figures of merit of ACMOMDs can be measured. A beamsplitter and

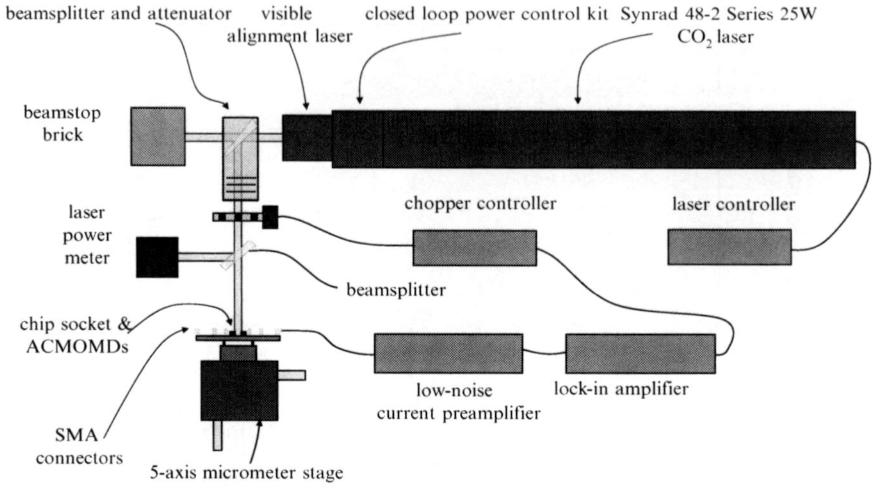

Fig. 3.34 Layout of experimental infrared testing arrangement. This setup provides for the infrared testing of ACMOMDs. By illuminating the devices with 10.6 μm radiation from a CO_2 laser and monitoring their response, the detection characteristics can be measured and compared to other detector technologies

attenuator is used to obtain the desired laser power. A mechanical chopper provides for square wave modulation of the irradiation, whereas another beam splitter and power meter allow for continuous monitoring of the beam power. The devices fabricated in this research are connected through titanium-gold leads and aluminum wirebonds to a LCC. The LCC is then placed into a chip socket that is soldered onto a printed circuit (PC) board. The devices are connected to the electrical sensing equipment through coaxial connectors on the PC board. A photograph of the PC board and the SMA connectors that provide an electrical connection for 20 devices is shown in Fig. 3.35.

A current preamplifier amplifies the current in the ACMOMD and passes a voltage signal to a lock-in amplifier that is synchronized with the reference frequency of the mechanical chopper and reads out the detector signal in volts. By dividing the displayed output voltage of the lock-in by the sensitivity of the preamplifier (V/A), the rms current at the detector can be calculated.

Figure 3.36 represents an electromagnetic wave incident on an ACMOMD. For this research, this incidence is linearly polarized 10.6 μm radiation from a CO_2 laser. The polarization angle of the incident radiation is represented by φ, where φ rotates in a plane parallel to the surface of the wafer.

The polarization of the incident radiation with the electric field parallel to the antenna axis ($\varphi = 0°$) is commonly referred to as p-polarization, whereas irradiation with the electric field perpendicular to the antenna axis ($\varphi = 90°$) is referred to as s-polarization. The angle of incidence is represented by θ.

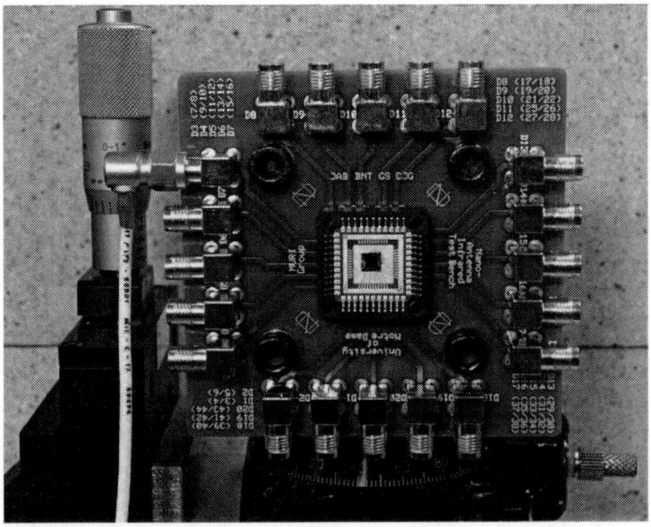

Fig. 3.35 PC board with SMA connectors, 44-pin LCC, and ACMOMDs. The wirebonded LCC is inserted in the chip socket, which is connected via PC board traces to SMA connectors. An SMA to BNC cable connects the devices to the current preamplifier and lock-in amplifier to measure the device output

3.4.2.1 IR Detector Characterization

There are many types of measurements that can be carried out to characterize the performance of the antenna-coupled MOM diode detectors fabricated for this research, which will be discussed in detail. Noise sources will be described and presented with an experimental analysis. The figures of merit that will allow for comparison to other IR detector technologies will be presented. Polarization dependence, antenna-length dependence, and angle of incidence dependence will also be presented.

The current response of an Al/AlO_x/Pt ACMOMD with respect to the infrared input power has been measured and is shown in Fig. 3.37. As the incident power of the laser is increased, the device current increases linearly. Hence, the detector functions as a square law detector; the response is proportional to the power of the incident radiation. This detector signal is measured with the polarization of the incident infrared radiation parallel to the antenna, referred to as p-polarization.

Figures of Merit

The responsivity, SNR, NEP, and normalized detectivity of antenna-coupled MOM diode detectors can be determined by measuring the response of the detector to infrared radiation normal to the sample. With the polarization of the electric field of

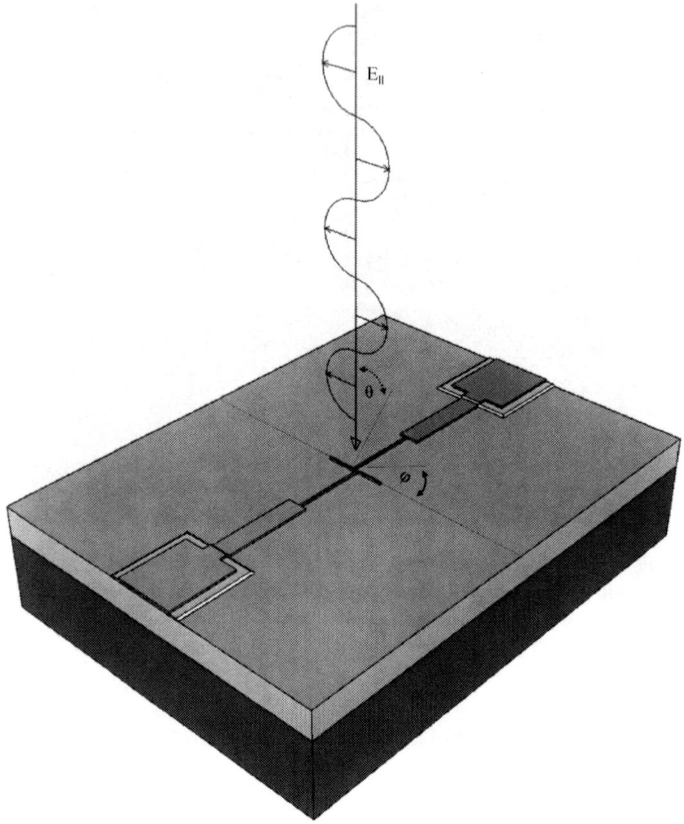

Fig. 3.36 Electromagnetic wave incident on antenna. When the electric field of the incident wave is parallel to the antenna ($\varphi = 0°$), it is referred to as p-polarization. S-polarization is when the electric field of the incident wave is perpendicular to the antenna ($\varphi = 90°$). Angle of incidence is represented by θ

the incident radiation parallel to the antenna axis, as shown in Fig. 3.36, the detector response is the highest. These figures of merit can be calculated according to the equations described in Sect. 3.1.3.

To determine these figures of merit, an effective area for the ACMOMD detectors must be established. Fumeaux et al. 1999) were able to determine the spatial response of a 3.1 μm dipole antenna-coupled MOM detector by deconvolution of the beam profile of the incident 10.6 μm CO_2 laser radiation. The antenna collecting area is represented as an ellipse. For a 3.1 μm dipole antenna, the full-width antenna response along the arms is approximately 12 μm that indicated the existence of a fringe field that extends approximately one dielectric wavelength beyond the physical ends of the dipole arms (Fumeaux et al. 1999). The axial ratio of the ellipse is $y_{ant}/x_{ant} = 0.54$, which agrees with the theoretical estimates presented by

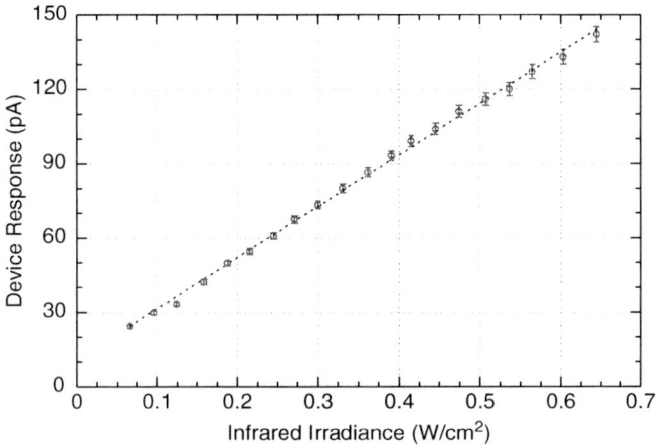

Fig. 3.37 Al/AlO$_x$/Pt ACMOMD device signal as a function of infrared input irradiance. Device current increases linearly with input power and hence the device functions as a square-law detector

Boreman et al. (1996). This ellipse, of approximate dimensions 12 × 6.5 μm, corresponds to an effective area equal to 61 μm^2. Since the geometry and substrate of the devices fabricated in this research are the same as those presented by G. Boreman, this effective area will be assumed for the ACMOMDs presented in this work.

Several figures of merit based on the measured characteristics have been calculated for these ACMOMDs. For a beam power of 62 mW, which corresponds to an input infrared irradiance of 498 mW/cm^2, a detected current SNR of 48.5 dB has been obtained for a measurement bandwidth of 10 Hz for Al/AlO$_x$/Pt devices fabricated with high oxygen exposure doses. Ambient light was found not to affect device response.

There are various sources of noise that contribute to the measured noise, due to the preamplifier and ACMOMD. The preamplifier noise is due to the Johnson noise of the feedback resistor in addition to the internal electronic noise of the amplifier, VNA. The noise of the ACMOMD has been determined to be mostly Johnson noise. These three sources of noise can be represented by the expression (Northrop 2005):

$$V = \sqrt{4kTR_{\mathrm{I}}\left(\frac{R_{\mathrm{F}}}{R_{\mathrm{I}}}\right)^2 B + 4kTR_{\mathrm{F}}B + V_{\mathrm{NA}}^2\left(1 + \frac{R_{\mathrm{F}}}{R_{\mathrm{I}}}\right)^2 B}$$

where R_{I} is the ACMOMD resistance, R_{F} is the feedback resistance, and B is the bandwidth in Hz. The first term is the Johnson noise of the input resistance, which takes into account the gain of the amplifier, the second term represents the Johnson noise of the feedback resistor, and the third term represents the noise of the operational amplifier. These terms correspond to the amplifier circuit shown in Fig. 3.38.

The output of the preamplifier V_{o} is passed to a second gain stage that amplifies the signal by a factor of 50. This signal is then passed to the lock-in amplifier. The

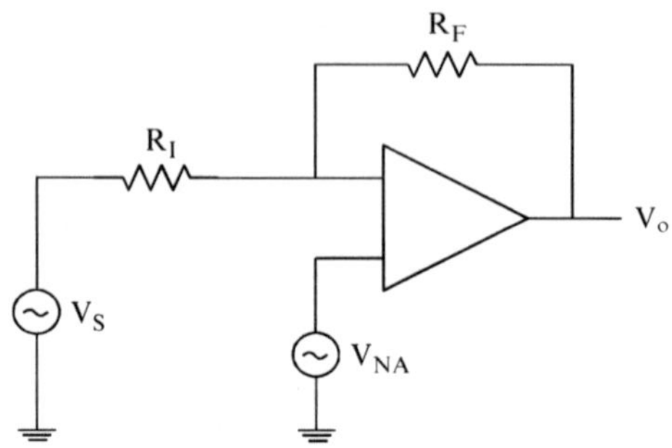

Fig. 3.38 Preamplifier circuit used to measure the signal and noise characteristics of ACMOMDs. The gain is determined by the ratio of resistances RF to RI. These resistances contribute Johnson noise and the amplifier, which is an AD743 operational amplifier, also contributes a noise figure

buffered Monitor Out output is connected to the spectrum analyzer input. The gain of this output is controlled by the input sensitivity of the lock-in: for input sensitivities of 2, 20, and 200 mV, the input gains are 2200, 220, and 210, respectively. Using metal film resistors with values of $R_{\mathrm{I}} = 1\,\mathrm{k}\Omega$, $R_{\mathrm{F}} = 100\,\mathrm{k}\Omega$, and using the 200 mV input sensitivity on the lock-in amplifier, the noise components were measured at a frequency of 1 kHz. The total noise value was measured to be approximately $470\,\mathrm{nV/Hz^{1/2}}$. The contribution from the Johnson noise in the feedback resistor noise is known to be $4kT\Delta fR$, which is $40.7\,\mathrm{nV/Hz^{1/2}}$. The Johnson noise contribution from the 1 kΩ resistor measured at the output is $407\,\mathrm{nV/Hz^{1/2}}$. These noise sources sum in quadrature, and since the total noise and Johnson noise contributions are known, the value of the operational amplifier noise can be solved, which gives $2.45\,\mathrm{nV/Hz^{1/2}}$. The operational amplifier noise, according to the Analog Devices AD743 Data Sheet, is approximately $2.5\,\mathrm{nV/Hz^{1/2}}$. Therefore, the expected value for the total noise is very close to the measured value. For large ACMOMD resistances, that is on the order of tens of kΩ to MΩs, the preamplifier noise is the dominant noise source. However, when ACMOMD resistances are on the order of approximately 1 kΩ, the measured noise value will increase.

Since the sensitivity of the current preamplifier is large, the noise generated by the lock-in amplifier and spectrum analyzer can be neglected. The measured noise leads to an estimated NEP of $1.15\,\mathrm{nW/Hz^{1/2}}$. This was calculated using the effective area of the antenna, a $6.5\,\mu\mathrm{m} \times 12\,\mu\mathrm{m}$ ellipse. Therefore, the normalized detectivity (D^*) for shadow evaporation ACMOMDs has been calculated as $2.15 \times 10^6\mathrm{cm \cdot Hz^{1/2} \cdot W^{-1}}$. This exceeds the performance of any previously reported MOM-based infrared detector (Abdel-Rahman 2004).

3.4.2.2 Polarization Dependence

The polarization dependence of a detector can lend insight into its operation mechanisms. For an antenna-coupled MOM diode, the response should be strongest for p-polarization, that is where the electric field of the incident radiation is parallel to the antenna axis ($\varphi = 0°$). For this polarization, the incident radiation induces waves that resonate in the antenna and are rectified by the MOM diode. A smaller response for s-polarization, that is where the incident electric field is perpendicular to the antenna axis ($\varphi = 90°$), is referred to as the polarization-independent contribution. For this orientation, the incident radiation does not induce any resonant current in the antenna. However, there still may be a small current present of thermal origin.

For the devices fabricated in this research, the polarization-independent signal is present regardless of the polarization of the incident radiation. The polarization-dependent response is represented by the signal response for p-polarization less that of the response for s-polarization. The response of these devices can be represented by the sum of a constant, thermally generated polarization-independent response, plus a cosine-squared polarization-dependent signal. The detected signal current $I(\varphi)$ can be expressed as:

$$I(q) = I_{\mathrm{ip}} + I_{\mathrm{p}}(\varphi) = I_{\mathrm{ip}} + I_{\mathrm{p}} \cos^2(\varphi_0 - \varphi)$$

where I_{ip} is the polarization-independent contribution, I_{p} is the polarization-dependent contribution, φ_0 is the location of the maximum polarization response, and φ is the angle between the antenna axis and the polarization direction (Fumeaux et al. 1998).

Figure 3.39 shows the polarization response of $Al/AlO_x/Pt$ ACMOMD.

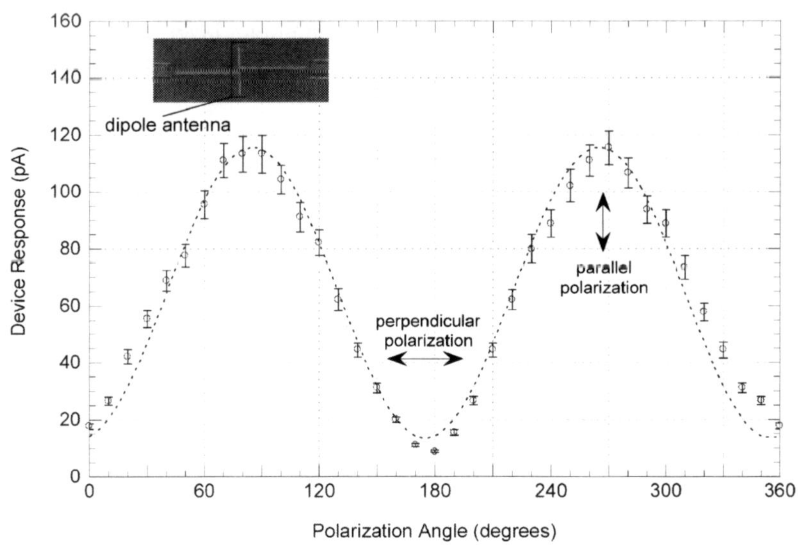

Fig. 3.39 Polarization response of an $Al/AlO_x/Pt$ ACMOMD (Bean et al. 2009). The electric field of the incident radiation is parallel to the antenna at 90° and 270°, where the maximum response was measured. The polarization ratio for this device, which is the maximum response divided by the minimum response, is about 10

As expected, the maximum signal was obtained when the electric field of the incident infrared radiation was parallel to the antenna (90°, 270°), whereas the minimum was measured with the electric field perpendicular to the antenna (0°, 180°). The data points represent the average response for each polarization angle, whereas the error bars represent the standard deviation of a time-averaged response for each polarization angle. The cross-polarization response is due to thermionic emission because of heating of the structure. Electrons with higher energies due to heating can gain enough energy to surmount or tunnel through the oxide barrier, as discussed in Sect. 3.2.3.

In agreement with antenna theory, the device response follows a cosine-squared dependence, denoted by the dotted line (Bean et al. 2009). The polarization ratio, which is the ratio between the maximum and minimum signal, is around 10. The variations in detector response are due in part to mode hopping in the CO_2 laser used for testing. In addition, small fluctuations in the angle of incidence, which is desired to be normal for polarization dependence measurements, can impact the measured signal. These errors can be caused if the sample is not affixed parallel to the LCC, the LCC is not seated properly in the chip socket, the chip socket is not mounted parallel to the PC board, or if the PC board is not parallel to the rotational stage. Every effort was made to minimize these errors by ensuring the beam was incident normal to the sample, but small fluctuations can cause the experimental errors shown in the measurements.

Tiwari et al. (2009) have investigated fabrication techniques that provide for an in situ oxide growth. In this process, the first antenna arm is patterned, developed, and aluminum is deposited on the sample. The second antenna arm is then patterned and developed. However, before the second metal evaporation, a Kaufman Ion Source is used to remove the native oxide of the first aluminum layer at the overlap area by argon bombardment. Oxygen can then be introduced to the chamber to reform an oxide on the first antenna arm and then the second metal evaporation can take place. By using an etch and regrowth method of the oxide barrier, control over diode characteristics, similar to those of shadow-evaporation devices shown in Fig. 3.33, can be created.

A representative polarization-dependence response of a two-lithography $Al/AlO_x/Pt$ ACMOMD fabricated by Tiwari is shown in Fig. 3.40.

3.4.2.3 Antenna Length Dependence

The response of an antenna-coupled detector depends on the antenna geometry and the frequency of the incident electromagnetic irradiation. The peak current at the center of a half-wavelength dipole occurs when the antenna length is equal to half the wavelength of the desired detection wavelength. For an antenna on a substrate, the length of the dipole corresponds to the equivalent substrate wavelength of the incident radiation. For dipole antennas fabricated on a silicon substrate with 1.5 μm silicon dioxide matching layers irradiated by 10.6 μm laser radiation, the optimum dipole length has been found to be 3.1 μm (Rakos 2006). For antennas

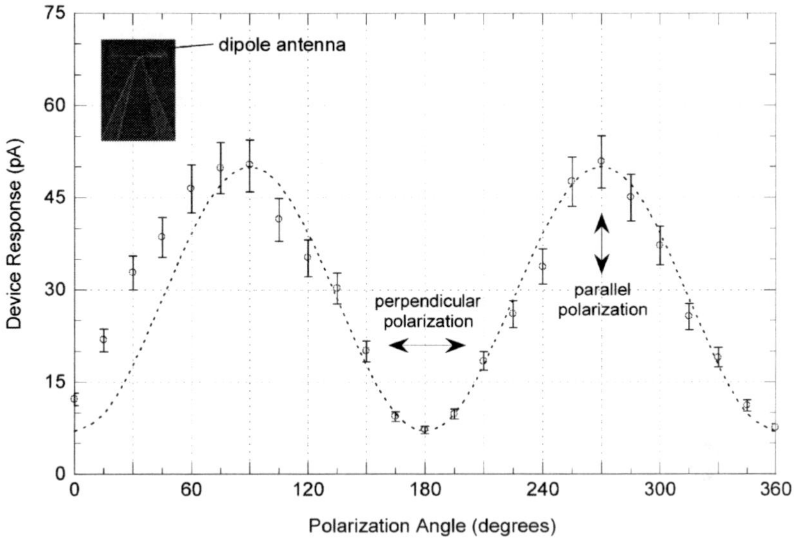

Fig. 3.40 Polarization response of two-step lithography Al/AlO$_x$/Pt ACMOMD (Tiwari et al. 2009). The electric field of the incident radiation is parallel to the antenna at 90° and 270°, where the maximum response was measured. The polarization ratio for this device, which is the maximum response divided by the minimum response, is about 7

of 3.1 μm length, the maximum signal response is for p-polarized 10.6 μm illumination and the minimum signal is found for s-polarized illumination. The ratio of the maximum signal to the minimum signal is the polarization ratio. Depending on the antenna length, incident 10.6 μm radiation will cause varying degrees of resonance. Along those lines, the polarization-dependent portion of the response will vary depending on length.

Figure 3.41 shows the polarization ratio that is the ratio of the maximum signal measured to the minimum level measured, as a function of antenna length for Al/AlO$_x$/Pt shadow-evaporation ACMOMDs. This is commonly shown as the relationship between the polarization-dependent response and antenna length, but polarization ratio is shown in an effort to normalize response from numerous detectors of the same length and minimize the impacts of mode-hopping of the CO_2 laser. A zero polarization-dependent response corresponds to a polarization ratio of one in this case, and therefore the theoretical $\sin^4(\pi L/2\lambda)$ dependence on antenna length (Wilke et al. 1994) still holds and is shown by the dotted line in the figure.

The first maximum, or resonance, occurs at the expected optimum antenna length, but there is some deviation from the expected result. For shorter antenna lengths, the electrical leads can have an effect on the detected polarization-dependent signal. This impact is assumed to be somewhat small, due to the fact that lead currents would be attenuated due to the Coleman effect since the leads are long, and the fact that the large lead widths used allow transverse currents, as discussed in Sects. 3.2.5 and 3.2.1, respectively. For the first minimum, the second

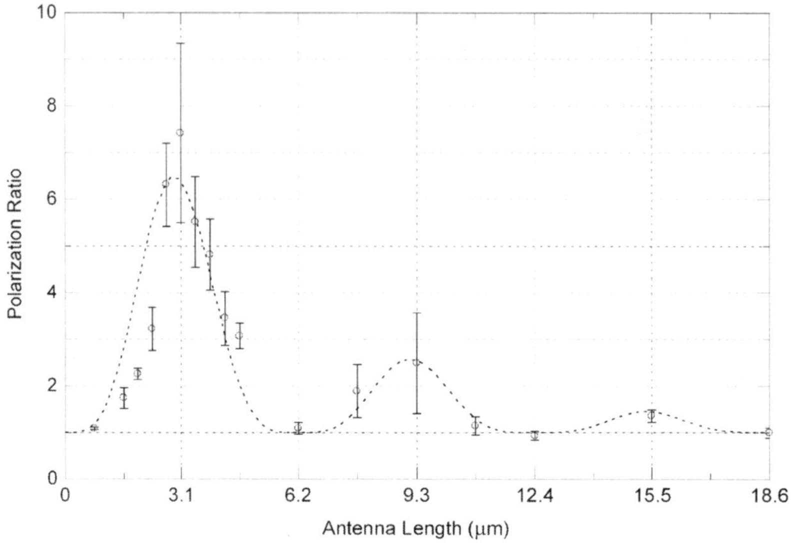

Fig. 3.41 Polarization ratio as a function of antenna length for $Al/AlO_x/Pt$ ACMOMDs. A polarization ratio of 1 means there is no polarization-dependent signal. The data points represent the average of no less than eight devices for each antenna length and the error bars represent the standard deviation of the response

maximum, second minimum, and third maximum, the measurements closely match the theory. However, beyond the third maximum, the attenuation of the antenna current is dominated by the Coleman effect. By fitting the data to a theoretical $\sin^4(\pi L/2\lambda)\exp\left(\Gamma_{exp}L/2\right)$ dependence (Wilke et al. 1994), the calculated experimental attenuation constant $\Gamma_{exp} = 0.40\,\mu m^{-1}$. This takes into account all forms of attenuation such as Coleman effect and surface attenuation. The theoretical attenuation due to Coleman effect $\Gamma_C = 1.48\,\mu m^{-1}$ and the theoretical surface attenuation constant is $\Gamma_{sr} = 22.7\,\mu m^{-1}$. Since the experimental attenuation constant is much smaller than the surface attenuation constant, it can be concluded that this imparts a negligible attenuation and that the experimental antenna current attenuation is due to the Coleman effect. This has also been confirmed by Wilke et al. (1994) and Fumeaux et al. (2000).

3.5 Comparison to Current Technologies

Dipole antenna-coupled MOM diode detectors have been fabricated to detect electromagnetic radiation in the thermal or LWIR band. Although there are several types of detectors currently commercially available, none provide for each and every project constraint detailed in Sect. 3.1.1. These include the ability to: function without cooling at room temperature, provide frequency selective and fast response in a compact $10 \times 10\,\mu m$ detector area, and offer CMOS compatible fabrication. Although this type of detector has been previously researched, this project is the first to apply one-step lithography, shadow-evaporation technique to fabricate ACMOMDs (Bean et al. 2009).

3.5.1 Comparison with Currently Available IR Detectors

The infrared response of the detectors fabricated for this research has been characterized; the results were presented in Chap. 4. The devices perform as expected in terms of polarization-dependent and antenna-length response. Table 3.4 provides an opportunity to compare the results from this research to current infrared detector technologies (Dereniak and Boreman 1996; Fumeaux et al. 2000; Bean et al. 2009).

From this evaluation, it can be concluded that of radiation-field detectors, those presented in this research provide the highest reported D^* at $2.15 \times 10^6 \mathrm{cm} \cdot \mathrm{Hz}^{1/2} \cdot \mathrm{W}^{-1}$. However, to be a commercially viable technology, normalized detectivity should be in the $10^8\,\mathrm{cm} \cdot \mathrm{Hz}^{1/2} \cdot \mathrm{W}^{-1}$ range. While the detectors provide for all of the constraints outlined in Sect. 3.1.1, the D^* value must be increased by roughly 50 times in order to be viable candidates for integration with commercial night-vision systems.

3.5.2 Integration with CMOS Imaging Chips

The ability to fabricate infrared detectors that respond to thermal IR radiation has been demonstrated in the preceding pages of this chapter. However, implementing these devices in an array, which might ultimately lead to the capability of producing

Table 3.4 Comparison of ACMOMDs to current IR detectors

Detector type	Area (mm^2)	Operating temperature (K)	Response time (ms)	D^* ($\mathrm{cm\,Hz}^{1/2}\mathrm{W}^{-1}$)
HgCdTe	~0.25	77	10^{-3}	1×10^{10}
Pyroelectric	0.78–63.6	300	10^{-4}	1×10^9
Thermistor	0.25–25	300	5	3×10^8
Thermopile	~5	208–343	5	7×10^8
Bolometer (Ge)	~5	0.3–2.0	0.5	3×10^8
Antenna-coupled bolometer (Nb)	0.0001	300	10^{-4}	6×10^5
Antenna-coupled MOM diode	0.0001	300	10^{-10}	1×10^6
Shadow-evaporation ACMOMD	0.0001	300	10^{-10}	2.15×10^6

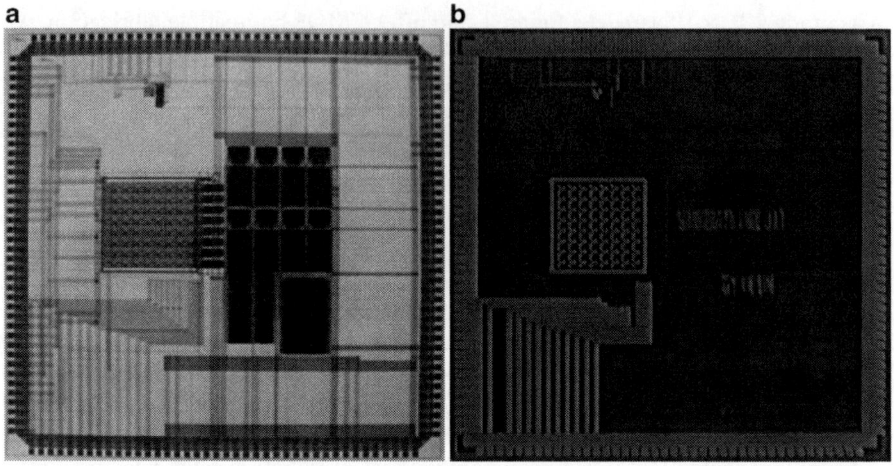

Fig. 3.42 (**a**) Layout and (**b**) optical micrograph of Eutecus Xenon-NC V1 CNN chip. This CMOS imaging chip contains an area where an 8 × 8 array of detectors can be implemented. ACMOMDs are chosen to provide for infrared detection

an image similar to the thermal IR image of Fig. 3.1, would prove commercial viability. This chapter discusses processes that can be used to integrate an 8 × 8 array of ACMOMDs with a CMOS imaging chip. The chip used in this case the Eutecus Xenon-NC V1 chip, shown in Fig. 3.42. The area left of the center of the chip, which is outlined by a square, denotes where ACMOMD infrared detectors developed, fabricated, and characterized for this project are to be integrated. The image in Fig. 3.42 is a screenshot of the layout of the CNN chip, whereas the image in B is an optical micrograph. The area in the box denotes the integration location for the ACMOMDs.

The SEM in Fig. 3.43 shows this integration area. This area is composed of 8 × 8 identical cells, each connected to amplification and computation circuitry. The ACMOMDs will be electrically connected between each of the 8 × 8, or 64 total, vias, which are the light gray squares, and the pad ring around the perimeter of the image. The inset image shows a magnified view of one cell, where the *via* is located at the center. One lead of the ACMOMD detector will be connected to the *via*, whereas the other connection will be made to the pad ring that is extended throughout the chip so that identical connections can be made for each ACMOMD.

The integration process is outlined in Fig. 3.44. This collection of images shows the entire 5 × 5 mm CNN chip, the 8 × 8 array integration area, and a single cell with integrated device, along with three SEMs showing the cell, completed device, and dipole antenna.

The last step in the process is wirebonding the completed CNN chip for testing with the Bi-I system. There are 168 pins on the CNN chip; the package used is a 181-pin ceramic package. An image of the completed CNN chip, integrated with ACMOMDs and wirebonded and ready for testing is shown in Fig. 3.45.

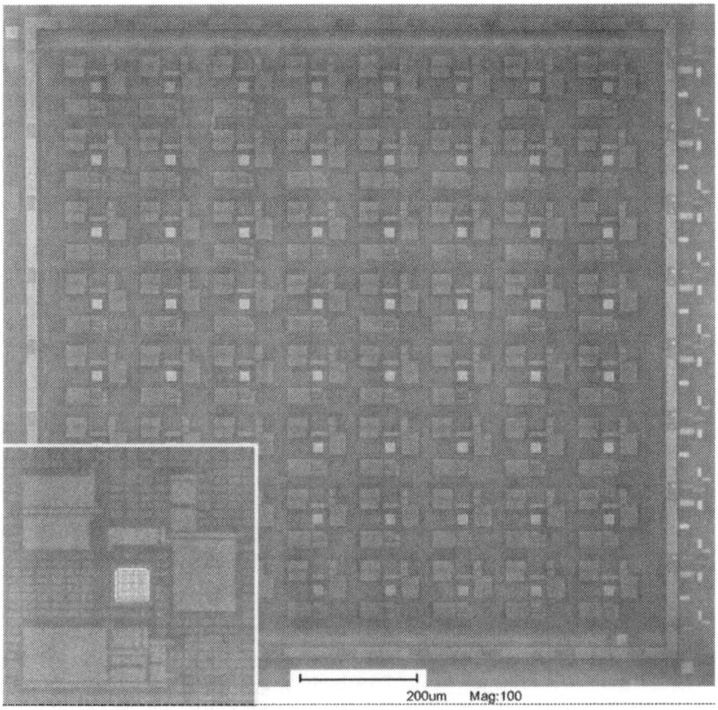

Fig. 3.43 Scanning electron micrograph of IR detector integration area on CNN. The inset image shows the integration area for one device; the light gray area in the *center* is a via contact, where one lead of the ACMOMD is connected

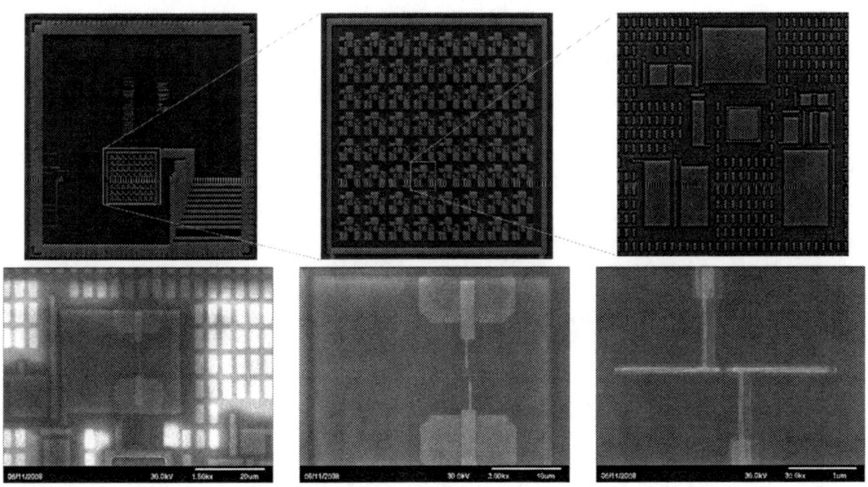

Fig. 3.44 Various magnifications of completed devices integrated with the CNN chip. The Ti/Pt contacts can be seen in the scanning electron micrographs with the ACMOMD integrated between the contacts

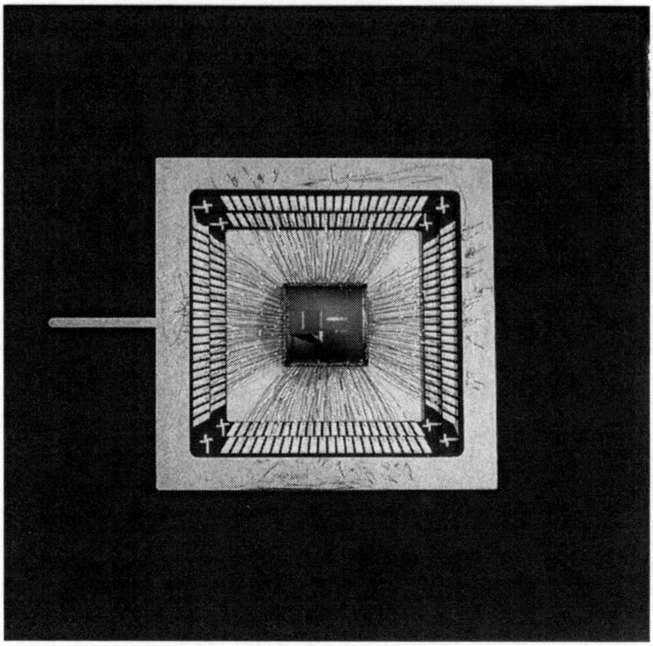

Fig. 3.45 Completed CNN chip after ACMOMD integration and wirebonding. 168 aluminum wirebonds connect the contacts on the edge of the CNN chip to the pins on the chip socket. Once a measurement test bench is finalized, the response of the array of ACMOMDs could be measured

The preceding steps could be used to integrate ACMOMDs with a CMOS vision chip. Numerous antenna geometries could be employed to provide for multi-spectral image acquisition. This would provide for a powerful image acquisition solution that could have endless possibilities.

References

Abdel-Rahman, M.R., Gonzalez, F.J., Boreman, G.D.: Antenna-Coupled Metal-Oxide-Metal Diodes for Dual-Band Detection at 92.5 GHz and 28 THz. Electron. Lett. **40**, 116–118 (2004)

Alda, J., Fumeaux, C., Gritz, M.A., et al.: Responsivity of Infrared Antenna-Coupled Microbolometers for Air-side and Substrate-side Illumination. Infrared Phys. Techn. **41**, 1–9 (2000)

Allen, C., Aramś, F., Wang, M., et al.: Infrared-to-Millimeter, Broadband, Solid State Bolometer Detectors. Appl. Optics **8**, 813–817 (1969)

Balanis, C.A.: Antenna Theory: Analysis and Design. John Wiley and Sons, Inc., Hoboken (2005)

Bean, J.A., Tiwari, B., Bernstein, G.H., Fay, P., Porod, W.: Thermal Infrared Detection Using Dipole Antenna-Coupled Metal-Oxide-Metal Diodes. J. Vac. Sci. Tech. B **27**, 11–14 (2009)

Beerman, H.P.: The Pyroelectric Detector of Infrared Radiation. IEEE Trans. Electron Devices **ED-16**, 554 (1969)

Bernstein, G.H., Hill, D.A., Wen-Ping, L.: New High-contrast Developers for Poly(methyl methacrylate) Resist. J. Appl. Phys. **71**, 4066–4075 (1992)

Block, W.H., Gaddy, O.L.: Thin Metal Film Room-Temperature IR Bolometers with Nanosecond Response Time. IEEE J. Quantum Electron. **QE-9**, 1044–1053 (1973)

Boreman, G.D., Dogariu, A., Christodoulou, C., et al.: Modulation Transfer Function of Antenna-coupled Infrared Detector Arrays. Appl. Opt. **35**, 6110–6114 (1996)

Boreman, G.D., Fumeaux, C., Herrmann, W., et al.: Tunable Polarization Response of a Planar Asymmetric-Spiral Infrared Antenna. Opt. Lett. **23**, 1912–1914 (1998)

Bradley, C.C., Edwards, G., Knight, D.J.E.: Absolute Measurement of Submillimetre and Far Infrared Laser Frequencies. Radio Electron. Eng. **42**, 321–327 (1972)

Bramley, P., Clark, S.: A Quantitative Model for the Thermocouple Effect Using Statistical and Quantum Mechanics. AIP Conf. Proc. **684**, 547–552 (2003)

Capper, P., Elliott, C.T.: Infrared Detectors and Emitters: Materials and Devices. Kluwer Academic, Norwell (2000)

Chong, N., Ahmed, H.: Antenna-Coupled Polycrystalline Silicon Air-Bridge Thermal Detector for Mid-Infrared Radiation. Appl. Phys. Lett. **71**, 1607–1609 (1997)

Chua, L.O., Yang, L.: Cellular Neural Networks: Theory. IEEE Trans. Circuits Syst. **35**, 1257–1272 (1988)

Chua, L.O., Yang, L.: Cellular Neural Networks: Applications. IEEE Trans. Circuits Syst. **35**, 1273–1290 (1988)

Chua, L.O., Roska, T.: Cellular Neural Networks and Visual Computing. Cambridge Press, Cambridge (2002)

Codreanu, I., Fumeaux, C., Spencer, D.F., et al.: Microstrip Antenna-Coupled Infrared Detector. Electron. Lett. **35**, 2166–2167 (1999)

Codreanu, I., Boreman, G.D.: Infrared Microstrip Dipole Antennas-FDTD Predictions Versus Experiment. Microw. Opt. Tech. Lett. **29**, 381–383 (2001)

Codreanu, L., Gonzalez, F.J., Boreman, G.D.: Detection Mechanisms in Microstrip Dipole Antenna-Coupled Infrared Detectors. Infrared Phys. Techn. **44**, 155–163 (2003)

Cohen-Solal, G., Riant, Y.: Epitaxial (CdHg)Te Infrared Photovoltaic Detectors. Appl. Phys. Lett. **19** 436–438 (1971)

Coleman, B.L.: Propagation of Electromagnetic Disturbances Along a Thin Wire in a Horizontally Stratified Medium. Philosoph. Mag. **41**, 276–288 (1950)

Corbeil, J.L., Lavrik, N.V., Rajic, S., et al.: "Self-leveling" Uncooled Microcantilever Thermal Detector. Appl. Phys. Lett. **81**, 1306–1308 (2002)

Daneu, V., Sokoloff, D., Sanchez, A., et al.: Extension of Laser Harmonic-Frequency Mixing Techniques into the 9um Region with an Infrared Metal-Metal Point-Contact Diode. Appl. Phys. Lett. **15**, 398–401 (1969)

Datskos, P.G., Lavrik, N.V., Rajic, S.: Performance of Uncooled Microcantilever Thermal Detectors. Rev. Sci. Instruments **75**, 1134–1148 (2004)

Dereniak, E.L., Boreman, G.D.: Infrared Detectors and Systems. John Wiley & Sons, Inc., New York (1996)

Diesing, D., Merschdorf, M., Thon, A., et al.: Identification of Multiphoton Induced Photocurrents in Metal-Insulator-Metal Junctions. Appl. Phys. B **B78**, 443–446, 2004

Dolan, G.J.: Offset Works for Lift-off Photoprocessing. Appl. Phys. Lett. **31**, 337–339 (1977)

Esfandiari, P., Bernstein, G., Fay, P., et al.: Tunable Antenna-Coupled Metal-Oxide-Metal (MOM) Uncooled IR Detector. Proc. SPIE – Int. Soc. Opt. Eng. **5783**, 470–482 (2005)

Faris, S.M., Gustafson, T.K., Wiesner, J.C.: Detection of Optical and Infrared Radiation with DC-Biased Electron-Tunneling Metal-Barrier-Metal Diodes. IEEE J. Quantum Electron. **QE-9**, 737–745 (1973)

Fastenau, J.M., Liu, W.K., Fang, X.M., et al.: Commercial Production of QWIP Wafers by Molecular Beam Epitaxy. Infrared Phys. Tech. **42**, 407–415 (2001)

Fisher, J.C., Giaever, I.: Tunneling Through Thin Insulating Layers. J. Appl. Phys. **32**, 172–177 (1961)

Fulton, T.A., Dolan, G.J.: Observation of Single-Electron Charging Effects in Small Tunnel Junctions. Phys. Rev. Lett. **59**, 109–112 (1987)

Fulton, T.A., Gammel, P.L., Bishop, D.J., et al.: Observation of Combined Josephson and Charging Effects in Small Tunnel Junction Circuits. Phys. Rev. Lett. **63**, 1307–1310 (1989)

Fumeaux, C., Herrmann, W., Kneubühl, F.K., et al.: Nanometer Thin-Film Ni-NiO-Ni Diodes for Detection and Mixing of 30 THz Radiation. Infrared Phys. Tech. **39**, 123–183 (1998)

Fumeaux, C., Boreman, G., Herrmann, W., Kneubühl, F., Rothuizen, H.: Spatial impulse response of lithographic infrared antenna. Applied Optics **38**, 37–46 (1999)

Fumeaux, C., Gritz, M.A., Codreanu, I., et al.: Measurement of the Resonant Lengths of Infrared Dipole Antennas. Infrared Phys. Tech. **41**, 271–281 (2000)

Gallagher, D.L., Adrian, M.L.: Two-Dimensional Drift Velocities from the IMAGE EUV Plasmaspheric Imager. J. Atmos. Sol.-Terr. Phys. **69**, 341–350 (2007)

George, S.M., Ott, A.W., Klaus, J.W.: Surface Chemistry for Atomic Layer Growth. J. Phys. Chem. **100**, 13121–13131 (1996)

Glass, A.M.: Investigation of the Electrical Properties of $Sr_{1-x}/Ba_x/Nb_2/O_6$ with Special Reference to Pyroelectric Detection. J. Appl. Phys. **40**, 4699–4713 (1969)

Gloos, K., Koppinen, P.J., Pekola, J.P.: Properties of Native Ultrathin Aluminium Oxide Tunnel Barriers. J. Phys.: Condens. Matter **15**, 1733–1746 (2003)

Gonzalez, F.J., Boreman, G.D.: Comparison of Dipole, Bowtie, Spiral, and Log-periodic IR Antennas. Infrared Phys. Tech. **46**, 418–428 (2005)

Green, S.I.: Point Contact M.O.M. Tunneling Detector Analysis. J. Appl. Phys. **42**, 1166–1169 (1971)

Gupta, H.M., Van Overstraeten, R.J.: Role of Trap States in the Insulator Region for MIM Characteristics. J. Appl. Phys. **46**, 2675–2682 (1975)

Gustafson, T.K., Bridges, T.J.: Radiation of Difference Frequencies Produced by Mixing in Metal-Barrier-Metal Diodes. Appl. Phys. Lett. **25**, 56–59 (1974)

Gustafson, T.K., Schmidt, R.V., Perucca, J.R.: Optical Detection in Thin-Film Metal-Oxide-Metal Diodes. Appl. Phys. Lett. **24**, 620–622 1974.

Hasnain, G., Arjavalingam, G., Dienes, A., et al.: Dispersion of Picosecond Pulses on Microstrip Transmission Lines. Proc. SPIE – Int. Soc. for Opt. Eng. **439**, 159–163 (1983)

Hegyi, B., Csurgay, A., Porod, W.: Investigation of the Nonlinearity Properties of the DC I-V Characteristics of Metal-Insulator-Metal (MIM) Tunnel Diodes with Double-Layer Insulators. J. Comp. Electron. **6**, 159–162 (2007)

Heiblum, M., Wang, S.Y., Gustafson, T.K., et al.: Edge-MOM Diode: An Integrated, Optical, Nonlinear Device. IEEE Trans. Electron Dev. **ED-24**, 1199 (1977)

Heiblum, M., Shihyuan, W., Whinnery, J.R., et al.: Characteristics of Integrated MOM Junctions at DC and at Optical Frequencies. IEEE J. Quant. Electron. **QE-14**, 159–169 (1978)

Kadlec, J., Gundlach, K.H.: Dependence of the Barrier Height on Insulator Thickness in Al-(Al-oxide)-Al Sandwiches. Solid State Comm. **16**, 621–623 (1975)

Kale, B.M.: Electron Tunneling Devices in Optics. Opt. Eng. **24**, 267–274 (1985)

Kovacs, G.T.A.: Bulk Micromachining of Silicon. Proc. IEEE **86**, 1536–1551 (1998)

Kwok, S.P., Haddad, G.I., Lobov, G.: Metal-Oxide-Metal (M-O-M) Detector. J. Appl. Phys. **42**, 554–563 (1971)

Lahiji, G.R., Wise, K.D.: A Batch-Fabricated Silicon Thermopile Infrared Detector. IEEE Trans. Electron Devices **ED-29**, 14–22 (1982)

Lang, S.B., Rice, L.H., Shaw, S.A.: Pyroelectric Effect in Barium Titanate Ceramic. J. Appl. Phys. **40**, 4335–4340 (1969)

Long, D.: Photovoltaic and Photoconductive Infrared Detectors. Opt. Infrared Detect. 101–147 (1977)

Lord, S.D.: A New Software Tool for Computing Earth's Atmospheric Transmission of Near- and Far-Infrared Radiation. NASA Tech. Memo. 103957 (1992)

Matsukura, Y., Nishino, H., Tanaka, H., et al.: Quantum Well Infrared Photodetectors (QWIP) with Selectively Re-Grown N-GaAs Plugs. Proc. SPIE **4369**, 481–488 (2001)

Mead, C.A.: Electron Transport Mechanisms in Thin Insulating Films. Phys. Rev. **128**, 2088–2093 (1962)

Michaelson, H.B.: The Work Function of the Elements and its Periodicity. J. Appl. Phys. **48**, 4729–4733 (1977)

Middlebrook, C.T., Zummo, G., Boreman, G.D.: Direct-Write Electron-Beam Lithography of an IR Antenna-Coupled Microbolometer Onto the Surface of a Hemispherical Lens. J. of Vac. Sci. & Tech. B **24**, 2566–2569 (2006)

Miyamoto, S., Kawashima, S., Shionoya, S.: Photo-Induced Infrared Absorption in ZnSe Single Crystals. J. Phys. Soc. Japan **24**, 1182 (1968)

Momida H, Hamada T, Ohno T: First-Principles study of Dielectric Properties of Amorphous High-k Materials. Jpn. J. Appl. Phys. **46**, 3255–3260 (2007)

Nagae, M.: Response Time of Metal-Insulatator-Metal Tunnel Junctions to Step Input Voltage. Jpn. J. Appl. Phys. **12**, 523–530 (1973)

Nelson, O.L., Anderson, D.E.: Potential Barrier Parameters in Thin-Film Al-Al_2O_3-Metal Diodes. J. Appl. Phys. **37**, 77–82 (1966)

Noda, A., Miyamoto, T., Murakami, S., et al.: A Dielectric Bolometer Mode of Infrared Sensor Using a New Ba ($Ti_{1-x}/Sn_x/O3$) Thin Film with a High Temperature Coefficient of Dielectric Constant. Integr. Ferroelectr. **49**, 305–314 (2002)

Northrop, R.B.: Introduction to Instrumentation and Measurements, Second Edition: Taylor & Francis, Boca Raton (2005)

Nossek, J.A., Seiler, G., Roska, T., et al.: Cellular Neural Networks: Theory and Circuit Design. Int. J. Circuit Theory Appl. **20**, 533–553 (1992)

Orlov, A.O., Amlani, I., Kummamuru, R.K., et al.: Experimental Demonstration of Clocked Single-Electron Switching in Quantum-Dot Cellular Automata. Appl. Phys. Lett. **77**, 295–297 (2000)

Ott, A.W., McCarley, K.C., Klaus, J.W., et al.: Atomic Layer Controlled Deposition of Al_2O_3 Films Using Binary Reaction Sequence Chemistry. App. Surface Sci. **107**, 128–136 (1996)

Pierret, R.: Advanced Semiconductor Fundamentals 2nd Edition. Prentice Hall, Upper Saddle River (2002)

Rakos, B.: Investigation of Metal-Oxide-Metal Structures for Optical Sensor Applications. Ph.D. Dissertation, University of Notre Dame, Notre Dame (2006)

Richards, R.K., Hutchinson, D.P., Bennett, C.A.: Room-Temperature QWIP Detection at 10 um. Proc. SPIE – Int. Soc. Opt. Eng. **4820**, 250–253 (2003)

Rogalski, A.: Infrared Detectors. Gordon and Breach, Amsterdam (2000)

Rutledge, D.B., Schwarz, S.E., Adams, A.T.: Infrared and Submillimetre Antennas. Infrared Phys. **18**, 713–729 (1978)

Sakuma, E., Evenson, K.M.: Characteristics of Tungsten-Nickel Point Contact Diodes Used as Laser Harmonic-Generator Mixers. IEEE J. Quant. Electron, **QE-10**, 599–603 (1974)

Sanchez, A., Davis, C.F., Jr., Liu, K.C., et al.: The MOM Tunneling Diode: Theoretical Estimate of its Performance at Microwave and Infrared Frequencies. J. Appl. Phys. **49**, 5270–5277 (1978)

Schwarz, S.E., Ulrich, B.T.: Antenna-Coupled Infrared Detectors. J. Appl. Phys. **48**, 1870–1873 (1977)

Simmons, J.G.: Electric Tunnel Effect Between Dissimilar Electrodes Separated by a Thin Insulating Film. J. Appl. Phys. **34**, 2581–2590 (1963)

Small, J.G., Elchinger, G.M., Javan, A., et al.: AC Electron Tunneling at Infrared Frequencies: Thin Film M-O-M Diode Structure with Broad-band Characteristics. Appl. Phys. Lett. **24**, 275–279 (1974)

Sokoloff, D.R., Sanchez, A., Osgod, R.M., et al.: Extension of Laser Harmonic-Frequency Mixing Into the 5-micrometer Regions. Appl. Phys. Lett. **17**, 257–259 (1970)

Summers, C.J., Zwerdling, S.: Material Characterization and Ultimate Performance Calculations of Compensated n-Type Silicon Bolometer Detectors at Liquid-Helium Temperatures. IEEE Trans. Microw. Theory Tech. **MTT-22**, 1009–1013 (1974)

Sun, Z.: Silicon-based Passives for Integrated Microwave and Infrared Applications. Ph.D. Dissertation, University of Notre Dame, Notre Dame (2006)

Thon, A., Merschdorf, M., Pfeiffer, W., et al.: Photon-Assisted Tunneling Versus Tunneling of Excited Electrons in Metal-Insulator-Metal Junctions. Appl. Phys. A (Mater. Sci. Process.) A78:189–199, 2004.

Tidrow, M.Z., Beck, W.A., Clark, W.W., et al.: Device Physics and Focal Plane Array Applications of QWIP and MCT. Proc. SPIE **3629**, 100–113 (1999)

Tiwari, B., Bean, J.A., Szakmany, G., et al.: Controlled Etching and Regrowth of Tunnel Oxide for Antenna-Coupled MOM Diodes. Submitted to J. Vac. Sci. Tech. B (2009)

Tucker, J.R., Millea, M.F.: Photon detection in nonlinear tunneling devices. Applied Phys. Lett. **33**, 611–613 (1978)

Twu, B.I., Schwarz, S.E.: Mechanism and Properties of Point-Contact Metal-Insulator-Metal Diode Detectors at 10.6 micrometers. Appl. Phys. Lett. **25**, 595–598 (1974)

Vanbesien, K., De Visschere, P., Smet, P.F., et al.: Electrical Properties of Al_2O_3 Films for TFEL-Devices Made With Sol-Gel Technology. Thin Solid Films **514**, 323–328 (2006)

Wang, S.Y., Izawa, T., Gustafson, T.K.: Coupling Characteristics of Thin-Film Metal-Oxide-Metal Diodes at 10.6 um. Appl. Phys. Lett. **27**, 481–483 (1975)

Wilke, I., Herrmann, W., Kneubuhl, F.K.: Integrated Nanostrip Dipole Antennas for Coherent 30 THz Infrared Radiation. Appl. Phys. B **B58**, 87–95 (1994)

Wilke, I., Oppliger, Y., Herrmann, W., et al.: Nanometer Thin-Film Ni-NiO-Ni Diodes for 30 THz Radiation. Appl. Phys. A **A58**, 329–341 (1994)

Yamashita O: Effect of Metal Electrode on Seebeck Coefficient of p- and n-type Si Thermoelectrics. J. Appl. Phys. **95**, 178–183 (2004)

Yasuoka, Y., Sakurada, T., Siu, D.P., et al.: Resistance Dependence of Detected Signals of MOM diodes. J. Appl. Phys. **50**, 5860–5864 (1979)

Yngvesson, S.: Microwave Semiconductor Devices. Kluwer Academic Publishers, Norwell (1991)

Chapter 4
Memristors: A New Nanoscale CNN Cell

Leon Chua

Abstract The circuit-theoretic foundation of the *memristor* and its generalizations to a lossless *memory capacitor* and a lossless *memory inductor* is presented along with the devices' constitutive relations. Their identifying fingerprints consist of a *pinched hysteresis loop* when plotted in the voltage versus current plane, voltage versus charge plane, and current versus flux plane, respectively. All three devices are *nonlinear* and their underlying physical mechanisms are expected to dominate and manifest their *memory character* as the device size scales below 20 nm, when electrons and ions are coupled strongly under intense electric and/or magnetic fields. While all three devices are ideal candidates for *nonvolatile nano memories*, their long-term significance lies in their enabling potentials for designing *nano CNNs*, and intelligent machines, with *learning* and *adaptive* capabilities. Even more fundamental is their nonlinear dynamics that underpins the biological basis of *life* itself, where *ion channels*, with their complex biochemical synaptic dynamics, are essentially *memristors*.

4.1 Introduction

The May 1 *Nature* paper (Strukov et al. 2008; Tour and He 2008) that unveils a working nano *memristor* device has generated immense interests among both nano device researchers and the memory-chip industry alike. As of June 1, 2008, Google had registered more than a million hits. This unprecedented interests was due in part to the high potential economic impact of the HP invention. Since the titanium-dioxide (TiO_2) HP memristor can in principle be scaled down to almost 1 nm[1] and is compatible with the current IC technology, some industry observers are predicting that the HP memristor device would eventually replace both flash memories and DRAMS. In the November 12, 2008 issue of *EE Times*, an article by

L. Chua (✉)
University of California, Berkeley, California, USA
e-mail: chua@eecs.berkeley.edu

[1] Private communications with Dr. Stan Williams from HP.

C. Baatar et al. (eds.), *Cellular Nanoscale Sensory Wave Computing*,
DOI 10.1007/978-1-4419-1011-0_4,

R. Colin Johnson, entitled "Is the U.S. falling behind in chip R & D," even went so far as predicting that "*HP Labs' memristors could make semiconductor memory obsolete*." Indeed, a PC with no booting times, and which remembers all data prior to unplugging the power, could become standard features within 5 years.

Notwithstanding the above cited excitements, we believe the long-term significance of the *memristor*, and its generalizations, lies in its potential as an enabling component for building *adaptive* and *intelligent nanoscale Cellular Neural Networks* (*CNN*) (Itoh and Chua 2008, 2009), as well as *brainlike* and *intelligent* machines that are endowed with *learning* and *self-programming* (Borghetti et al. 2009) capabilities. For example, memristors could mimic *synaptic plasticity* and *long-term memory* mechanisms, such as LTP (Bliss et al. 2003), and distributed cellular memories in human brains (Kandel 2006; Dudai 1989). In particular, since the memristor is a *dynamic nonlinear* device, it can provide *dynamic* and *self-tuning couplings* between locally active CNN cells. In terms of the *standard* CNN cells (Chua 1998; Chua and Roska 2002), memristors would make it possible to tune the coefficients of both the feedback template **A** and the feedforward template **B**, via on-chip adaptive and learning rules. They could also be used to design *field-programmable nonvolatile* CNN chips for executing high-volume custom-designed dynamic motion recognition tasks.

We also believe that memristors[2] are responsible for the highly nonlinear electrical conduction dynamics in *ion channels* and many *bioelectric* phenomena. For example, the *potassium conductance* and *sodium conductance* referred to by Hodgkin and Huxley in their Nobel prize-winning paper (Hodgkin and Huxley 1952) as *time-varying conductances* are conceptually *wrong* from a nonlinear circuit-theoretic perspective. Like many other *anomalous* phenomena (Mauro 1961) reported in the biophysics literature over the last century (Cole 1941, 1947), they are in fact all memristors. The reported anomalies were the direct consequence of mistaken identities. In the case of the Hodgkin-Huxley circuit model, we will see in Sect. 4.3.3 that the *time-varying potassium conductance* is nothing but a first-order memristor and the time-varying sodium conductance is nothing but a second-order memristor. We believe that a deep understanding of many biochemical and biophysical mechanisms of the brain at the molecular level is possible only if their underpinning memristive mechanisms are fully uncovered and analyzed incisively.

4.2 Background Information on Memristors

From the circuit-theoretic point of view, the three basic two-terminal circuit elements are defined in terms of a relationship between two of the four fundamental circuit variables, namely, the *current* i, the *voltage* v, the *charge* q, and the *flux* φ.

[2] Since a *memristive system* (Chua and Kang 1976) is a generalization of the memristor and behaves qualitatively like a memristor, we will henceforth use the name memristor to includes both the *ideal* memristor defined in Chua (1971) and its current and future generalizations (Chua and Kang 1976; Di Ventra et al. 2009).

Of the six possible combinations of these four variables, five have led to well-known relationships (Chua 1969). Two of these relationships are already related by their *definitions*, namely[3], $q(t) \triangleq \int_{-\infty}^{t} i(\tau)\mathrm{d}\tau$ and $\varphi(t) \triangleq \int_{-\infty}^{t} v(\tau)\mathrm{d}\tau$. Three other relationships are given, respectively, by axiomatic definitions of the three classical circuit elements, namely, the *resistor* (defined by a constitutive relation between the *voltage* v and the *current* i), the *inductor* (defined by a constitutive relation between the *flux* φ and the *current* i), and the *capacitor* (defined by a constitutive relation between the *charge* q and the *voltage* v). Only one relationship remains undefined, namely, between the *flux* φ and the *charge* q. From the logical as well as axiomatic points of view, it is necessary for the sake of *completeness* in logic, and *symmetry* principle to postulate the existence of a *fourth basic two-terminal* circuit element that is characterized by a φ versus q constitutive relation. This element was called the *memristor* because it behaves somewhat like linear *resistor with memory*. The symbol of the memristor and a hypothetical $\varphi - q$ curve are shown in Fig. 4.1.

By definition, a *memristor* is characterized by a *constitutive relation* $f(\varphi, q)=0$. It is said to be *charge-controlled* (respectively, *flux-controlled*) if this relation can be expressed as a single-valued function $\varphi = \varphi(q)$ of the charge q (respectively, $q = q(\varphi)$ of the flux φ). The voltage across a *charge-controlled* memristor is obtained by differentiating both sides of $\varphi = \varphi(q)$ and invoking the chain rule, namely,

$$\boxed{\text{Ideal Current-Controlled Memristor}} \qquad v = M(q)i \tag{4.1}$$

where

$$M(q) \triangleq \frac{\mathrm{d}\varphi(q)}{\mathrm{d}q}, \text{ Ohms} \tag{4.2}$$

Similarly, the current of a *flux-controlled* memristor is given by

$$\boxed{\text{Ideal Voltage-Controlled Memristor}} \qquad i = W(\varphi)v \tag{4.3}$$

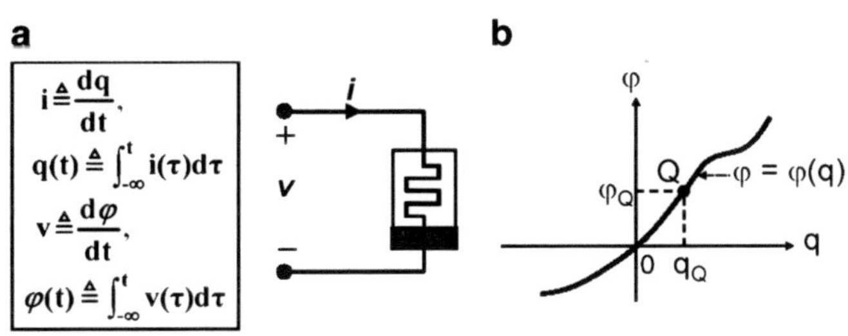

Fig. 4.1 Symbol of the memristor (**a**), and a typical φ versus q constitutive relation (**b**)

[3] The reader should interpret the *charge* q and the *flux* φ as *names* given by the definitions in Fig. 4.1a; they need not be associated with a *physical charge*, or *flux*.

where

$$W(\varphi) \triangleq \frac{dq(\varphi)}{d\varphi}, \text{ Siemens} \tag{4.4}$$

Observe that $M(q_Q) = 1/W(\varphi_Q)$ is just the *slope* of the $\varphi - q$ curve at an *operating point* $Q(q_Q, \varphi_Q)$. Since $M(q)$ has the unit (Ohm) of *resistance*, it is called the *small-signal memristance* of the memristor at the oprating point $\mathbf{Q}(q = q_Q)$. Similarly, the function $W(\varphi)$ is called the *small-signal memductance* of the memristor at the operating point $\mathbf{Q}(\varphi = \varphi_Q)$ because it has the unit (Siemens) of *conductance*. Observe that the value of the *small-signal* memristance (respectively, memductance) at any time t_0 depends on the *time integral* of the memristor current (respectively, voltage) from $t = -\infty$ to $t = t_0$. Hence, while the memristor behaves like an ordinary resistor at a given instant of time t_0, its resistance (respectively, conductance) depends on the complete past history $(-\infty < t \leq t_0)$ of the memristor current (respectively, voltage). This observation justifies our choice of the name *memory resistor*, or *memristor*. It is interesting to observe that once the memristor voltage $v(t)$, or current $i(t)$, is specified, the memristor behaves like a *linear time-varying resistor*. This observation is what misled Hodgkin and Huxley into erroneously identifying the potassium and sodium circuit elements in their model as a *time-varying conductance*.[4]

In the very special case where the memristor $\varphi - q$ curve is a straight line, we obtain $M(q) = R$, or $W(\varphi) = G$, and the memristor reduces to a *linear time invariant resistor*. Hence, there is no point introducing a linear memristor in linear network theory.[5] Since the memristor cannot be synthesized by any combination of *passive* two-terminal nonlinear resistors, inductors, and capacitors, it follows that the *memristor* must be considered as the *fourth basic circuit element*. The four pairs of circuit variables defining the four basic circuit elements are depicted in the "*Basic Circuit-Element Quartet*" in Fig. 4.2a. Its striking resemblance to Aristotle's four basic building blocks of matter (Mainzer 2007), depicted in Fig. 4.2b via four Platonic solids, *circa* 350 BC, is truly remarkable.

The following passivity criterion specifies what class of memristors could exist in a pure "device form" without internal power supplies.

[4] To understand the nature of this serious conceptual mistake, observe that any nonlinear device described by a dynamical system equation $i = f(x, v)$, $\dot{x} = g(\mathbf{x}, v)$, where $f(\mathbf{x}, 0) = 0$, would behave like a *time-varying conductance* on recasting the first equation into the form $i = \mathrm{G}(\mathbf{x}, v)v$, where the conductance $G(\mathbf{x}, v) \triangleq (f(\mathbf{x}, v))/v$ is assumed to be well-defined. Unfortunately, such a time-varying conductance does not behave like a frequency-independent conductance even under small-signal sinusoidal excitations.

[5] Since research in circuit theory in the past has been dominated by circuits made of *linear* elements, it is not surprising that the concept of the memristor never arose there. Neither is it surprising that this element has not been invented in a device form until the HP breakthrough 37 years later because the *memristive phenomenon* seems to become dominant only at the *nanoscale*. Besides, it is somewhat "unnatural" to associate charge with flux-linkage. Moreover, since the memristor is intrinsically both *nonlinear* and *dynamic*, the necessity to design a $\varphi - q$ *curve tracer* had precluded the slim possibility of an accidental discovery.

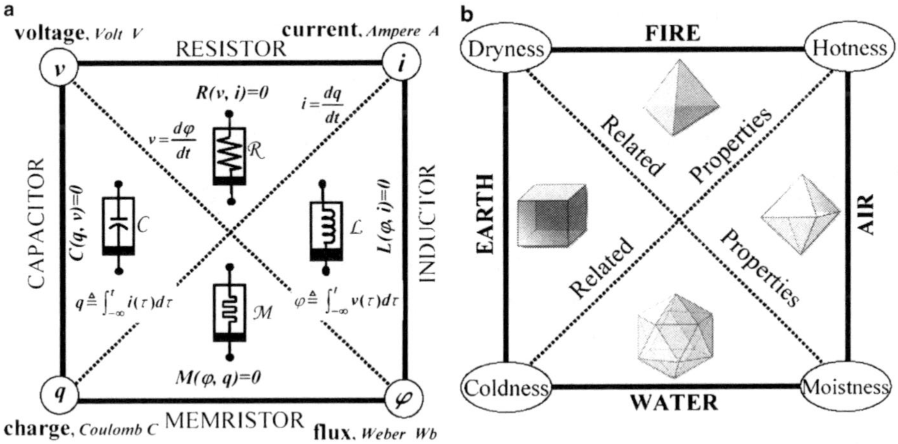

Fig. 4.2 A mnemonic diagram depicting four basic circuit elements in (**a**) and four building blocks of matter in (**b**)

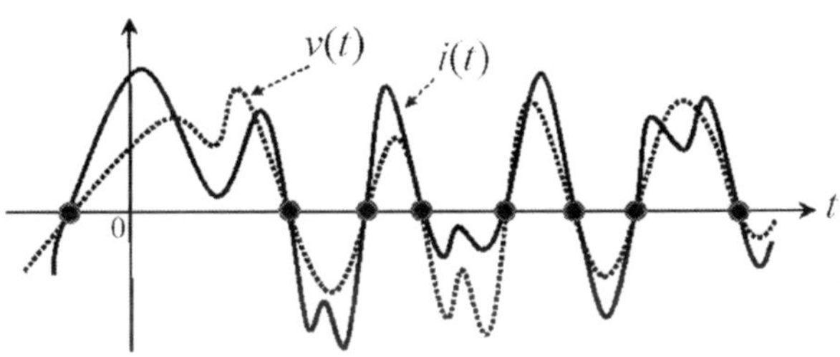

Fig. 4.3 Any associated pair of voltage $v(t)$ and current $i(t)$ of a strictly passive memristor must have *identical zero* crossings

Theorem 4.1. *Passivity Criterion* (Chua 1971)

A memristor characterized by a differentiable charge-controlled $\varphi = \varphi(q)$ [respectively, flux-controlled $q = q(\varphi)$], constitutive relation is *passive* if, and only if, its *small-signal* memristance $M(q)$ [respectively, memductance $W(\varphi)$] is nonnegative, that is, $M(q) \geq 0$ (respectively, $W(\varphi) \geq 0$).

Equations (4.1) and (4.3) imply that for a *strictly passive* [i.e., $M(q) > 0$ and $W(\varphi) > 0$] memristor, we always have

$$v(t) = 0 \Leftrightarrow i(t) = 0 \tag{4.5}$$

In other words, both $v(t)$ and $i(t)$ of a strictly passive memristor must have *identical zero crossings*, as illustrated in Fig. 4.3. It follows from Fig. 4.3 that any periodic voltage and current waveforms of a memristor must have *zero phase shifts*. The above properties translate into the following "signature" of a memristor:

Theorem 4.2. *Pinched Hysteresis Loop Fingerprint*

The loci (Lissajous figure) *of any corresponding pair of periodic voltage and current waveforms of a strictly passive memristor in the i versus v plane must pass through the origin, and be restricted to only the first and third quadrants.*

A typical memristive i versus v loci (Lissajous figure) under a sinusoidal voltage excitation $v = A \sin \omega t$ is shown in Fig. 4.4, henceforth called a *pinched hysteresis loop*.

An important phenomenon exhibited by all memristor pinched hysteresis loops is that the area enclosed inside each lobe *shrinks* as the frequency ω increases, and in fact converges to a straight line through the origin as $\omega \to \infty$, as depicted in Fig. 4.5. This high-frequency lobe-area shrinking phenomenon follows from the observation that

$$\varphi(t) = \varphi_0 + \int_{t_0}^{t} A \sin(\omega\tau)\mathrm{d}\tau = \varphi_0 + \frac{A}{\omega}\cos \omega t \to \varphi_0, \text{ as } \omega \to \infty \tag{4.6}$$

resulting in the memductance (and hence also the memristance) trending to a constant:

$$W(\varphi(t)) \to W(\varphi_0), \text{ as } \omega \to \infty \tag{4.7}$$

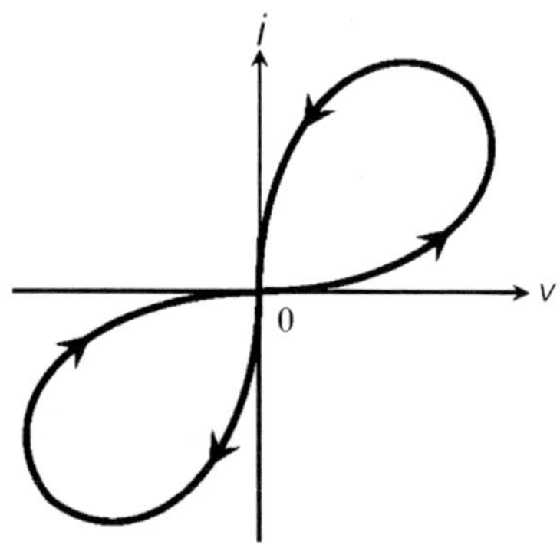

Fig. 4.4 The fingerprint of a strictly passive memristor is that its current versus voltage Lissajous figure corresponding to any periodic excitation similar to a sine wave is a *double-valued "pinched" hysteresis loop*

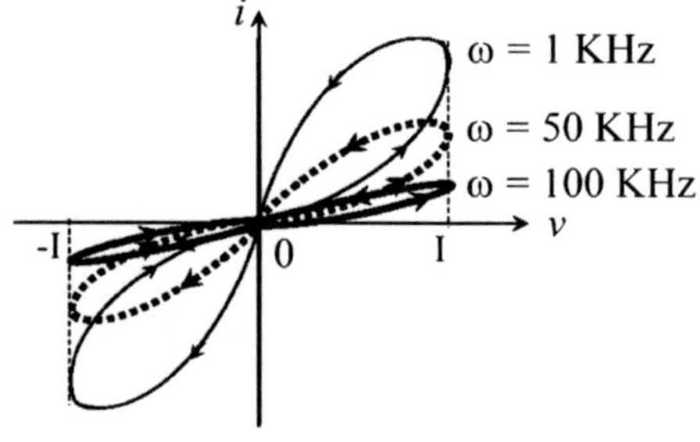

Fig. 4.5 The area enclosed within each lobe of the pinched hysteresis loop of a memristor *decreases* as the *frequency increases*, for the same *amplitude* of sinusoidal voltage input waveforms

This means that the physical mechanism responsible for the memductance (respectively, memristance) modulation exhibits some *inertia to motion*, and can not respond fast enough as the frequency becomes too high.

Figure 4.3 and Theorem 4.2 imply that the *instantaneous power*

$$p(t) \triangleq v(t)\,i(t) \tag{4.8}$$

entering a strictly passive memristor is always *nonnegative* for all times. Since no power is returned, the power is completely dissipated as *heat*. This *no-energy discharge property* implies the following fundamental property, which distinguishes memristors from capacitors and inductors.

Theorem 4.3. *No-energy storage property.*
A passive memristor cannot store energy.

A *memristor* prototype was built and demonstrated experimentally in Chua (1971) as a proof of principle. However, that memristor was made form transistors and op amps, which are *active* elements requiring a power supply. Is it possible to build a physical memristor without a power supply?[6] If so, what restrictions must be exhibited by the memristor constitutive relation $f(\varphi, q) = 0$? The answer is given by our next theorem.

Theorem 4.4. *Memristor passivity condition.*

The $\varphi = \varphi(q)$ [respectively, $q = q(\varphi)$] constitutive relation of all physically *realizable passive memristors* must be a *monotone-increasing* function.

Proof. This fundamental theorem was proved in Chua (1971). The method of proof consists of showing that any *nonmonotonic* $\varphi = \varphi(q)$ [respectively, $q = q(\varphi)$] curve must have at least one point Q having a *negative slope*. By applying a sufficiently small *current* $i(t)$ [respectively, *voltage* $v(t)$] signal about Q at $t \geq 0$ so that the dynamics remain within the negative *memristance* $M(q) < 0$ (respectively, *memductance* $W(\varphi) < 0$) region of the $\varphi - q$ (respectively, $q - \varphi$) curve, we obtain $p(t) = v(t)\,i(t) < 0$ for all $t \geq 0$. This implies the memristor is a power source, which is absurd.

4.2.1 HP Memristor

The HP memristor unveiled in Strukov et al. (2008) is a two-terminal device made from a thin film of TiO_2, sandwiched between two platinum electrodes, as shown in Fig. 4.6.

[6] A memristor that functions without a power supply is said to be a *passive* circuit element (Chua 1969).

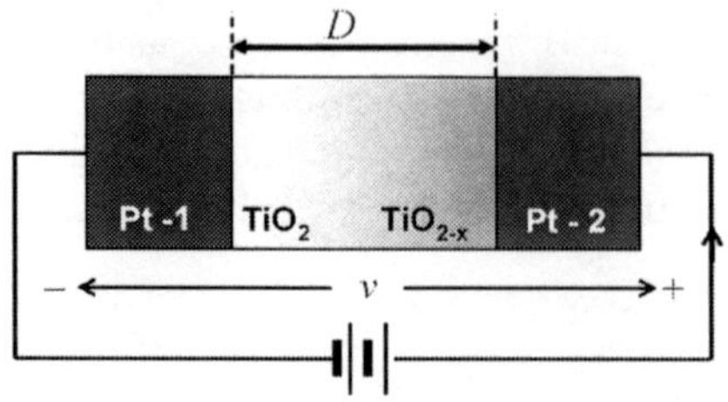

Fig. 4.6 The working HP memristor is made by sandwiching a thin film of TiO_2 of film width D between two platinum electrodes, where the TiO_2 on the right is doped with *oxygen vacancies* resulting in a nonuniform distribution of *mobile + 2-charged dopants*

Fig. 4.7 The φ versus q constitutive relation of the TiO_2 HP memristor

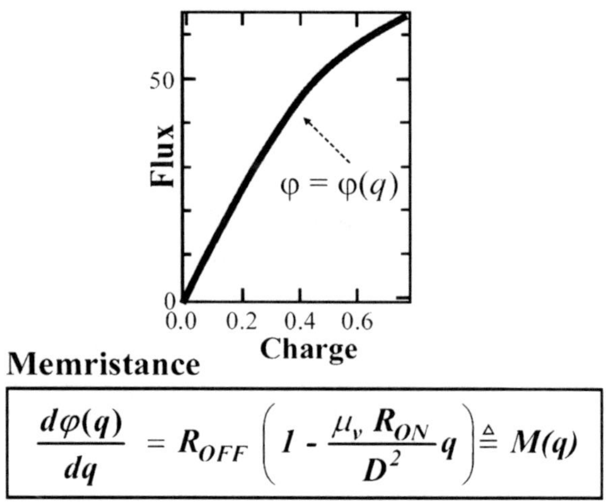

Although TiO_2 is a semiconductor, it can be made to conduct by removing some oxygens near the positive electrode, thereby creating positively charged dopants that drift toward the negative electrode when a voltage is applied as in Fig. 4.6 (Williams 2008). The *exact* equation derived by HP for this ideal memristor can be recast in the form

$$\varphi = \alpha_1 q - \alpha_2 q^2 \tag{4.9}$$

where α_1, and α_2 are device parameters.

Differentiating both sides of Eq. (4.9), we obtain

$$v = M(q)i \tag{4.10}$$

where

$$M(q) \triangleq R_{\text{OFF}}\left(1 - \frac{\mu_{\text{v}} R_{\text{ON}}}{D^2} q\right) \tag{4.11}$$

is the *memristance*, R_{OFF}, R_{ON}, μ_{v} are device parameters, and D is the width of the TiO_2 thin film. The $\varphi = \varphi(q)$ constitutive relation of this ideal memristor is given in Strukov et al. (2008) and redrawn in Fig. 4.7. The typical voltage and current

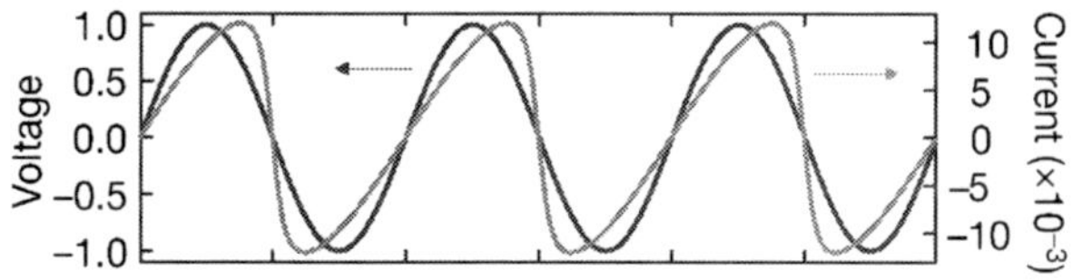

Fig. 4.8 The current waveform of the HP TiO_2 memristor has identical *zero crossings* as the sinusoidal input voltage

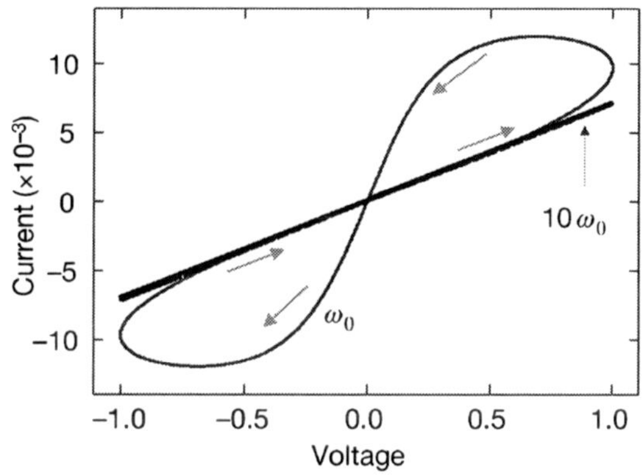

Fig. 4.9 The pinched hysteresis loop corresponding to the voltage and current waveforms depicted in Fig. 4.8 for $\omega = \omega_0$ and $\omega = 10\omega_0$. The latter shrinks to a very thin loop almost indistinguishable from a straight line

waveforms shown in Fig. 4.8 exhibit the *zero-crossing fingerprint of a memristor*. Observe that the current waveform in Fig. 4.8 does not exhibit the *quarter-wave* symmetry of the *sinusoidal* voltage. This loss of symmetry will manifest itself as a *double-valued* hysteresis. The corresponding v–i Lissajous figure shown in Fig. 4.9 is a *pinched hysteresis* loop at frequency ω_0, which shrinks to a very thin loop (resembling a straight line) at high frequencies (10 ω_0 in this case).

Figure 4.10b shows a more complicated pinched hysteresis loop (Strukov et al. 2008) made of three consecutive lobes 1, 2, and 3 in the first quadrant, and lobes 4, 5, 6 in the third quadrant. This pinched hysteresis loop is generated by the voltage and current waveforms shown in Fig. 4.10a.

4.2.2 How to Read and Write Memory States

In conventional electronic circuits, the dc operating points of the active devices (e.g., tunnel diodes, transistors, etc) are set by a *dc* power supply. When the power is switched off, all memory states disappeared. In a memristor, the operating point is set by a narrow *voltage pulse* $v_s(t)$ as depicted in Fig. 4.11. Observe that the operating point Q remains unchanged even though $v_s(t) = 0$ for $t > t_0 + \Delta$ because Q is defined by the memristor *flux* $\varphi(t) = \varphi_Q = E\Delta$, and not by the memristor voltage $v(t) = v_Q = 0$ for $t > t_0 + \Delta$. By tuning the *pulse height* E (or pulse width Δ),

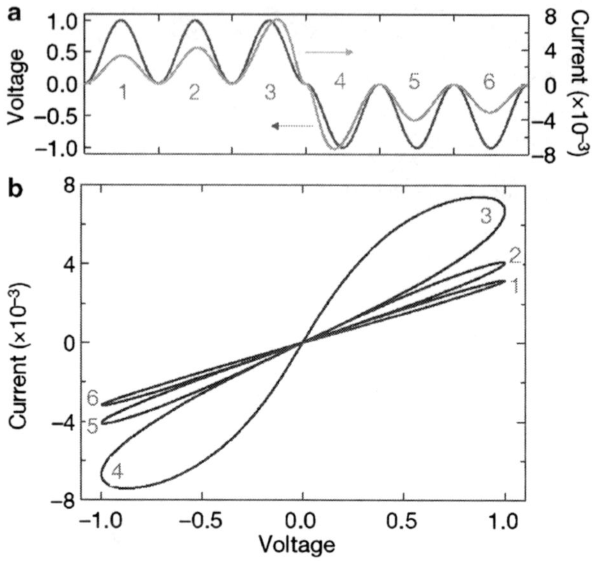

Fig. 4.10 A six-lobe pinched hysteresis loop derived from the HP TiO_2 memristor when driven by a six-pulse periodic voltage waveform defined by $v = v_0 \sin^2 \omega t$ for $0 \le t \le 3$, and $v = -v_0 \sin^2 \omega t$ for $3 \le t \le 6$, as shown in (**a**). The corresponding Lissajous figure in (**b**) is complex but nevertheless *pinched* at the origin, as expected. Here, the upper three lobes labeled 1, 2, and 3 in (**b**) are associated with the first three consecutive positive voltage pulses during the time interval $0 \le t \le 3$ in (**a**). Similarly, the lower three lobes 4, 5, and 6 correspond to the three consecutive negative voltage pulses during the time interval $3 \le t \le 6$

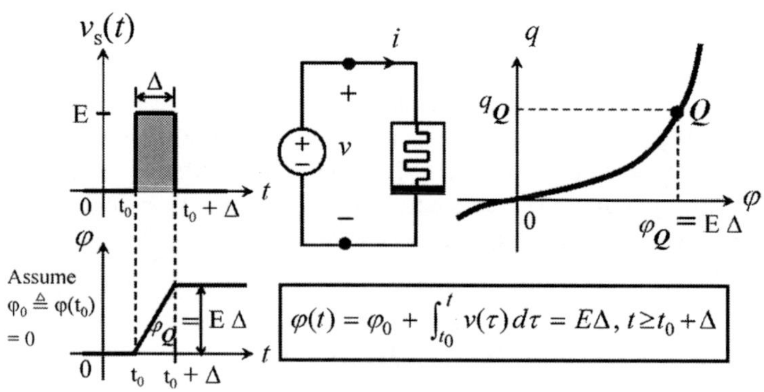

Fig. 4.11 The memristor dc operating point Q is set by a *voltage pulse*

any point on the $q = q(\varphi)$ curve in Fig. 4.11 can be chosen as the operating point Q, whose slope

$$W(Q) \triangleq \left. \frac{\mathrm{d}q(\varphi)}{\mathrm{d}\varphi} \right|_{\varphi=\varphi_Q} \tag{4.12}$$

is the *memductance* at Q. Since the slope associated with the $q-\varphi$ curve in Fig. 4.11 is continuously tunable, this memristor can be used as a *nonvolatile analog* memory by *biasing* it at an operating point Q, whose small-signal *conductance* can be chosen over a broad continuous range of conductance values.

Since the $q-\varphi$ curve is a *monotone-increasing* function (passivity condition), we can always choose an operating point $\boxed{0}$ in a *high-resistance* region, and an operating point $\boxed{1}$ in a *low-resistance* region that is sufficiently far from $\boxed{0}$, and use the memristor as a *binary memory*. The state of the memristor memory at any time can be easily written, or read, as illustrated below.

Example 4.1. Writing the memristor memory state

Consider the memristor circuit shown in Fig. 4.12 where the initial condition is chosen to be $\varphi(0) = 0$ to avoid clutter. The flux $\varphi(t)$ obtained by integrating $v_s(t)$ is a ramp that saturates at $\varphi = \Delta\varphi = (\Delta E)\,(\Delta t)$ for $t \geq \Delta t$. For the $q-\varphi$ curve shown in Fig. 4.12, the state **0** (at $\varphi_Q = 1$) is set by choosing $\Delta E = 1/\Delta t$ volts. Similarly, the state **1** (at $\varphi_Q = 3$) is set by choosing $\Delta E = 3/\Delta t$ volts.

Example 4.2. Reading the memristor memory state

Consider the memristor circuit shown in Fig. 4.13. The stored memory state can be determined by applying a symmetrical dual polarity pulse, such as the "square" doublet shown in this figure. A small alternating voltage pulse is chosen so that the

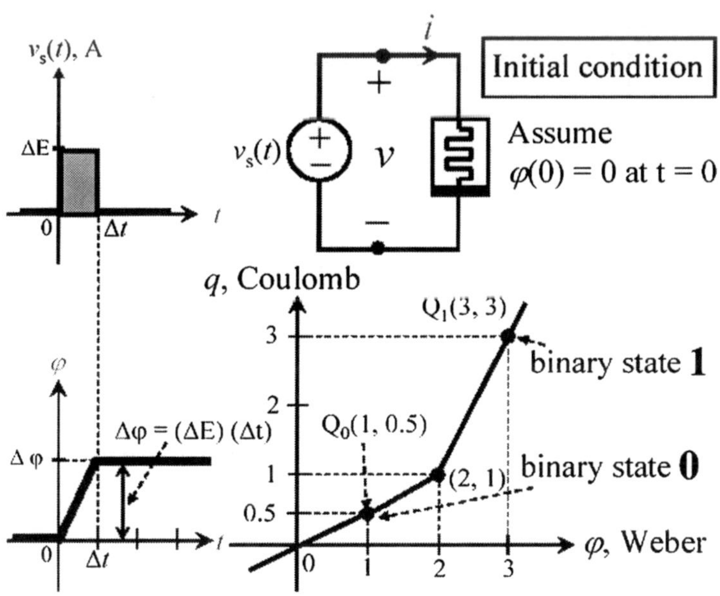

Fig. 4.12 For zero initial flux $\varphi(0) = 0$, the two binary states **0** and *1* can be set by choosing $\Delta E = 1/\Delta t$ and $\Delta E = 3/\Delta t$ volts, respectively, for fixed pulse width Δt

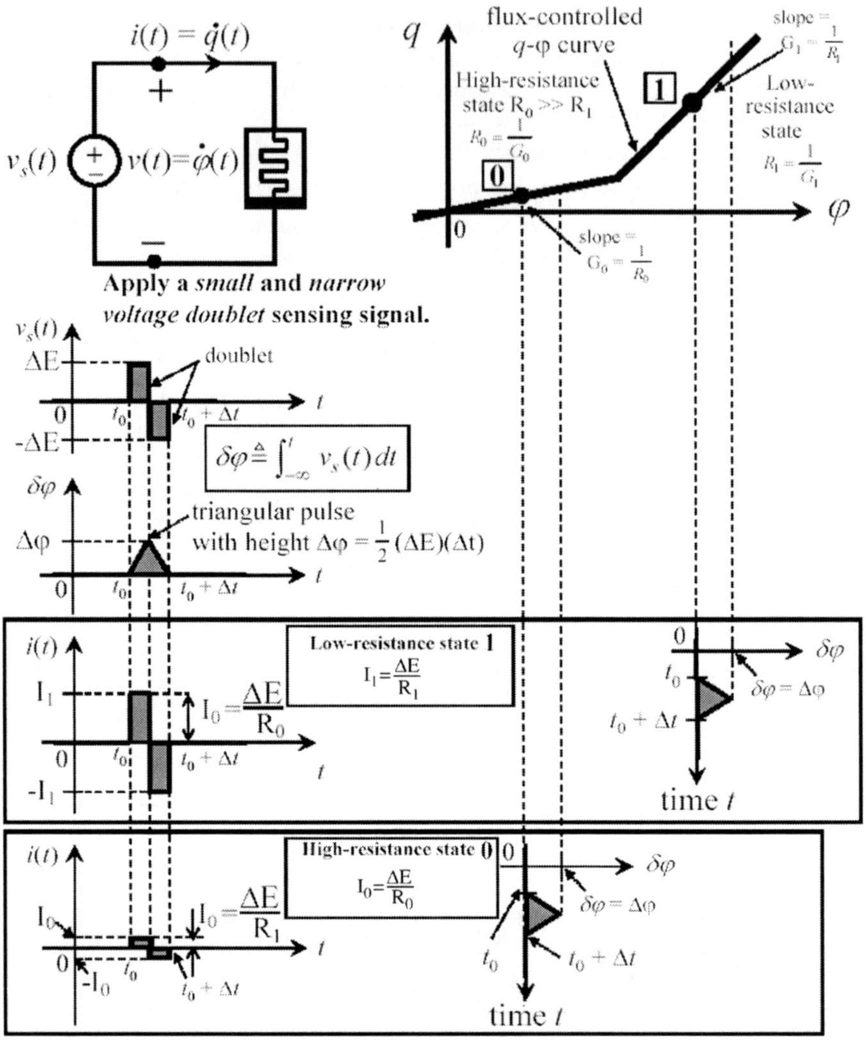

Fig. 4.13 A *large* read-out current doublet indicates state $\boxed{1}$. Conversely, a *small* read-out current doublet indicates state $\boxed{0}$

"interrogating" pulse $v_s(t)$ does not perturb the location of the operating point, as evidenced from the triangle waveform $\delta\varphi$ that returns to *zero* after the brief interrogating time interval Δt. Now, depending on the memory state, we will elicit either a large *current doublet* at $\boxed{1}$ whose the memristance $R_1 = 1/G_1$ is *small*, or a much smaller current doublet at $\boxed{0}$ whose memristance $R_0 = 1/G_0$ is much larger.

4.3 Memristive Devices and Systems

The HP memristor is an ingenious and inspiring invention (Williams 2008) destined for future textbooks. But it is *not* a *discovery* because *the memristive fingerprint* uncovered in the preceding section is a generic *phenomenon of nature*, both physical and nonphysical (e.g., ecological, social, economics, etc), that lurks behind many totally unrelated systems, both man-made and otherwise. For example, Figure 4.14 presents a sample of 12 unrelated 2-terminal electrical devices that share a common intrinsic characteristic, namely, they all exhibit a *pinched hysteresis loop* in the *v–i plane*. They were all mistaken for a nonlinear resistor endowed with some weird parasitic hysteresis. They are in fact all memristive devices – a natural generalization of the memristor (Chua and Kang 1976) – defined by one of the following two equivalent "dual" representations:

Current-Controlled Memristor	$v = M(\boldsymbol{x},\, i)\, i$ $\dot{x} = \boldsymbol{f}(\boldsymbol{x},\, i)$	(4.13)

Voltage-Controlled Memristor	$i = W(\boldsymbol{x},\, v)\, v$ $\dot{x} = \boldsymbol{f}(\boldsymbol{x},\, v)$	(4.14)

where

$$\boldsymbol{x} \triangleq [x_1 \; x_2 \, x_3 \; \cdots \; x_n] \tag{4.15}$$

denotes "n" *state variables* associated with the internal *device* mechanisms that determine the device's time evolution under all possible *input* signals [$i(t)$ for current-controlled representation, and $v(t)$ for voltage-controlled representation].

In the special case where we choose $n = 1$, $x = q$, $M(\boldsymbol{x},\, i) = M(x)$, and $\mathbf{f}(\boldsymbol{x}, i) = i$, we obtain $\dot{q} = i$, $v = M(q)i$, which is Eq. (4.1). Similarly, if we choose $n = 1$, $x = \varphi$, $W(\boldsymbol{x}, v) = W(x)$, and $\mathbf{f}(\boldsymbol{x}, v) = v$, we would obtain $\dot{\varphi} = v$, and $i = W(\varphi)\, v$, which is Eq. (4.3). Let us consider an example of each representation that generalizes the *ideal* memristor equation while preserving its pinched hysteresis fingerprint.

Example 4.3. Incandescent lamp bulbs

The resistance of the *filament* of the incandescent bulb, shown in Fig. 4.15(a), is *not* a constant, but changes with temperature in accordance with the heat balance equation (Cunningham 1951):

$$v = (R_0 T)i \tag{4.16}$$

$$\dot{T} = aTi^2 - b\left(T^4 - T_0^4\right) \tag{4.17}$$

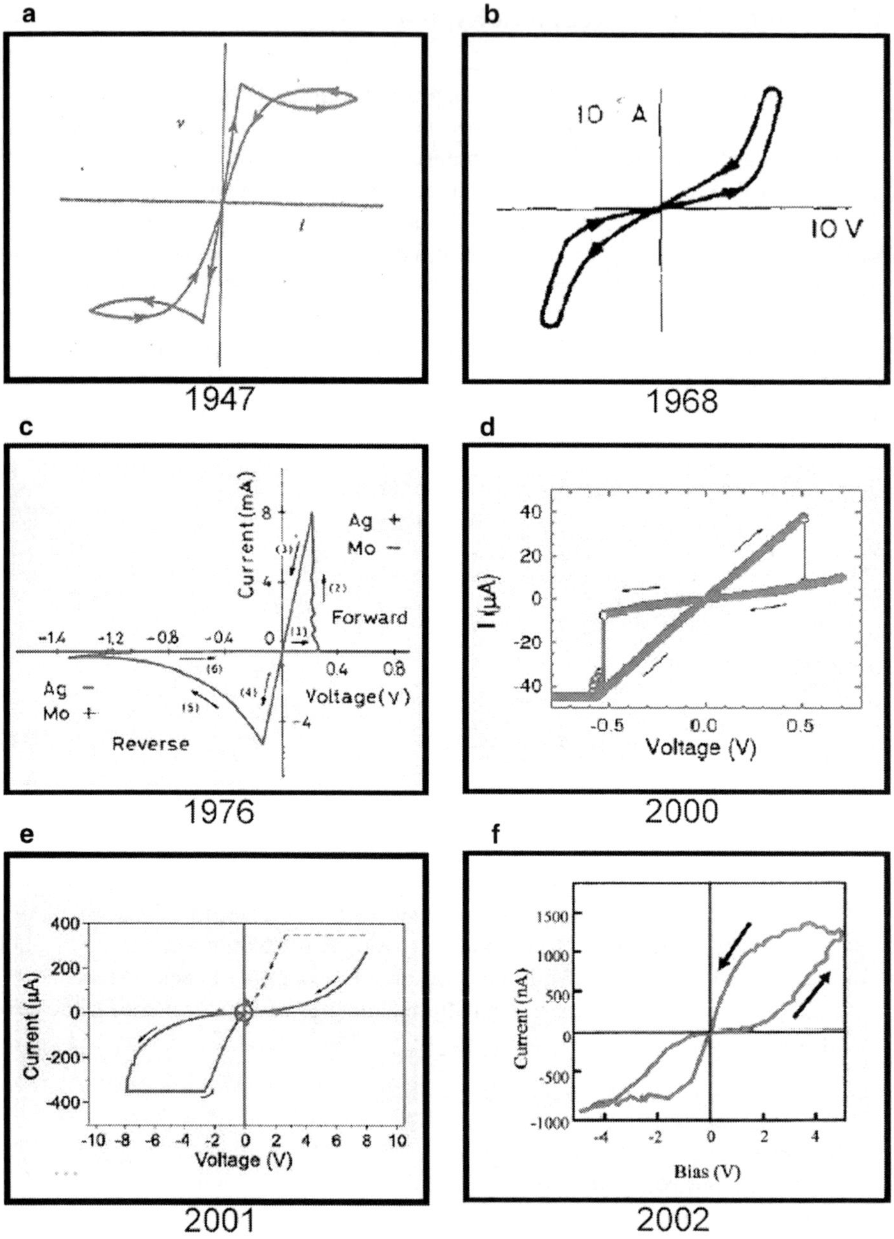

Fig. 4.14 Examples of two-terminal electrical devices whose voltage-current characteristics are *pinched* hysteresis loops: (**a**) Francis (1947), (**b**) Argall (1968), (**c**) Hirose and Hirose (1976), (**d**) Beck et al. (2000), (**e**) Rossel et al. (2001), (**f**) Duan et al. (2002), (**g**) Sluis (2003), (**h**) Seo and Lee (2004), (**i**) Oka and Nagaosa (2005), (**j**) Sawa et al. (2006), (**k**) Schindler (2007), (**l**) Dong et al. (2008)

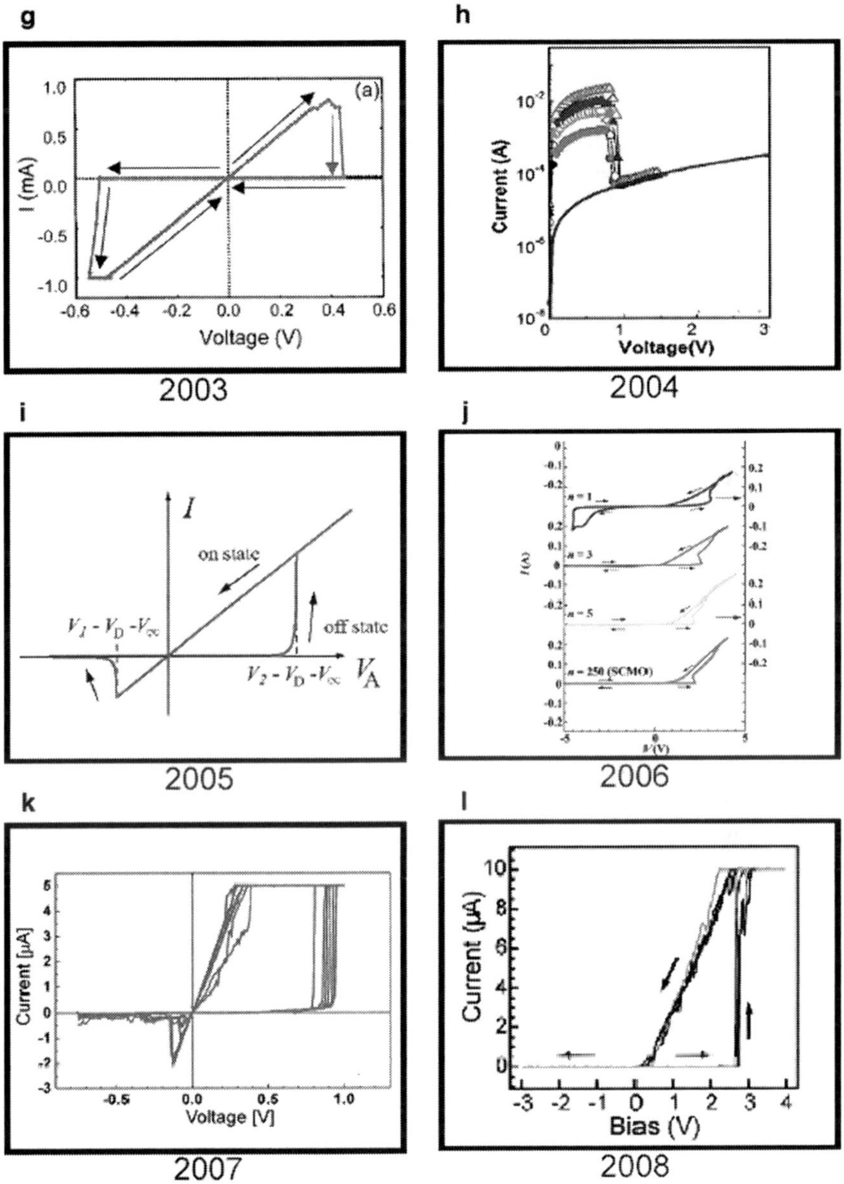

Fig. 4.14 (continued)

where T and T_0 denote the *filament* temperature and the *ambient* temperature, respectively, and where a and b are constants depending on the filament material.

Equations (4.16) and (4.17) is an example of a current-controlled memristor with $n = 1$, $\boldsymbol{x} = x \triangleq T$, $\boldsymbol{M}(\boldsymbol{x}, i) \triangleq R_0 T$, and $\boldsymbol{f}(\boldsymbol{x}, i) \triangleq aTi^2 - b(T^4 - T_0^4)$.

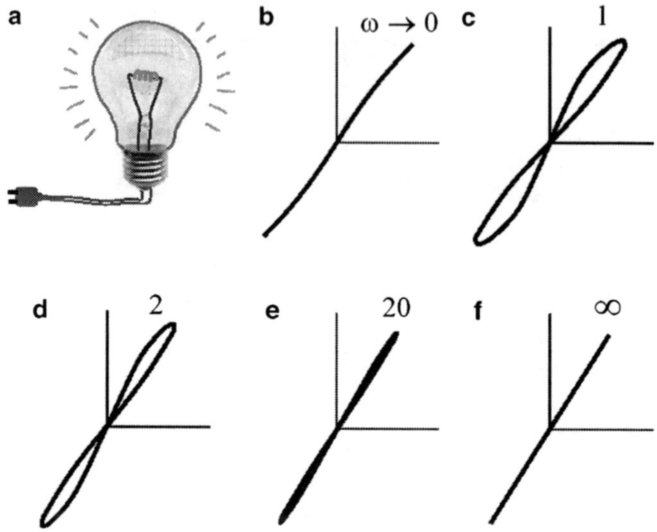

Fig. 4.15 The current versus voltage Lissajous figure of an incandescent lamp bulb under sinusoidal voltage inputs is a pinched hysteresis loop whose lobe area varies with frequency ω

The current versus voltage characteristics of a tungsten filament bulb under dc and sinusoidal voltage excitations were measured over a broad range of frequencies in Cunningham (1952) and replotted in Figs. 4.15a–f.

Observe that at dc, the tungsten lamp behaves just like a *nonlinear* resistor with a *single-valued* $v - i$ curve, as predicted in Chua and Kang (1976). For this example, we can actually derive an *implicit* equation of the *dc v–i* curve by setting $\dot{T} = 0$ in Eq. (4.17) to obtain

$$g(T) \triangleq \frac{T^4 - T_0^4}{T} = \frac{ai^2}{b} \tag{4.18}$$

Solving Eq. (4.18) numerically for T, we obtain

$$T = g^{-1}\left(\frac{ai^2}{b}\right) \tag{4.19}$$

Substituting Eq. (4.19) for T in Eq. (4.16), we can plot the nonlinear *dc v–i curve* numerically via the equation

$$v = \left[R_0 g^{-1}\left(\frac{ai^2}{b}\right)\right] i \triangleq h(i) \tag{4.20}$$

An example of this *dc v–i* curve is shown in Fig. 4.15b.

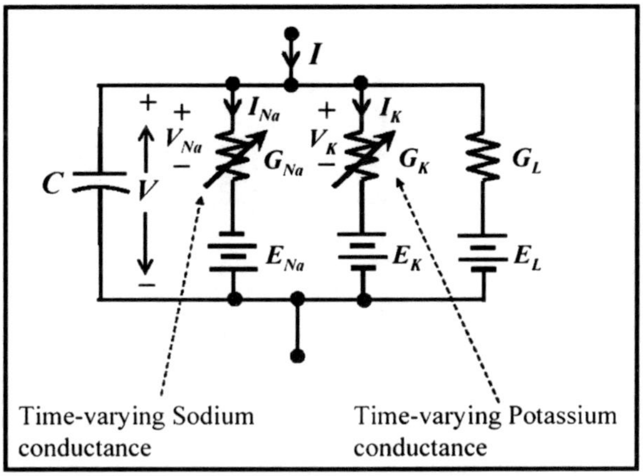

Fig. 4.16 The Hodgkin–Huxley nerve membrane circuit model

Observe that as the frequency ω of the sinusoidal voltage input signal increases, the area inside both lobes of the pinched hysteresis loop shrinks continuously and tends to a straight line as $\omega \to \infty$, as predicted.

Example 4.4. Hodgkin–Huxley Time-Varying Conductances

In the celebrated nerve membrane model shown in Fig. 4.16, Hodgkin and Huxley had mistakenly identified the *potassium conductance* G_K and the *sodium conductance* G_{Na} as *time-varying conductances*, as highlighted in their circuit diagram by two oblique arrows superimposed on the standard resistor symbols. Notwithstanding the fact that this model has been widely used in biological *system-level* computer simulations, at a electrophysiological level, the time-varying conductances have led to numerous anomalies (Cole 1941, 1947, 1971; Mauro 1961).

Many attempts to uncover the molecular origin of *the small-signal inductive* impedances manifested in experimental measurements had failed because the *inductive* components predicted from the small-signal impedance analysis of G_K and G_{Na} are mere illusions resulting from Hodgkin–Huxley's incorrect identifications of G_K and G_{Na} as time-varying conductances. The fact that they are memristive was first pointed out in Chua and Kang (1976) in a *nonbiological* journal and had remained obscure until the recent publication in *Nature* (Strukov et al. 2008). Indeed, the Hodgkin–Huxley model should be revised as shown in Fig. 4.17, where the two time-varying conductances G_K and G_{Na} are replaced by memristors, defined in Sects. 4.3.1 and 4.3.2, respectively.

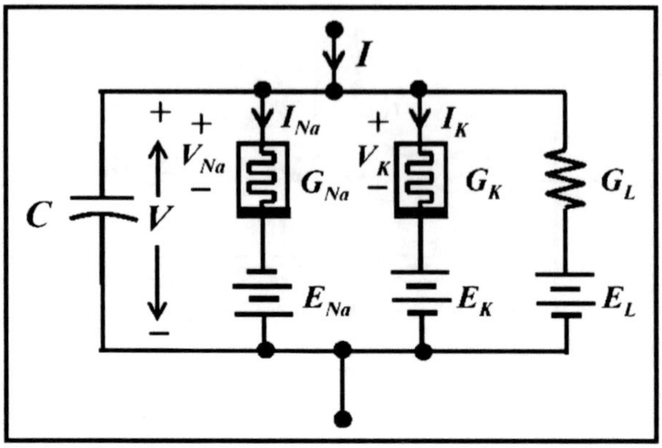

Fig. 4.17 Memristive Hodgkin–Huxley model with G_{K} and G_{Na} replaced by memristors

4.3.1 *Potassium Memristor*

The equations given in Hodgkin and Huxley (1952) for the *potassium time-varying conductance* G_{K} are recast below in the form of a first-order voltage-controlled memristor as defined in Eq. (4.14).

$$i_{\mathrm{K}} = \bar{g}_{\mathrm{K}} n^4 v_{\mathrm{K}} \triangleq G_{\mathrm{K}}(n) v_{\mathrm{K}} \tag{4.21}$$

$$\frac{\mathrm{d}n}{\mathrm{d}t} = \frac{0.01(v_{\mathrm{K}} + E_{\mathrm{K}} + 10)}{\exp[(v_{\mathrm{K}} + E_{\mathrm{K}} + 10)/10] - 1}(1-n) - 0.125 \exp\left(\frac{v_{\mathrm{K}} + E_{\mathrm{K}}}{80}\right) n \triangleq f(n, v_{\mathrm{K}}) \tag{4.22}$$

where $\bar{g}_{\mathrm{K}}$ and $\bar{E}_{\mathrm{K}}$ are constants.

Here, the number of state variables is equal to *one*, $\boldsymbol{x} \triangleq n$, $W(\boldsymbol{x}, v) \triangleq \bar{g}_{\mathrm{K}} n^4 \triangleq G_{\mathrm{K}}(n)$, $\boldsymbol{f}(\boldsymbol{x}, v) \triangleq f(n, v_{\mathrm{K}})$, $v = v_{\mathrm{K}}$, and $i = i_{\mathrm{K}}$. Clearly, $\mathrm{G_K}$ is a *first-order voltage-controlled memristor* whose dynamics is characterized by a single state variable $\boldsymbol{x} \triangleq n$.

4.3.2 *Sodium Memristor*

The equations given in Hodgkin and Huxley (1952) for the sodium time-varying conductance $\mathrm{G_{Na}}$ are recast below in the form of a second-order *voltage-controlled memristor* as defined in Eq. (4.14).

$$i_{\mathrm{Na}} = \bar{g}_{\mathrm{Na}} m^3 h v_{\mathrm{Na}} \triangleq G_{\mathrm{Na}}(m, h) v_{\mathrm{Na}} \tag{4.23}$$

$$\frac{dm}{dt} = \frac{0.1(v_{Na}-E_{Na}+25)}{\exp[(v_{Na}-E_{Na}+25)/10]-1}(1-m) - 4\exp\left(\frac{v_{Na}-E_{Na}}{18}\right)m \triangleq f(m, v_{Na})$$
$$\frac{dh}{dt} = 0.07\exp\left(\frac{v_{Na}-E_{Na}}{20}\right)(1-h) - \frac{1}{\exp[(v_{Na}-E_{Na}+30)/10]+1}h \triangleq f_2(h, v_{Na}) \tag{4.24}$$

where $\bar{g}_K$ and $\bar{E}_K$ are constants.

Here, the number of state variables is equal to *two*, $\boldsymbol{x} \triangleq [m, h]$, $W(\boldsymbol{x}, v) \triangleq \bar{g}_{Na} m^3 h \triangleq G_{Na}(m, h)$, $\boldsymbol{f}(\boldsymbol{x}, v) \triangleq f\,[f_1(m, v_{Na}),\ f_2(h, v_{Na})]$, $v = v_K$, and $i = i_K$. Clearly, G_{Na} is a *second-order voltage-controlled memristor* whose dynamics is characterized by *two state variables* $\boldsymbol{x} \triangleq [x_1\ x_2]$, where $x_1 \triangleq m$ and $x_2 \triangleq h$.

It is important to observe that neither Eq. (4.21) nor Eq. (4.23) contains the *time-variable t explicitly* in the potassium conductance $G_K(n)$ and the sodium conductance $G_{Na}(m,\ h)$, respectively. In other words, both G_K and G_{Na} are well-defined *time-invariant memristors*!

We now present a special case of Eq. (4.14) that can be exploited for applications as a *nonvolatile memory* in the sense that the *memristance* (respectively, *memductance*) is preserved even after the power is switched off.

Theorem 4.5. *Nonvolatile memory property*

If $\boldsymbol{f}(\boldsymbol{x}, i) = \boldsymbol{f}(i)$ does not depend on $\boldsymbol{x}$ and $\boldsymbol{f}(0) = \boldsymbol{0}$ (respectively, $\boldsymbol{f}(\boldsymbol{x},\ v) = \boldsymbol{f}(v)$ does not depend on $\boldsymbol{x}$ and $\boldsymbol{f}(0) = \boldsymbol{0}$), then the *memristance* $M(\boldsymbol{x}(t))$ (respectively, *memductance* $W(\boldsymbol{x}(t))$) at $t = T$ remains unchanged for all times $t > T$ when the power is switched off at $t = T$, that is, $i(t) = 0$ (respectively, $v(t) = 0$) for $t \geq T$.

Proof. By hypotheses, Eqs. (4.13) and (4.14) assume the following special form:

Current-Controlled representation	Voltage-Controlled representation
$v = M(\boldsymbol{x}, i)i$ $\dot{x} = \boldsymbol{f}(i)$ where $\boldsymbol{f}(0) = \boldsymbol{0}$	$i = W(\boldsymbol{x}, v)v$ $\dot{x} = \boldsymbol{f}(v)$ where $\boldsymbol{f}(0) = \boldsymbol{0}$

(4.25)

It follows from Eq. (4.25) and $\boldsymbol{f}(0) = \boldsymbol{0}$ that

$\boldsymbol{x}(t) = \int_{-\infty}^{t} \boldsymbol{f}(i(\tau))d\tau,$ for $t \leq T$ $= x(T),$ for $t > T$ Hence, $M(\boldsymbol{x}(t), i(t)) = M(\boldsymbol{x}(T), 0),$ for $t \geq T$	$\boldsymbol{x}(t) = \int_{-\infty}^{t} \boldsymbol{f}(v(\tau))d\tau,$ for $t \leq T$ $= x(T),$ for $t > T$ Hence, $W(\boldsymbol{x}(t),\ v(t)) = W(\boldsymbol{x}(T), 0),$ for $t \geq T$

(4.26)

4.4 Lossless Nonvolatile Memory Circuit Elements

As *nonvolatile* memories, *memristors* do not consume power when idle. It does dissipate a little heat whenever it is being "written" or "read." In other words, like resistors, *memristors* are *not lossless*. We will now introduce two new "dual" *nonvolatile* memory *lossless* circuit elements.

4.4.1 Memory Capacitor

A *memory capacitor*, or *memcapacitor* for short, is a two-terminal circuit element defined by a *constitutive relation* $\sigma = \sigma(\varphi)$ between the *flux* $\varphi(t) \triangleq \int_{-\infty}^{t} v(\tau)\mathrm{d}\tau$ and the *integrated charge* $\sigma(t) \triangleq \int_{-\infty}^{t} q(\tau)\mathrm{d}\tau$.

Our symbol of the memcapacitor is shown in Fig. 4.18a. A hypothetical σ versus φ constitutive relation $\sigma = \sigma(\varphi)$ is shown in Fig. 4.18b.

We can obtain a relationship between the *charge* and the *voltage* of a *memcapacitor* by differentiating both sides of its constitutive relation

$$\sigma = \sigma(\varphi) \tag{4.27}$$

to obtain

$$\underbrace{\frac{\mathrm{d}\sigma}{\mathrm{d}t}}_{q} = \frac{\mathrm{d}\sigma(\varphi)}{\mathrm{d}t} = \underbrace{\frac{\mathrm{d}\sigma(\varphi)}{\mathrm{d}\varphi}}_{C(\varphi)} \underbrace{\frac{\mathrm{d}\varphi}{\mathrm{d}t}}_{v} \tag{4.28}$$

Hence, we can recast the constitutive relation $\sigma = \sigma(\varphi)$ of a *memcapacitor* into the following *equivalent* form reminiscent of a *linear capacitor*:

Ideal Voltage-Controlled Memcapacitor

$$q = C(\varphi)\, v \tag{4.29}$$

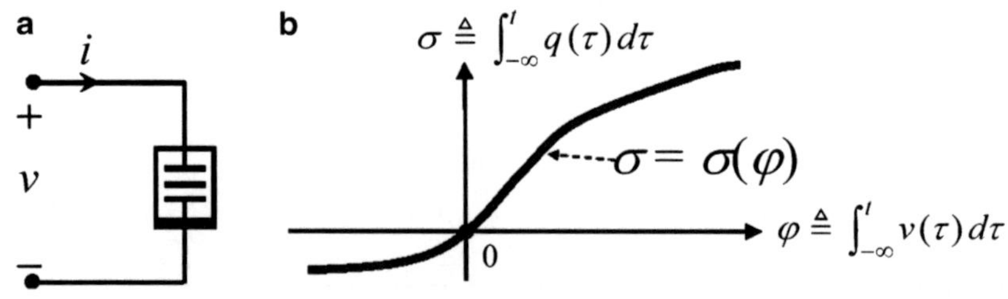

Fig. 4.18 (**a**) Symbol of memcapacitor. (**b**) Hypothetical $\sigma - \varphi$ characteristic curve of a memcapacitor

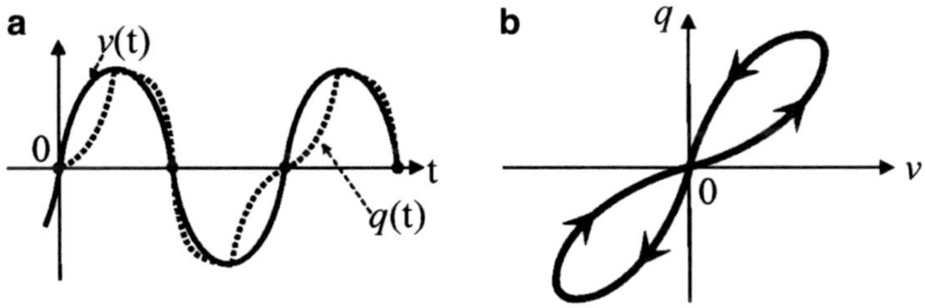

Fig. 4.19 (**a**) An associated pair of periodic waveforms $q(t)$ and $v(t)$. (**b**) The corresponding *Lissajous figure* is a *pinched* hysteresis loop

where

$$C(\varphi) \triangleq \frac{d\sigma(\varphi)}{d\varphi}, \quad \text{Farads} \tag{4.30}$$

is called the *memcapacitance*.

Observe that Eq. (4.29) can be interpreted as *a flux-controlled capacitor*. If we plot a typical pair of periodic waveforms $q(t)$ and $v(t)$ associated with a memcapacitor under a sinusoidal voltage excitation, as depicted in Fig. 4.19a, we would obtain a *pinched* hysteresis-loop *Lissajous figure* in the q versus v plane because, except for a change of symbols, the dynamical equations (4.27), (4.29), and (4.30) are exactly identical to the constitutive relation $\varphi = \varphi(q)$ in Fig. 4.1b, $v = M(q)i$ in Eq. (4.1), and $M(q) \triangleq (d\varphi(q))/dq$ in Eq. (4.2), respectively, defining a memristor. It follows that the following theorem also holds true:

Theorem 4.6. *Memcapacitor passivity condition*

The $\sigma = \sigma(\varphi)$ constitutive relation *of all physically realizable passive memcapacitors* is a *monotone-increasing function*.

Theorem 4.7. *Lossless memcapacitance property*

Every *passive memcapacitor* is *lossless* in the sense that the total net area $\int_0^T v(t)i(t)dt = \int_{-\infty}^T v(q(t))dq(t)$ enclosed by the two oppositely oriented pinched hysteresis lobes under sinusoidal excitation is zero over each period T.

Proof. For simplicity, let us assume that the $\sigma = \sigma(\varphi)$ characteristic curve is a piecewise-linear function with a positive slope for all segments. Since each linear region in this case is equivalent to a *linear positive capacitor*, the memcapacitor is lossless within each linear region. By decomposing any periodic input signal over corresponding piecewise-linear intervals, the total net area enclosed by the pinched *hysteresis* loop when replotted in the q versus v plane must likewise sum to zero over each period T. Hence, the *memcapacitor* is lossless.

Let us now illustrate how to *write* and *read* a memory state on a memcapacitor.

Example 4.5. Writing memcapacitor memory state

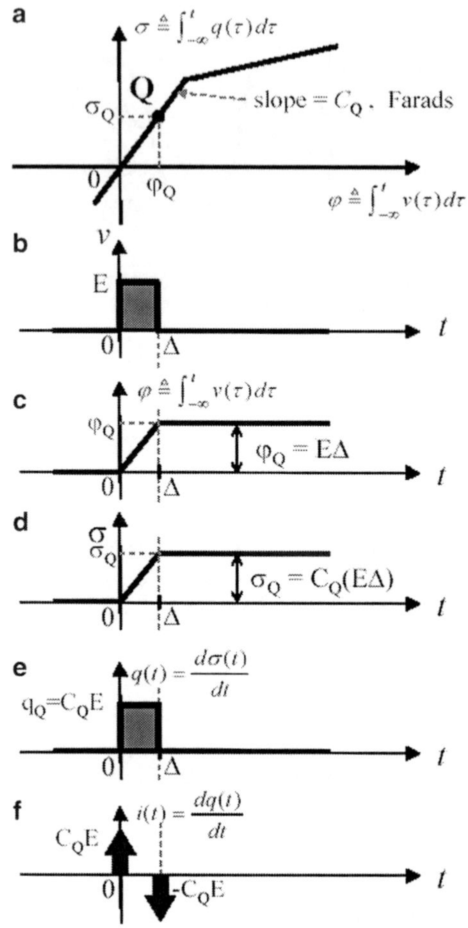

Fig. 4.20 Waveforms associated with a "Write" voltage pulse for biasing at Q on the *memcapacitor* $\sigma - \varphi$ curve

Consider the hypothetical memcapacitor $\sigma - \varphi$ curve shown in Fig. 4.20a. Suppose we wish to bias it at the operating point $Q(\varphi = \varphi_Q)$. This can be easily set by applying the small narrow voltage pulse shown in Fig. 4.20b such that $\varphi_Q = E\Delta$, assuming $\varphi(0) = 0$, where E is the pulse height and Δ is the pulse width of the "*writing*" pulse. The corresponding *flux* $\varphi(t)$, *integrated charge* $\sigma(t)$, *charge* $q(t)$, and *current* $i(t)$ are shown in Figs. 4.20c, d, e, and f, respectively. The bold double arrow symbol shown in Fig. 4.20f denotes a *current doublet* composed of a pair of sign-alternating impulses.[7]

Example 4.6. Reading memcapacitor memory state

[7] In a physical circuit, the square voltage pulse in Fig. 4.20b will have *non-zero rise* and *fall* times. In this case, the corresponding *current doublet* in Fig. 4.20f will consist of two very narrow but smooth current pulses of opposite polarity.

To determine the memory state of a *two-state memcapacitor*, simply apply a small alternating voltage pulse, such as the *triangle-shape* voltage doublet shown in Fig. 4.21a. The reason for choosing an *alternating* voltage pulse is to prevent the operating point Q from drifting with the corresponding *flux* $\varphi(t)$, as illustrated in Fig. 4.21b. Observe that after the short "sensing time interval" 2Δ where $\varphi(t)$ increases by $\Delta\varphi = 1/2(\Delta E)(\Delta)$, the flux returns to its original value $\varphi = \varphi_Q$. The

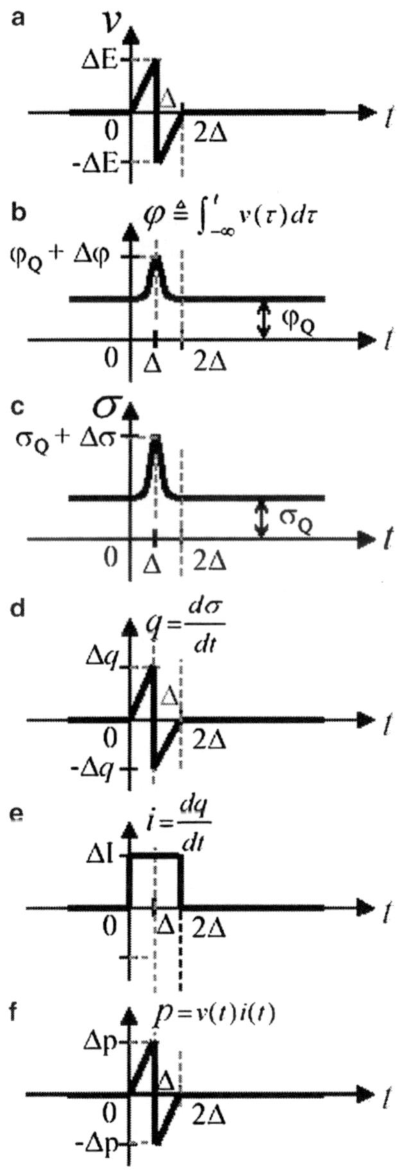

Fig. 4.21 Waveforms associated with an alternating "Read" voltage pulse composed of two short and narrow "triangle" pulses of opposite polarity

corresponding waveforms of $\sigma(t)$, $q(t)$, and $i(t)$ are shown in Figs. 4.21c, d, and e, respectively, where $\Delta\sigma = C_Q(\Delta\varphi), \Delta q = C_Q(\Delta E)$, and $\Delta I = C_Q(\Delta E)/\Delta$. Observe that since ΔI is proportional to the slope C_Q of the $\sigma - \varphi$ curve at the operating point Q, the "strength" ΔI of the "sensed" current pulse can be used to identify the memory state.

Observe also that the waveform of the *instantaneous power* $p(t) = v(t)i(t)$ depicted in Fig. 4.21f mimics that of $v(t)$ except for the scaling constant $\Delta p = (\Delta E)(\Delta I)$. Hence, the *total energy* dissipated in the memcapacitor over the reading period 2Δ is given by

$$\Delta W = \int_0^{2\Delta} p(t)\mathrm{d}t = 0 \tag{4.31}$$

This shows that the *memcapacitor* is lossless, as predicted.

Just as in the memristor theory presented in the preceding section, we end this section with a generalized definition of a memcapacitor via the constitutive relation

Voltage-Controlled Memcapacitor	$q = C(\boldsymbol{x}, v)v$ (4.32) $\dot{\boldsymbol{x}} = f(\boldsymbol{x}, v)$ (4.33)

where

$$x \triangleq [x_1\ x_2\ \cdots\ x_n]$$

are *state variables* that determine the internal *dynamics* of the corresponding *physical memcapacitor*.

4.4.2 Memory Inductor

Applying the circuit *duality* principle (Chua 1969), we define a memory inductor, or *meminductor* for short, by a *constitutive relation* $\rho = \rho(q)$ between the *charge* $q(t) \triangleq \int_{-\infty}^{t} i(\tau)\mathrm{d}\tau$ and the *integrated flux* $\rho(t) \triangleq \int_{-\infty}^{t} \varphi(\tau)\mathrm{d}\tau$.

Our symbol of the meminductor is shown Fig. 4.22a. A hypothetical constitutive relation $\rho = \rho(q)$ of a *meminductor* is shown in Fig. 4.22b.

We can obtain a relationship between the *flux* and the *current* of a meminductor by differentiating both sides of the constitutive relation

$$\rho = \rho(q) \tag{4.34}$$

to obtain

$$\underbrace{\frac{d\rho}{dt}}_{\varphi} = \frac{d\rho(q)}{dt} = \underbrace{\frac{d\rho(q)}{dq}}_{L(q)}\ \underbrace{\frac{dq}{dt}}_{i} \tag{4.35}$$

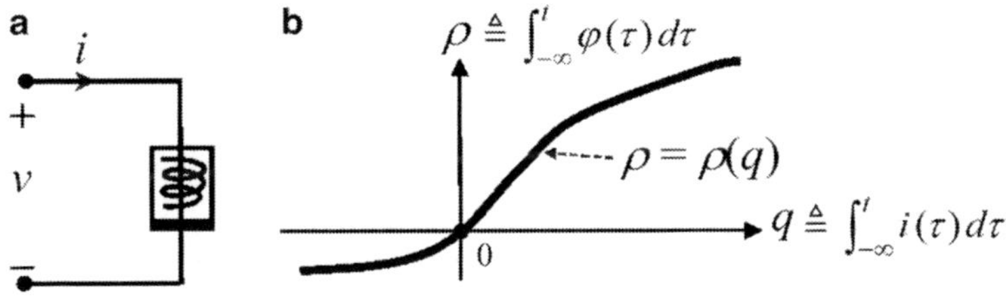

Fig. 4.22 (**a**) Symbol of meminductor. (**b**) Hypotetical $\rho - q$ characteristic curve of a meminductor

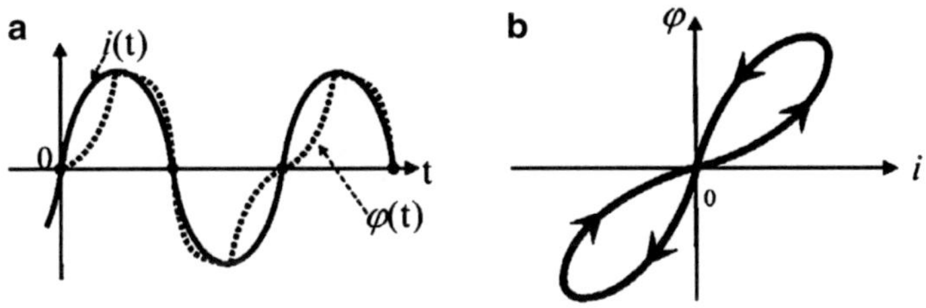

Fig. 4.23 (**a**) An associated pair of periodic waveforms $\varphi(t)$ and $i(t)$. (**b**) The corresponding *Lissajous figure* is a *pinched* hysteresis loop

We can recast the constitutive relation $\rho = \rho(q)$ of a *meminductor* into the following equivalent form reminiscent of a *linear inductor*:

Ideal Current-Controlled Meminductor

$$\varphi = L(q)\, i \tag{4.36}$$

where

$$L(q) \triangleq \frac{\mathrm{d}\rho}{\mathrm{d}q}, \quad \text{Henrys} \tag{4.37}$$

is called the *meminductance*.

Observe that Eq. (4.36) can be interpreted as a *charge-controlled* inductor. The "dual" pinched hysteresis loop associated with a meminductor is shown in Fig. 4.23.

The "duals" of Theorems 4.5 and 4.6 can be formulated by simply substituting ρ for σ and φ for q:

Theorem 4.8. *Meminductor passivity condition*

The $\rho = \rho(\varphi)$ constitutive relation *of all physically realizable passive meminductors is a monotone-increasing function.*

Theorem 4.9. *Lossless meminductance property*

Every *passive meminductance* is *lossless*.

Let us now illustrate how to *write* and *read* a memory state on a meminductor.

Example 4.7. Writing meminductor memory state

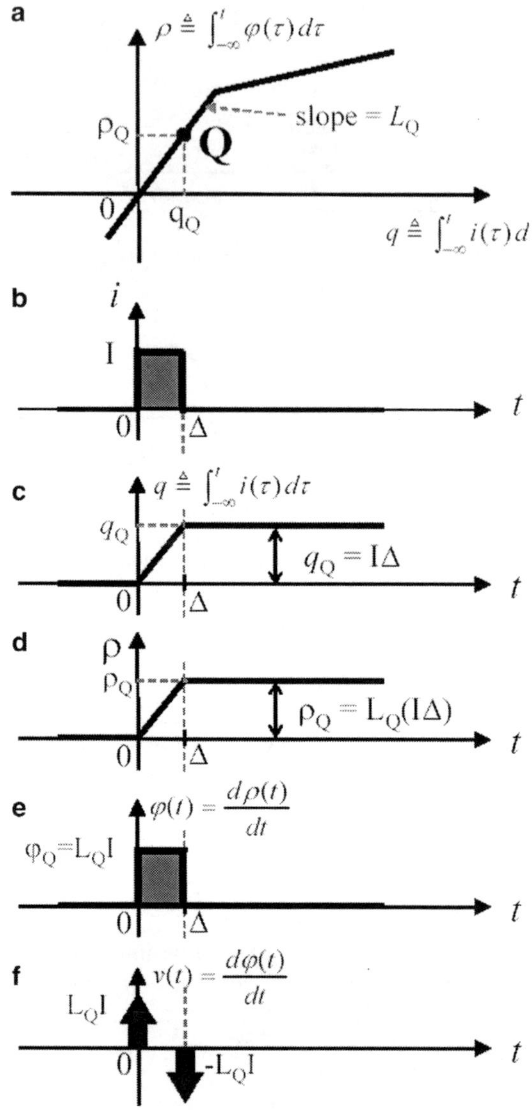

Fig. 4.24 Waveforms associated with a "write" current pulse for biasing at Q

Consider the hypothetical *meminductor* $\rho - q$ curve shown in Fig. 4.24(a). Suppose we wish to bias it at the operating point $Q(q = q_Q)$. This can be easily set by applying the small narrow current pulse shown if Fig. 4.24b such that $q_Q = I\Delta$, where I is the pulse height and Δ is the pulse width. The corresponding *charge* $q(t)$, *integrated flux* $\rho(t)$, *flux* $\varphi(t)$, and *voltage* $v(t)$ are shown in Fig. 4.24c, d, e, and f, respectively. The bold double-arrow symbol shown in Fig. 4.24f denotes a *voltage doublet* composed of a pair of sign-alternating impulses.

Example 4.8. Reading meminductor memory state.

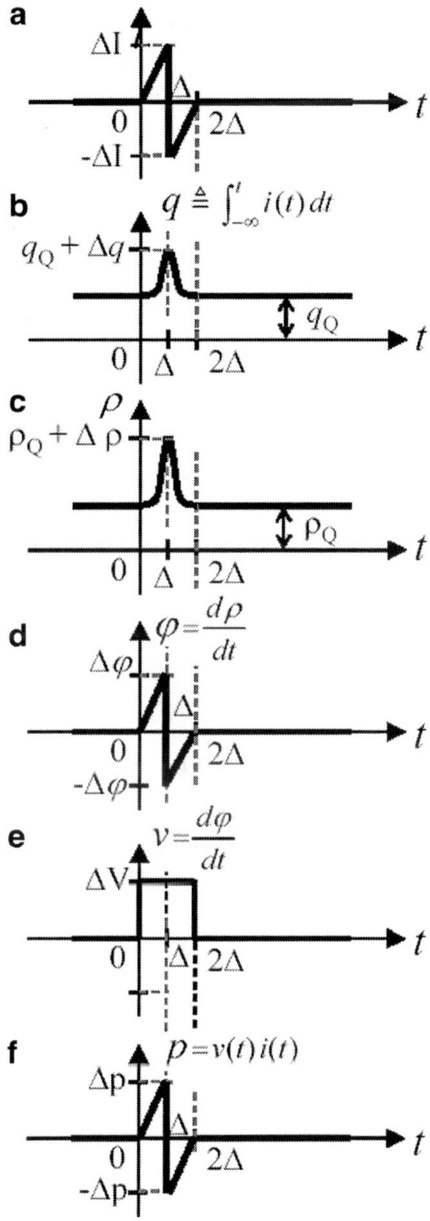

Fig. 4.25 Waveforms associated with an alternating current pulse composed of two short "triangle" pulses of apposite polarity

To determine the memory state of a two-state meminductor, simply apply a small alternating current pulse, such as the triangle-shape *current doublet* shown in Fig. 4.25a. The waveforms "dual" to those of Fig. 4.21 are shown in Fig. 4.25, obtained by simply substituting i for v in (a), q for φ in (b) ρ for σ in (c), φ for q

in (d), and v for i in (e). Again, Fig. 4.25f shows that Eq. (4.31) holds. Hence, the *meminductor* is *lossless*, as predicted.

The "dual" generalized constitutive relation of a meminductor is given by

Current-Controlled Meminductor	$\varphi = L(\boldsymbol{x}, i)i$ (4.38)
	$\dot{x} = f(\boldsymbol{x}, i)$ (4.39)

where

$$\boldsymbol{x} \triangleq [x_1 x_2 \cdots x_n]$$

are *state variables* that determine the internal *dynamics* of the corresponding *physical meminductor*.

References

Argall F (1968) Switching pheonomena in Titanium oxide thin films. Solid State Electron, Vol. 11:535–541

Beck A, Bednorz J G, Gerber C,Rossel C, Widmer D (2000) Reproducible switching effect in thin oxide films for memory applications. Appl. Phys. Letters, Vol. 77, No. 1

Bliss T, Collingridge G, Morris R (2003) LTP—Long-Term Potenlication. Oxford, New York

Borghetti J, Li Z, Straznicky J, Li X, Ohlberg D A A, Wu W, Stewart D, Williams R S (2009) A hybrid nanomemristor/transistor logic circuit capable of self-programming. PNAS. doi: 10.1073/pnas.0806642106

Chua L O (1969) Introduction to Nonlinear Network Theory. McGraw-Hill, New York

Chua L O (1971) Memristor-The missing circuit element. IEEE Trans. Circuit Theory, Vol. CT-18, No. 5

Chua L O (1998) CNN: A Paradigm for Complexity. World Scientific

Chua L O, Kang S M (1976) Memristive devices and systems. Proc. IEEE, Vol. 64, No. 2

Chua L O, Roska T (2002) Cellular neural networks and visual computing. Cambridge

Cunningham W J (1952) Incandescent lamp bulbs in voltage stabilizers. J. Appl Phys, Vol. 23, No. 6:658–662

Cole K S (1941) Rectification and inductance in the squid giant axon. J. Gen Physiol. Vol. 25:29–51

Cole K S (1947) Four lectures in Biophysics, Rio de Janeiro, Universidade do Brasil

Cole K S (1972) Membranes, Ions and Impulses. University of California Press, Berkeley

Di Ventra M, Pershin Y V, Chua L O (2009) Circuit elements with memory: memristors, memcapacitors and meminductors, *Proceedings of the IEEE*, Vol. 97, No. 10, 2009 (in press)

Dong Y, Yu G, McAlpine M C, Lu W, Lieber C M (2008) Si/a-Si core/shell nanowires as nonvolatile crossbar switches. Nano Letters, Vol. 8, No. 2:386

Duan X, Huang Y, Lieber C M (2002) Nonvolatile memory and programmable logic from molecule-gated nanowires. Nano Letters, Vol. 2, No. 5:487

Dudai Y (1989) The Neurobiology of Memory. Oxford, New York

F. Argall F (1968) Switching phenomena in titanium oxide thin films. Appl. Phys. Letters, Vol. 11:535–551.

Francis V J (1947) Fundamentals of Discharge Tube Circuits. Methuen & Co. London

Hirose Y, Hirose H (1976) Polarity-dependent memory switching and behavior of Ag dendrite in Ag-photodoped amorphous As_2S_3 films. J. Appl. Phys., Vol. 47, No. 6:2767

Hodgkin A L, Huxley A F (1952) A quantitative description of membrane current and its application to conduction in nerve. J. Phys. Vol. 117:500–544

Itoh M, Chua L O (2008) Memristor oscillators. Int. J. Bifur. Chaos, Vol. 18, No.11:3183–3206

Itoh M Chua L O (2009) Memristor cellular automata. Int. J. Bifur. Chaos, Vol. 19, No. 12: in press
Johnson R C (2008) Will memristors prove irresistible. EE Times issue 1538, August 18:30–32
Kandel E R (2006) In Search of Memory. Norton, New York
Mainzer K (2007) Thinking in complexity. Fifth edition, Springer, Berlin
Mauro A (1961) Anomalous impedance, a phenomenological property of time-variant resistance an analytic review. Biophysical J, Vol. 1:353–372
Oka T and Nagaosa N (2005) Interfaces of correlated electron systems: proposed mechanism for colossal electroresistance. Physical Review Letters, Vol. 95:266403
Rossel C, Meijer G I, Bremaud D, Widmer D (2001) Electrical current distribution across a metal–insulator–metal structure during bistable switching. J. Appl. Phys., Vol. 90, No. 6:2892
Sawa A, Fujii T, Kawasaki M, Tokura Y (2006) Interface resistance switching at a few nanometer thick perovskite manganite active layers. Appl. Phys. Letters, Vol. 88:232112
Schindler C, Thermadam S C P, Waser R, Kozicki M N (2007) Bipolar and Unipolar Resistive Switching in Cu-Doped SiO_2. Electron Devices, IEEE Trans. on, Vol. 54, No. 10:2762
Seo S, Lee M J et al (2004) Reproducible resistance switching in polycrystalline NiO films. Appl. Phys. Letters, Vol. 85, No. 23:5655
Sluis van der P (2003) Non-volatile memory cells based on $Zn_xCd_{1-x}S$ ferroelectric Schottky diodes. Appl. Phys. Letters, Vol. 82, No. 23:4089
Strukov D B, Snider G S, Stewart D R, Williams R S (2008) The missing memristor found. Nature, 453, No. 7191:80–83
Tour J M, He T (2008) The fourth element. Nature, Vol. 453, No. 7191:42–43
Williams R S (2008) How we found the missing memristor. IEEE Spectrum, Vol. 45, No. 12:28–35

Chapter 5
Circuit Models of Nanoscale Devices

Árpád I. Csurgay and Wolfgang Porod

Abstract On the nanoscale, equivalent circuit models are not scale invariant. An ideal equivalent circuit can be a valid model of a device at the macro or even microscale, but it might not reveal even the qualitative properties of the same device during downscaling. We illustrate the consequences of downscaling to the nanoscale with an example, the nanoscale capacitor. The circuit models combine four groups of state variables: (1) classical mechanical, (2) classical electromagnetic, (3) quantum mechanical, and (4) quantum electromagnetic. In general, a quantum-classical equivalent circuit is combined from four coupled "subcircuits," representing the classical mechanical dynamics of the nuclei, the classical dynamics of the electromagnetic field, the quantum wave-dynamics of the electrons, and the QFT dynamics of photons. The modeling procedure should determine the state-variables of the four subcircuits and their couplings.

Two examples illustrate the quantum-classical models. The first combines the mechanical dynamics of the nuclei with the quantum wave behavior of the electrons. The second illustrates an application of the nanocapacitor as a nonlinear infrared sensor.

5.1 Introduction

Engineering design has been, and is deeply rooted in physics, and the problems raised by the theory of design are mathematical problems. However, the challenges engineers face are fundamentally different from those that physicists and mathematicians face. Engineers are called upon to invent, design, and build artificial objects that do not exist in nature on their own. Engineers build *machines* from *components*. Engineering is about *syntheses of complex machines from simple*

Á.I. Csurgay (✉) and W. Porod
Faculty of Information Technology, Pázmány Péter Catholic University, Budapest
and
Center for Nano Science and Technology, University of Notre Dame, Notre Dame IN 46556

C. Baatar et al. (eds.), *Cellular Nanoscale Sensory Wave Computing*,
DOI 10.1007/978-1-4419-1011-0_5,

components. These components communicate with each other through their interfaces, e.g., through terminals or ports. Geometrical and chronometrical similarities, as well as physical conservation relations, such as conservation of charge, energy, and momentum, result in equivalent or at least approximately equivalent terminal and port behavior in the case of many internally different physical systems. This common terminal behavior in a specific experimental framework can be represented by an ideal mathematically defined component model. This discovery led to the emergence of the notion of *equivalent circuits*. Two circuits with only finite number of accessible terminals were called to be equivalent if in a given experimental frame no measurements at the accessible terminals could discover differences between them (Helmholtz 1853; Thévenin 1883; Mayer 1926; Norton 1926).

When engineers are challenged to design and build nanoscale artifacts with nanotechnologies, they have to rely on the laws of nature valid at the nanoscale. (We are talking about nano if at least one of the dimensions of an artifact is below ~100 nm.) Nature is not scale invariant, and behavior of matter on the nanoscale has its peculiarities.

On the nanoscale

1. From among the four fundamental forces of nature (gravity, electromagnetism, weak and strong nuclear forces), the electromagnetic interaction is the only dominant one. Gravity is negligible, and the nuclear forces act only inside nuclei where at the huge energy of that ladder nanoscale objects do not exist anymore. The Coulomb force between a proton and an electron is $F_C = e^2/4\pi\varepsilon_0 r^2$, where e is the charge of the proton, r is the distance between them, and $\varepsilon_0 = 8.85 \cdot 10^{-12}$ As/Vm. The gravitational force is $F_G = \gamma \cdot m_e m_p/r^2$, where $\gamma = 6.67 \times 10^{-11}$ Nm^2/kg^2, m_e and m_p are the mass of the electron and the proton, respectively. The ratio of the Coulomb and gravitational forces is $F_C/F_G \cong 0.7 \times 10^{40}$, i.e., the Coulomb force is $\sim 10^{40}$ times stronger than the gravitational force.
2. Vacuum is not "empty"; effects of vacuum fluctuations can become significant, even comparable with Coulomb forces (see, e.g., the Casimir effect);
3. The "quantum ladder" characterizes the state of objects under specific external conditions. The quantum ladder classifies a material system into hierarchic levels (steps on the ladder), whose distances apart are conditioned by size-energy relations: there is a threshold activation energy for each successive step on the ladder, below which it should be considered as "inert" (Weisskopf 1970);
4. All electrons, protons, neutrons are identical, and even atoms and molecules show identity in interactions, as long as they stay on a "step" of the quantum ladder;
5. Wave nature of electrons cannot be neglected, and below ~5 nm electronic quantum phenomena have a dominant sway.

Equivalent circuit models are not scale invariant. An ideal equivalent circuit can be a valid model of a device at the macro or even microscale, but it might not reveal even the qualitative properties of the same device during downscaling.

We illustrate the consequences of downscaling to nanoscale with an example, the nanoscale capacitor. At the macroscale, two metallic contacts separated by an insulating layer behave as a simple capacitor, and the charge is a unique function of the exciting voltage, and no conductive, only displacement current flows through the capacitor.

If the insulator becomes thin enough, the wave nature of the electron gets a dominant sway, and quantum tunneling begins to occur; now, not only displacement but also conductive current flows, and the structure behaves as a nonlinear tunneling diode (Esaki 1958; Scanlan 1966; Lent and Kirkner 1990). If the capacitance is small enough, and the two electrodes are made from metals with different work functions, the quantum tunnel diode can serve as an infrared detector, or as a mixer (Sanchez et al. 1978). The nanocapacitor becomes a nonlinear metal-oxide-metal (MOM) diode.

The capacitance of structures composed of thin oxide layers and metallic nanoparticles can be a few attofarad. In these small capacitors the effect of a single electron's charge can become significant, because the capacitance between two metallic plates, of a diameter of, e.g., 10 nm and a distance of 1 nm, is only 0.7 aF ($1\,\text{aF} = 10^{-18}$ farad); thus, if a single electron is added to the charge of the capacitor, there is a 220 mV voltage drop. The probability of tunneling depends exponentially on the voltage; thus, if the voltage of a junction drops, there is an exponential decrease in the probability of tunneling. The next electron's tunneling will be blocked by the Coulomb force of the former tunneling electron. The Coulomb force of a tunneling electron in metal–insulator–metal structures is utilized in SETs (single-electron transistors), which combines quantum tunneling of a single electron with the voltage drop caused by the tunneling electron itself, i.e., with the Coulomb blockade (Grabert and Devoret 1992). SET circuits have been suggested, and methods to design SET circuits have been developed (Likharev 1999; Hoekstra 2007).

Tunneling combined with Coulomb-force interactions provided the framework for the first field-coupled integrated-circuit concept, the quantum-dot cellular automata (QCA) (Lent et al. 1993).

It was discovered that a simple MIM diode can emit visible light (Tsui 1964, Jain et al. 1978). As long as an electron is inside a conductor, the electron gas shields the Coulomb field of the electron, and the interaction is so weak that it is unable to produce any observable effect. In the case of tunneling, when the electron is outside the cathode electrode, its Coulomb field is not screened any more by the other electrons. Thus, its long-range Coulomb field polarizes the surface charge density at the other electrode, and surface charges appear which prevent the field from penetrating inside the anode electrode. When the electron is absorbed by the anode, its Coulomb field becomes shielded again; thus, the surface polarization disappears. Tunneling electrons excite time-varying surface charge oscillations, thus generating surface plasmon waves. An MIM diode can function not only as a detector but also as a light emitting device. Surface plasmon waves can be generated on gold and silver nanoparticles, and on submicron waveguides by tunneling or by attenuated total reflection (ATR) of laser light. Integrated optical circuits have been envisioned, making use of the dynamics of plasmons (Maier 2001; Csurgay and Porod 2004).

Changing the size of a device and the experimental frame, we can cross a threshold on the quantum ladder. If we pull out the insulator material, thus there remain only two parallel neutral metal discs. Even if there is no voltage, and no charge, there still exists a strong pressure which pushes the two metal plates toward each other. It turns out that this strong pressure is caused by vacuum fluctuations.

In free space (where there are no charged particles at all), permanent electromagnetic fluctuations can be observed, which are explained as the consequence of the energy-time uncertainty principle of quantum field theory (QFT). The electric and magnetic field vectors are vibrating as fields of electromagnetic harmonic oscillators. On average, the fields cancel out, the expectation value of the E and B fields are zero, and in this sense the vacuum is "empty." However, the expectation value of the vacuum energy is not zero, it is equal to the zero-point energy of a harmonic oscillator, namely it is $\hbar\omega/2$. Significant pressure can be observed, caused by vacuum fluctuations, such as the Casimir effect which can play a significant role on the nanoscale, e.g., in nano-electromechanical sensors and in nanoactuators.

On the nanoscale, the very same object, a device built from two parallel metal discs, can be a NEMS, an oscillator, a detector tunnel diode, a simple capacitor, or a combination of them. Vacuum fluctuations can never be stopped, only neglected. In general, mechanical, quantum mechanical, electromagnetic and quantum electromagnetic dynamics have to be combined to understand the dynamics of nanodevices. The Born–Oppenheimer approximation helps in the separation of mechanical and quantum mechanical dynamics, to separate phonons and photons. In a nanodevice, the classical mechanical dynamics of the nuclei can never be separated from the quantum mechanical dynamics of the electrons. The modeling of coupling between the mechanical and quantum mechanical state variables can be approximated by the Hellmann-Feynman Theorem.

The circuit models combine four groups of state variables: (1) classical mechanical, (2) classical electromagnetic, (3) quantum mechanical, and (4) quantum electromagnetic. In general, a quantum–classical equivalent circuit is combined from four coupled "subcircuits," representing the classical mechanical dynamics of the nuclei, the classical dynamics of the electromagnetic field, the quantum wave dynamics of the electrons, and the QFT dynamics of photons. The modeling procedure should determine the state variables of the four subcircuits and their couplings.

5.2 Vacuum Fluctuations in Nanocircuits

In vacuum, there are forces between electrically neutral and highly conductive metal particles. They are manifestations of quantum fluctuations. The boundary conditions imposed on the electromagnetic fields lead to a spatial redistribution of the mode density with respect to free space, creating a spatial gradient of the zero-point energy density and hence a net force between the metals.

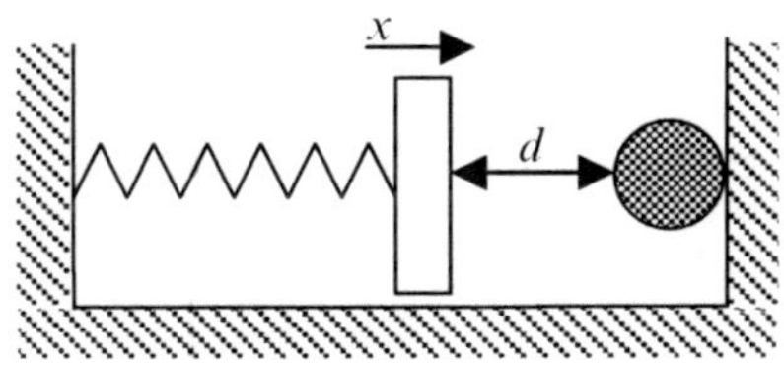

Fig. 5.1 Nonlinear Casimir oscillator. The equilibrium position of the plate in the absence of the Casimir force, is chosen to be 40 nm. The classical mechanical spring constant is 0.02 N m^{-1}

Between two parallel plates the force is attractive and assumes the form $F_{\text{Cas}} = -\frac{\pi^2 \hbar c}{240} \frac{A}{d^4}$ where A is the area of the particles and d is their distance. If one of the interacting surfaces is spherical, that modifies the force $F_{\text{Cas}} = -\frac{\pi^3 \hbar c}{360} \frac{R}{d^3}$, where R is the radius of the sphere and d is the distance between the plate and the sphere. The force and its sign can be controlled by tailoring the shapes of the interacting surfaces. In electromechanical systems of size smaller than 100 nm, a coupled classical mechanical and QFT dynamics can realize complex dynamical systems.

The Casimir force can be significant. Let us compare the Casimir and Coulomb forces between the plates of a capacitor. The Coulomb force is $F_{\text{C}} = Q \cdot E/2$, where the charge of the capacitor $Q = C \cdot V = (\varepsilon_0 A/d) \cdot V$ and $E = V/d$, i.e., $F_{\text{C}} = \varepsilon_0 A V^2/d^2$. The ratio is

$$\frac{F_{\text{Cas}}}{F_{\text{C}}} = -\frac{\pi^2 \hbar c}{240} \frac{A}{d^4} \cdot \frac{d^2}{\varepsilon_0 A V^2} \cong 146 \frac{1}{(d^{nm})^2 \, V^2}.$$

If the distance between the plates is in the order of a few nanometers, the force caused by the vacuum fluctuation is on the order of the Coulomb force caused by a few volts (at 1 nm it is ~12 V, at 10 nm it is ~1.2 V). In the modeling of MEMS and NEMS devices, vacuum fluctuations, i.e., the effect of QFT cannot be neglected.

The first experimental observation of bistability and hysteresis caused by the effect of QFT was published by Capasso et al. (2001). Figure 5.1 show a simple model of their oscillator which consists of a movable metallic plate subjected to the restoring force of a spring and the force arising from vacuum fluctuation between the plate and a fixed metallic sphere.

This nonlinear classical mechanical oscillator has been embedded in an electronic circuit, and was used as a MEMS component (Capasso et al. 2001).

5.3 Mixed Quantum Classical Electromechanical Models

Molecular dynamics can be approximately modeled by quantum–classical models. The simplest case is the quantum–classical molecular dynamics (QCMD) of two interacting particles, one of them moves as a classical particle, the other one behaves as a quantum mechanical object. This assumption is useful in case of a simple molecule of two masses which differ significantly, therefore, the heavier particle of mass M can be modeled classically while the lighter one of mass m remains a quantum particle. The quantum particle is described by a wave function $\Psi(\mathbf{r}, t)$ which obeys Schrödinger's equation

$$\mathrm{j}\hbar\Psi_q(\mathbf{r}, t) = \left(-\frac{\hbar^2}{2m} + V(\mathbf{r}, \boldsymbol{\mu})\right)\Bigg|_{\boldsymbol{\mu}=\mathbf{q}(t)} \cdot \Psi_q(\mathbf{r}, t)$$

with a parameterized potential which depends on the position $\mathbf{q}(t)$ of the classical particle, making the potential time-dependent. The location of the classical particle is a solution of a classical Hamiltonian equation of motion,

$$M\dot{\mathbf{q}} = \mathbf{p}, \quad \dot{\mathbf{p}} = -\nabla_q U,$$

in which the time-dependent potential $U(\mathbf{q})$ is given as the original classical potential $V(\mathbf{r},\mathbf{q})$, weighted with the probability of finding the quantum particle

$$U(\mathbf{q}) = \int V(\mathbf{r}, \mathbf{q}) \cdot |\Psi(\mathbf{r}, t)|^2 \, \mathrm{d}V.$$

The forces in classical equations of motion $\dot{\mathbf{p}} = -\nabla_q U$ are the so-called Hellmann–Feynman forces (Hellmann 1937; Feynman 1939)

$$\nabla_q U = \langle \Psi, \nabla V \cdot \Psi \rangle .$$

The Schrödinger's equation can be replaced by its density-matrix representation, the Liouville-Neumann equation. An arsenal of efficient simulation tools has been developed. It turns out to be feasible to combine classical molecular dynamics with the simultaneous evaluation of the forces using quantum density functional theory.

Quantum–classical models have been developed for integrated circuits composed of Coulomb coupled nanodevices (Csurgay and Porod 2001).

As schematically shown in Fig. 5.2, the individual device (molecule) is dissipatively coupled to a heat bath, and it is exposed to external forces, such as clocking circuitry, and it couples to its neighbors through electric fields. We have shown that the electronic (or magnetic) state at time t of any open quantum system can be

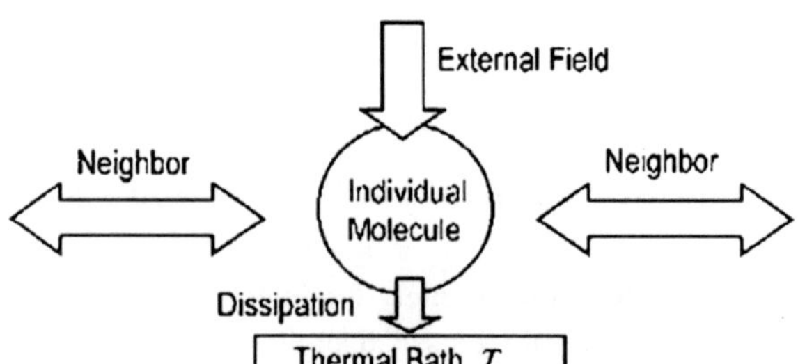

Fig. 5.2 A nanodevice (molecule) coupled to its neighbors and excited by external field

described by a state vector, the so-called coherence vector λ(t), which represents the Hermitian density matrix of the system. For the case of a two-level system, the coherence vector has three components, which corresponds to a 2-by-2 density matrix.

The electronic dynamics of such a nanostructure may be described by quantum Markovian master equations of finite-state systems. This model describes the dynamics of a device as the irreversible evolution of an open quantum system coupled to a reservoir (heat bath). This coupling to the environment introduces damping terms in the dynamic equations, which then take the general form of:

$$\hbar \frac{\mathrm{d}\lambda(t)}{\mathrm{d}t} = \Omega\lambda(t) + R\lambda(t) + k$$

Here, Ω is the Bloch matrix of the corresponding conservative (nondissipative) quantum system, and R and k are the damping matrix and vector, respectively. The details can be found in Csurgay and Porod (2001). Note that both Ω and R depend on the coherence vector of the open system itself, as well as on the coherence vector of the coupled neighboring systems.

The mixed quantum–classical equations describe the time evolution of the state of the nanodevice. The coherence vector determines the electronic evolution within the framework of a density-matrix description, and all experimentally observable quantities are related to its components. For the case of a two-state system, the third component of the vector $\lambda_3(t)$ determines the electronic charge configuration.

Notice that the above (ordinary differential) equation resemble circuit dynamics. The equations for the various components of $\lambda(t)$ can be interpreted as the state equations of a nonlinear circuit with state variables λ_1, λ_2, and λ_3. The various terms in the coupled equations can be viewed as nonlinear resistors, capacitors, inductors, and controlled sources. This is schematically shown in Fig. 5.3 above for the case of two-state nanostructure with a 3-dimensional electronic state vector λ, and one degree of freedom for nuclear vibration.

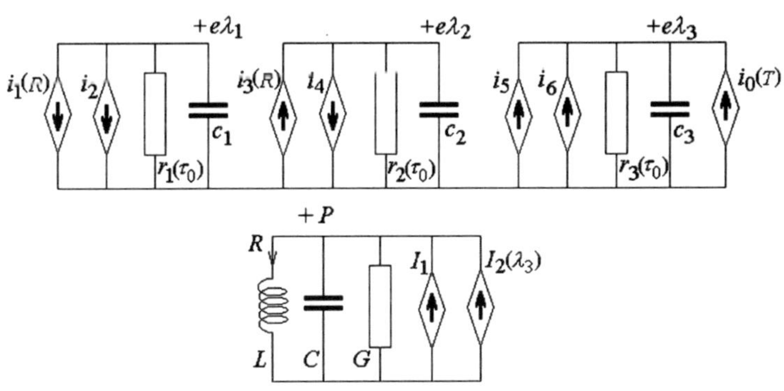

Fig. 5.3 Equivalent-circuit representation of the mixed quantum-classical dynamics for a 2-state nanostructure with one-dimensional nuclear vibration. Note that the nonlinearity is represented by the nonlinear controlled sources

We assume that the individual devices or molecules in an array are fixed in space, and that the electronic dynamics takes place inside each individual molecule (no inter-molecular charge transfer). We also assume that the molecules are far enough apart from each other that the overlap between their wave functions can be ignored. We can then identify sets of private electrons and Hamilton operators as belonging to each molecule. Intermolecular forces due to field coupling are relatively weak and their effects can be considered as perturbations.

To model the Coulombic interactions between individual molecules, we need to be able to describe the way in which charge is distributed inside each molecule. It is well known that Coulomb interactions between charges localized inside spheres can be specified by the interactions between multipoles (point charges, dipoles, quadrupoles, octopoles, etc.) representing the charge distribution inside the isolated sphere surrounding a molecule.

In this way, the time-varying Coulomb field of an individual molecule can be represented by multipoles at fixed positions with time-varying multipole moments. If the dynamics of a molecule with its time-varying electronic charges are known, then the potential at the site of the neighbor can be determined (and thus the interaction energies). This allows us to model the effects of the neighbors on any individual molecule in the array.

For the equivalent circuit model, the effect of the neighbors is represented by controlled sources, which are dependent upon the state variable that describes the charge configuration (λ_3 for the case of a 2-state device).

Much before the modeling of the mixed electronic-mechanical dynamics, a purely electronic Coulomb-coupled architecture was proposed and demonstrated: the QCA concept (Lent et al. 1993). The Notre Dame proposal was based on a cell which contains five quantum dots. In the ideal case, this cell is occupied by two electrons. The electrons are allowed to "jump" between the individual dots in a cell by the mechanism of quantum mechanical tunneling. Based upon the emerging technology of quantum-dot fabrication, the Notre Dame NanoDevices group has proposed a scheme for computing with cells of coupled quantum dots (Porod et al. 1999), which has been termed "quantum-dot cellular automata" (QCA) (Lent et al. 1993; Amlani et al. 1999; Toth et al. 1996; Porod et al. 1999, 2003; Snider et al. 1999). For a review, see Chapter 6 in *Handbook of Nanoscience, Engineering, and Technology* entitled *Nanoelectronic Circuit Architectures* (Porod 2007).

5.4 Circuit Model of a Double-Band Infrared Sensor

A proposed double-band sensor is shown in Fig. 5.4 (Matyi 2004). Two coupled nanoantennas are lithographically fabricated on a substrate covered with a reflector top metal layer (Fig. 5.4b). Two MOM diodes are providing two rectified DC voltages (VDC1; VDC2). We intended to design the antenna geometry to meet the double-band requirements.

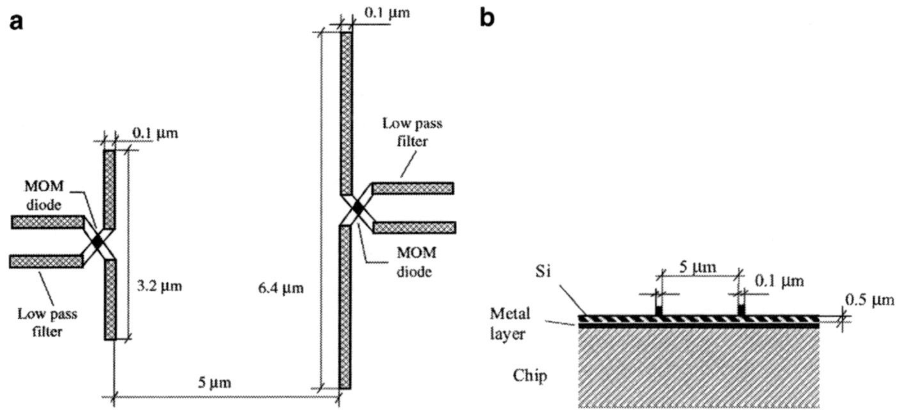

Fig. 5.4 Double-band sensor. (**a**) Layout of the double-band sensor; and (**b**) cross section of the double-band sensor. The metal layer serves as a reflector for the two coupled antennas

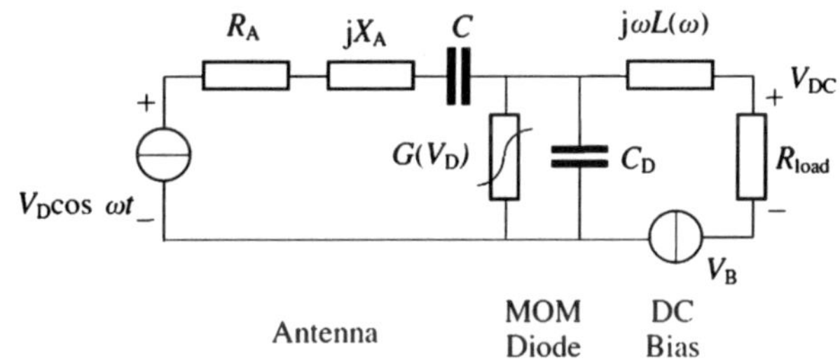

Fig. 5.5 Single-band nanoantenna-MOM diode sensor and its equivalent circuit

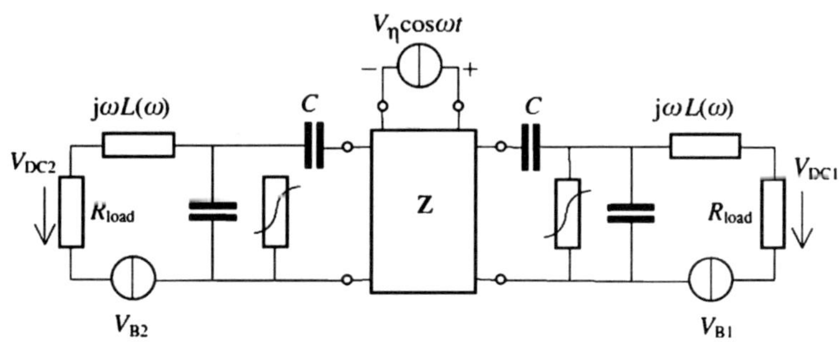

Fig. 5.6 Equivalent circuit of the double-band infrared sensor

The circuit model shown in Fig 5.5 (Sanchez et al. 1978) has been extended for this case. Figure 5.6 shows the equivalent circuit of the double-band sensor. The two MOM diodes are biased independently ($VB1$; $VB2$), capacitors C separate the high-frequency circuits from the DC currents, the large inductor reactance $j\omega L\,(\omega)$

serves as a lowpass filter, thus high-frequency currents do not flow in the direction of the loads. The three-port Z couples the incident radiation to the diodes. If the antennas were far from each other, diode 1 would see just the radiation resistance and reactance of antenna 1, and the same were true for diode 2. In general the two antennas are coupled, and this effect is represented with Z.

The arrangement behaves as a double-band sensor for 12 ± 2 THz (Band 1) and 20 ± 2.5 THz (Band 2).

References

Amlani I., Orlov AO, Toth G, Bernstein GH, Lent CS, Snider GL (1999) Digital logic gate using quantum-dot cellular automata. *Science* 284:289–291

Capasso F, Munday JN, Iannuzzi D, Chan HB (2007) IEEE J Select Top Quantum Electr 13: 400–414

Chan HB, Aksyuk VA, Kleiman RN, Bishop DJ, Capasso F (2001) Nonlinear micro-mechanical Casimir oscillator. Phys Rev Lett 87(21):211801–04

Csaba G, Csurgay AI, Porod W (2001) Computing architecture composed of next-neighbor-coupled optically pumped nanodevices. Int J Circuit Theory Appl 29: 73-91

Csurgay AI, Porod W, Lent CS (2000) Signal processing with near-neighbor-coupled time-varying quantum-dot arrays. IEEE Transact Circuits Syst I, 1212–1223

Csurgay AI, Porod W, Rakos B (2003) signal processing by pulse-driven molecular arrays. Int J Circuit Theory Appl 31(1):55–66

Csurgay AI, Porod W (2001) Equivalent circuit representation of Coulomb-coupled nanoscale devices: modelling, simulations and reliability. Int J Circuit Theory Appl 29(1):3–35

Csurgay AI, Porod W (2004) Surface plasmon waves in nanoelectronic circuits. Int J Circuit Theory Appl 32:339–361

Esaki L (1958) New phenomenon in narrow germanium p–n junctions. Phys Rev 109:603–604

Feynman RP (1939) Forces in Molecules, Phys Rev 56:340–343

Grabert H, Devoret MH (1992) Single electron tunneling—Coulomb blockade in nanostructures. NATO-ASI Series B-294. Plenum Press, New York

Hellmann H (1937) Einführung in die Quantenchemie, F. Deuticke, Leipzig

Helmholtz H (1853) Über einige Gesetze der Verteilung elektrischer Ströme in körperlichen Leitern mit Anwendung auf die thierisch-elektrischen Versuche. Annalen der Physik 89(6):211–233

Hoekstra J (2007) Toward a circuit theory for metallic single-electron tunneling devices. Int J Circuit Theory Appl 35(3):213–238

Jain RK, Wagner S, Olson DH (1978) Stable room-temperature light emission from metal–insulator–metal junctions. Appl Phys Lett 32(1):62–64

Lent CS, Kirkner DJ (1990) The quantum transmitting boundary method. J Appl Phys 67: 6353–6359

Lent CS, Tougaw PD, Porod W, Bernstein GH (1993) Quantum cellular automata. Nanotechnology 4:49–57

Likharev KL (1999) Single-electron devices and their applications. Proc IEEE 87(4):606–632

Maier SA et al. (2001) Plasmonics – a route to nanoscale optical devices, Advanced Materials, 13(19): 1501–1505

Matyi G (2004) Nanoantennas for uncooled, double-band, CMOS compatible, high-speed infrared sensors. Int J Circuit Theory Appl 32:425–430

Mayer HF (1926) Über das Ersatzschema der Verstärkerröhre. Telegraphen- und Fernsprech-Technik 15:335–337

Norton EL (1926) Design of finite networks for uniform frequency characteristic. Technical Report TM26–0–1860, Bell Laboratories

Porod W (2007) Nanoelectronic circuit architectures. Chapter 6 in: Goddard WA, Brenner DW, Lyshewski SE, Iafrate GJ (eds) Handbook of Nanoscience, Engineering, and Technology, CRC Press, Boca Raton

Porod W, Csaba G, Csurgay AI (2003) The role of field coupling in nano-scale cellular nonlinear networks. Int J Neural Syst 13(6):387–395

Porod W, Lent CS, Bernstein GH, Orlov AO, Amlani I, Snider GL, Merz JL (1999) Quantum-dot cellular automata: computing with coupled quantum dots. Int J Electron 86(5):549–590

Sanchez A, Davis CF, Liu KC, Javan A (1978) The MOM tunneling diode: theoretical estimate of its performance at microwave and infrared frequencies. J Appl Phys 49:155–163

Scanlan JO (1966) Analysis and synthesis of tunnel diode circuits. Wiley, London

Snider GL, Orlov AO, Amlani I, Zuo X, Bernstein GH, Lent CS, Merz JL, Porod W (1999) Quantum-dot cellular automata: review and recent experiments (invited). J Appl Phys 85(8):4283–4285

Stone AJ (1997) The theory of intermolecular forces. Clarendon Press, Oxford

Thévenin L (1883) Sur les conditions de sensibilité du pont de Wheatstone. Annales T'elégraphiques 10:225–234

Thompson A Wasshuber C (2000) Design of single-electron systems through artificial evolution. Int J Circuit Theory Appl 28(6):585–599

Toth G, Lent CS, Tougaw PD, Brazhnik Y, Weng W, Porod W, Liu RW, Huang Y-F (1996) Quantum cellular neural networks. Superlatt Microstruct 20:473–477

Tsui DC (1969) Observation of surface Plasmon excitation by tunneling electrons in GaAs-Pb tunnel junctions, Phys Rev Lett 22(7):293–295

Weisskopf VF (1970) Physics in the 20th century. Science 168:923–930

Chapter 6
A CMOS Vision System On-Chip with Multi-Core, Cellular Sensory-Processing Front-End

Angel Rodríguez-Vázquez, Rafael Domínguez-Castro, Francisco Jiménez-Garrido, Sergio Morillas, Alberto García, Cayetana Utrera, Ma. Dolores Pardo, Juan Listan, and Rafael Romay

Abstract This chapter describes a vision-system-on-chip (VSoC) capable of doing: image acquisition, image processing through on-chip embedded structures, and generation of pertinent reaction commands at thousands frame-per-second rate. The chip employs a distributed processing architecture with a pre-processing stage consisting of an array of programmable sensory-processing cells, and a post-processing stage consisting of a digital microprocessor. The pre-processing stage operates as a retina-like sensor front-end. It performs parallel processing of the images captured by the sensors which are embedded together with the processors. This early processing serves to extract image features relevant to the intended tasks. The front-end incorporates also smart read-out structures which are conceived to transmit only these relevant features, thus precluding full gray-scale frames to be coded and transmitted. The chip is capable to close action–reaction loops based on the analysis of visual flow at rates above 1,000 F/s with power budget below 1 W peak. Also, the incorporation of processors close to the sensors enables signal-dependent, local adaptation of the sensor gains and hence high-dynamic range signal acquisition.

6.1 Introduction

The Strategic Research Agenda of the European Nano-electronics Initiative Advisory Council SRA-ENIAC (ENIAC 2007), as well as the International Technology Roadmap For Semiconductors (ITRS) (International Technology Roadmap for

A. Rodríguez-Vázquez (✉), R. Domínguez-Castro, F. Jiménez-Garrido, S. Morillas, A. García, C. Utrera, Ma.D. Pardo, J. Listan, and R. Romay
AnaFocus (Innovaciones Microelectrnicas S.L.), Avda Isaac Newton, Pabelln de Italia, Ático, Parque Tecnolgico Isla de la Cartuja, 41092-Sevilla (Spain)
e-mail: angel@imse.cnm.es

A. Rodríguez-Vázquez and R. Domínguez-Castro
IMSE-CNM/CSIC and Universidad de Sevilla, Parque Tecnológico Isla de la Cartuja, 41092-Sevilla, Spain
e-mail: angel.rodriguez-vazquez@anafocus.com

C. Baatar et al. (eds.), *Cellular Nanoscale Sensory Wave Computing*, 129
DOI 10.1007/978-1-4419-1011-0_6,

Semiconductors (ITRS) 2007), highlight the gap between the potential of enabling IC technologies, on the one hand, and the actual capabilities of systems designed by using these technologies, on the other hand. Systems with many-billion devices can be implemented. However, *special-purpose*, dedicated architectures are needed to reach *performance levels* matched to these levels of complexity. The SRA-ENIAC acknowledges this fact and states the needs of devising *new concepts and architectures* of *smart electronic systems* capable of interacting with the environment and closing sensing–processing–actuating loops. It also identifies the key role of *applications* for the specification and driving of technology developments.

Vision systems, and more generally systems intended to handle massive set of topographical data are among the most challenging of the application drives mentioned by ENIAC's SRA. The design of *imaging systems* (sensors + readout + data conversion + controller + drivers) on CMOS chips has been making good progress during the last decade (El Gamal and Eltoukhy 2005). The main design target for CMOS imaging chips is *reproducing images* with given *accuracy* and *speed*. The target for *vision systems* is different. Similar to imagers, they have 2D light intensity maps as inputs. Also, they may output images for monitoring purposes. However, their primary *outputs* are not images, but *reaction commands*. For instance, these commands may be needed to discard defective parts following visual inspection in a production line; or to trigger evasive maneuver following the visual detection of looming objects moving into collision course toward a vehicle; or to align unmanned aerial vehicles while landing in a platform following the signaling provided by a set of light beacons; or to trigger alert mechanisms if suspicious events are detected into a scene subjected to video surveillance; just to mention some examples.

Vision applications require to complete the full "sense → process → analyze → make decision" cycle. It involves *large amount of data*, especially in applications where high frame-rate is essential. Making a real-time decision also requires *low latency* from the system, which makes the analysis of the large input data set even more demanding.

The industrial state-of-the-art considers vision systems as "seeing computers" or "computers that see." This vision (now in the metaphoric meaning of the word) is reflected on the architecture typically used for them, namely: an *imager* (image sensor) to acquire and digitize the sensory data and a *host processor* to handle this huge amount of raw data. Such brute-force approach does completely ignore the specifics of the data, the ways how interesting pieces of information emerge from the data, and hence results in largely inefficient systems. Consider for instance the application of finding defective parts in a production line where the parts may be placed with different orientations, corresponding to up to 360° rotations. Current vision technologies can hardly go above 10 F/s (frames per second), even by using a low-resolution front-end sensor with only 128×128 pixels (Cognex Ltd.).

Not only conventional computer architectures are inadequate. Conventional algorithmic solutions used in these architectures are also inadequate. This fact has been highlighted in a very recent paper published in *Vision System Design* (Devaraj et al. 2008). It states that brute force pattern matching, the conventional approach adopted by many system developers, is not the right tool in many applications. Instead, sic, "*a majority of smart camera applications can be solved using only a small number of image processing algorithms that can be learned quickly and used very effectively*" (Devaraj et al. 2008). Interestingly enough these simple algorithms (thresholds, blob analysis, edge detection, average intensity, binary operators, etc.) can be mapped down onto dedicated, processor architectures composed of simple processors with mostly local interactions – the sort of architectures addressed by this chapter.

Unconventional architectures and implementations for smart imaging chips (imagers with some embedded intelligence) and vision-dedicated chips have been reported elsewhere. For example AER silicon retina chips (Delbruck and Lichsteiner 2006), optical flow sensors (Green et al. 2004), visual depth sensors (Philipp et al. 2006), etc. These devices include many remarkable architectural concepts and optimized circuitry and are very efficient in some *specific early-vision* tasks. Also, during the last few years the authors have relied on the concept of *visual cellular microprocessors* (Chua and Roska 2002; Roska and Rodríguez-Vázquez 2001) and have devised different programmable *general-purpose* early-vision chips based on this concept (Rodríguez-Vázquez et al. 2004; Liñán et al. 2004; Carmona et al. 2003). However, all these chips are not *autonomous* systems; i.e., they must combine with off-chip controllers and processors for completing *medium-* and *high-level* vision tasks. This chapter reports a complete, autonomous vision-system-on-chip (VSoC) called Eye-RIS_v2.1. It is composed by two *multi-core* stages, namely:

- A *pre-processing* stage consisting of an array of mixed-signal sensing-processing cores; one per pixel. These cores are interconnected to realize spatial operations (such as linear convolutions, diffusions, etc.) on input images. Each pixel also contains memories for storage of intermediate processing results and control circuits for data and task scheduling. Filtering in time is achieved through data scheduling and memories. Nonlinear operations (such as thresholding, mathematical morphology, etc.) are realized through data-dependent scheduling and adaptation.
- A *post-processing* stage. It is a 32-bit RISC micro-Processor running at 100 MHz. This micro-processor is a silicon-hardened version of the ALTERA's NIOS-II μP which was initially conceived and released by ALTERA only for FPGA implementation.

Interactions between stages are handled by an embedded controller. The chip also embeds a 256 kB memory for program and data storage. Figure 6.1 shows the floor-plan of the chip with the pre-processing (called Q-Eye), post-processing, and memory sections labeled. It also shows a chip microphotograph and the external aspect of the vision system built with this chip – called Eye-RIS_v2.1 (AnaFocus Ltd.).

6.2 Architectural Concept of the Eye-RIS System

Eye-RIS systems are targeted to complete vision tasks at very high speed. For instance, to segment moving objects within a scene, compute their speeds and trajectories and provide this information to a control system which tracks the objects.

Fig. 6.1 The Eye-RIS_v2.1: VSoC floorplan; microphotograph; and packaged, stand-alone vision system

This is hard task for conventional vision systems composed of a front-end sensor followed by a DSP. These architectures operate on a frame-by-frame basis. The front-end sensor must captures all frames, one by one, at the required speed; then it must read all pixel date per each frame; convert and codify all these data into a digital format; and drive a digital processor with the resulting data flow. High-speed applications require large frame rates (well above the standard video rate) and each frame is composed of a large 2D set of data. Hence, since the digital processor must analyze a huge amount of information, either sophisticated processor architectures are employed or real-time operation becomes unfeasible.

The bottleneck of these conventional architectures is found in their frame-based operation. Reading and downloading complete frames is needed for applications whose target is reproducing full images (imaging applications), but not for vision applications. In these latter applications whole images are not important. Only those image features which convey the information required to complete the required vision tasks are. For instance, in tracking applications only the locations and speeds of the relevant objects are important. Hence, why to read-out, convert/codify, and transmit full image frames? By doing so we are wasting precious resources in

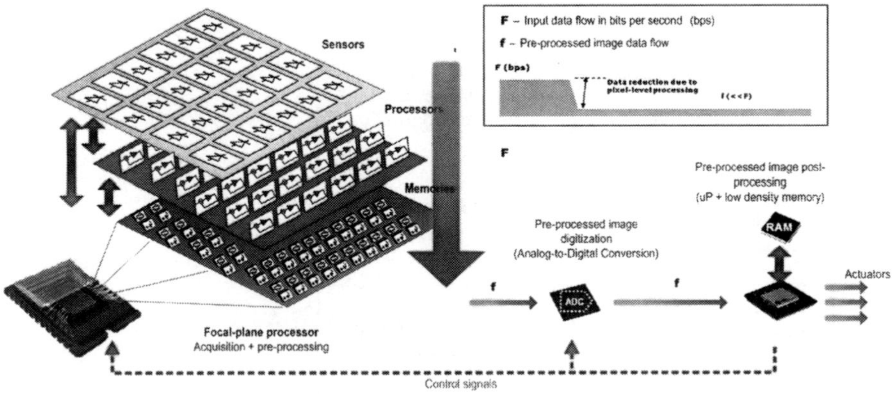

Fig. 6.2 Conceptual architecture of the Eye-RIS_v2.1 VSoC. Processing is distributed within two main stages. The second stage uses a conventional digital processor. The first one uses a multi-core cellular sensor-processor where each core embeds sensing, processing, and memory resources

handling useless information (pixel data) and overloading the DSP with such useless information. In the Eye-RIS architecture this problem is overcame by incorporating processing in the sensory front-end, as illustrated by Fig. 6.2.

The idea underlying the architecture of Fig. 6.2 is distributing the tasks among different cores and, more specifically performing a significant part of the processing at a *front-end* section consisting of simple, tightly coupled programmable processing cores. This front-end section, conceptually depicted as a multi-layer one in Fig. 6.2, is realized on-chip as a *multi-functional structure* with all conceptual layers implemented within a common semiconductor substrate. Relevant features of the incoming frames are extracted by this sensory-processing front-end, and only these relevant features are converted, codified, and transmitted for further analysis by the DSP.

Figure 6.3 illustrates the overall architectural target with reference to a conceptual representation of a vision processing chain (Russ 1992). The figure includes several processing steps and shows that the amount of data decreases as information travels along the chain, namely:

- At initial steps, the number of data is huge and many of the data are redundant and hence useless to the purposes of reaction prompting.
- As information flows across the processing chain and abstract features are extracted from the incoming images, the number of data decreases.

In conventional vision architectures the border between sensors and processors is placed at a point where the amount of data is large. However, in the Eye-RIS architecture this border is located at a point where the amount of data is small. Assume for illustration purposes that we target to tracking objects moving at 40 m/s into a scene. It requires capturing and analyzing images at 2,000 F/s rate. At the outcome of the capture/analyze process, the only pertinent data is the predicted position of

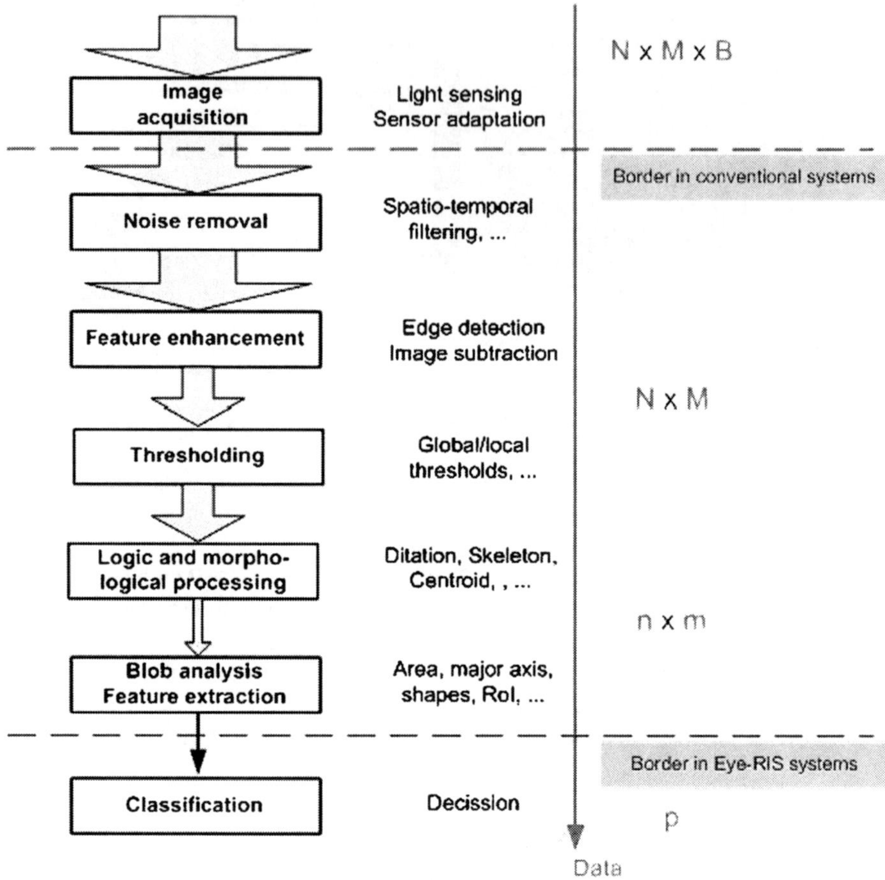

Fig. 6.3 Processing chain of vision. As data evolve from the sensor interface (raw data), the amount of data decreases and the abstraction level increases. N represents the number of rows, M the number of columns, and B the number of bits used per pixel data: $n < N$; $m < M$, and $p < (n, m)$. The Eye-RIS_v2.1 architecture map this layered data structure by using processing strategies fitted to each step in the chain

the objects. This is actually the only information driven to the digital processor. But to extract this information the following tasks must be completed:

- Image acquisition
- Low-pass filtering
- Activity detection
- Motion estimation
- Object tracking
- Loop control
- Position prediction

In the Eye-RIS system of Fig. 6.1, this is achieved by the so-called Q-Eye focal-plane processor (AnaFocus Ltd.).

6.3 The Eye-RIS Chip

The Eye-RIS chip is targeted to complete medium complexity vision tasks (segmentation, adaptation, tracking, movement estimation, feature analysis, etc.) at rates above 1,000 F/s with moderate power consumption (<1 W peak power). Such speed is required to close action–reaction loops for different applications.

As already mentioned, the architecture fits to the specifics of the vision processing chain. Parallel-processing is used during pre-processing to efficiently handle the vast data of early vision. Address-event reading is used at the link between pre- and post-processing stages to transmit only pertinent data. It reduces the computational payload of the post-processor and precludes the allocation of memory resources for redundant data.

Figure 6.4 shows the VSoC architecture. The pre-processor (area at the bottom left in the figure) is instrumental for performance. It computes the analog data capturedby the embedded sensors and employs digital signals to control processing function, parameters, and internal data flow. Digital codes are interchanged with the μP through the controller. Analog data are passed on to the μP by means of a battery of analog-to-digital converters (8 bit at 50 MHz).

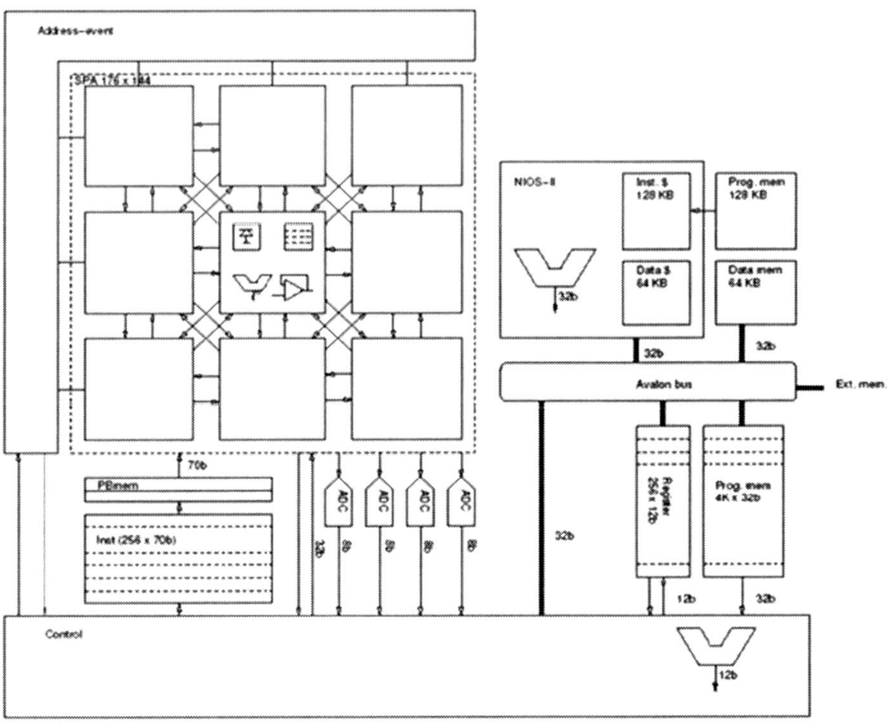

Fig. 6.4 Eye-RIS_v2.1 chip architecture

The pre-processing stage consists of an array of *multi-functional* pixels (see Fig. 6.2) including photo-sensors, processing circuits, and memories. These functions include: 2D image sensing, image processing (a sequence of programmable space and time-domain operations are executed in the 2D processor array); 2D memorization of both analog and digital data; 2D data-dependent task scheduling, control and timing, input/output operations; and storage of user-selectable instructions (programs) and parameter configurations. Pixels interchange information with their neighbors to realize a variety of operations such as:

- Linear convolutions with programmable masks
- Time- and signal-controlled diffusions (by means of an embedded resistive grid)
- Image arithmetics
- Signal-dependent data scheduling
- Gray-scale to binary transformation
- Logic operation on binary images
- Mathematical morphology on binary images

Power management strategies are used for optimum power consumption depending on the application demands (frame-rate, computational payload, ...). Most of the pre-processor blocks and the analog reference buffers used for biasing have independent power up/down signals and programmable bias and hence speed. Also, a temperature sensor and a temperature correction loop are embedded on-chip to preclude the impact of heating on optical sensors and analog memories. Other calibration techniques are embedded for robustness enhancement. Specifically, analog block offsets are stored in static (nonvolatile) digital memories and automatic calibration is performed by dedicated state machines which control in-loop A/D converters.

6.4 The Eye-RIS' Front-End: The Q-Eye

The name Q-Eye applies to the sensory-processing front-end of the Eye-RIS chip. The Q-Eye significantly differs from its antecessors; namely the ACE front-end chips (Rodríguez-Vázquez et al. 2004). Main drawbacks of the ACE chips were lack of robustness, large power consumption, and reduced cell density. The Q-Eye overcomes these drawbacks by making significant changes at both architectural and circuital design level. At the outcome, the cell density is increased by about 6.5 times and the power consumption is largely reduced. Also, these improvements are not detrimental of the functionality embedded per pixel; rather on the contrary, the Q-Eye pixels include co-processing structures which are not found in the ACE chips.

Figure 6.5 shows a floor-plan of the Q-Eye. The external interface is completely digital and synchronous. It is composed of a 32-bit data bus for image I/O and two additional buses, namely a 10-bit data bus and 12-bit address bus. These latter buses are employed to program a digital control system which contains 256 control words of 60-bit and individual register for analog references and miscellaneous configura-

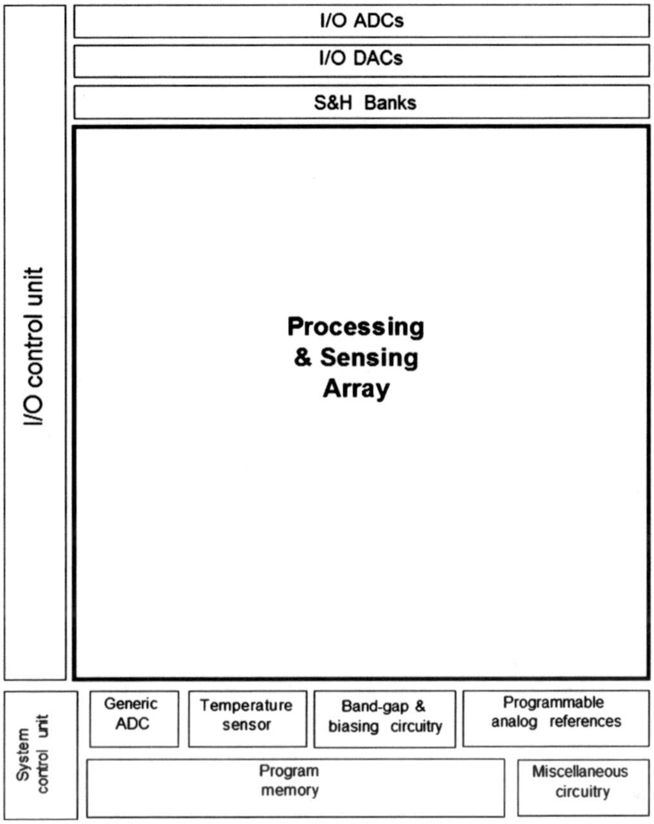

Fig. 6.5 Floor-plan of the Q-Eye front-end

tion. This system controls the array of processing-sensing cells, on one hand, and the I/O control unit which handles all basic I/O process, on the other. The I/O interface can operate in three modes:

- Loading–downloading of gray-scale images
- Loading–downloading of binary images (codified as 1-bit per pixel)
- Address-event mode

Gray-scale values are coded into digital form by on-chip 8-bits A/D Flash converters, and decoded by on-chip 8-bits D/A resistors string converters.

To the purpose of improved *power management,* and hence reduced power consumption, most of the processing blocks in the Q-Eye and the analog reference buffers used for biasing have independent power up/down signals Also, the operation speed of most blocks is programmable. Thus, the chip can be tuned to process either very high frame rates or low frame rates with optimum power consumption for each configuration.

Robustness enhancement is achieved through improved calibration techniques. In the ACE16k chip, offsets were stored into analog memories which experienced significant degradation especially at high temperatures. Instead, offsets in the Q-Eye are stored in static (nonvolatile) digital memories. Automatic calibration in the Q-Eye is performed by dedicated state machines which control in-loop A/D converters. Also, a temperature sensor and a temperature controlled correction loop are embedded in the Q-Eye to preclude the impact of junction temperature increases onto optical sensors and analog memories.

The core of the Q-Eye front end consists of an array of programmable cells. Each cell is capable to store several gray-scale and black-and-white pixel data and to perform arithmetic operations involving data of the neighbor pixels (spatial filtering) and past samples of the pixel data (temporal filtering). Figure 6.6 shows two alternative block diagrams for the pixel of the Q-Eye. As for the ACE chips, basic analog processing operations among pixels are *linear convolutions* with programmable masks. However, the Q-Eye does not employ transconductance multipliers (as it happens in the ACE chips) but a multiplier–accumulator circuit unit (MAC) which processes neighbor pixels into an *algorithmic sequence*. Despite this sequential operation, computation times are similar since no calibration of the transconductors is needed.

The area saving reported by the absence of spatially replicated structures (i.e., the transconductance multipliers employed for linear convolutions (Rodríguez-Vázquez et al. 2004)) enables the incorporation at the Q-Eye pixel of functions which are not found in the ACE chips. Overall the following tasks can be realized at the front-end:

- Pixel-wise "cosmetic operations": Each pixel is transformed independently of its neighbors, and remains on the same location.
- Generalized convolutions: Each pixel is transformed as a combination of the pixels within its neighborhood:
  - Linear convolution kernels
  - Morphological operators
  - Nonlinear operations, anisotropic diffusion
- Movements: pixels are moved to a different position. Movements can be decomposed into shifts.
- Image-wise operations: Pixels in different images in the same or different locations are combined (either linearly or nonlinearly)

6.5 The Eye-RIS Chip in Operation

Some main features of the Eye-RIS_v2.1 chip include:

- 176×144 pixel count. It suffices for many factory automation applications.
- Active sensing area of $5.4 \times 4.4\,\text{mm}^2$ which allows for 1/2 in. optical format.

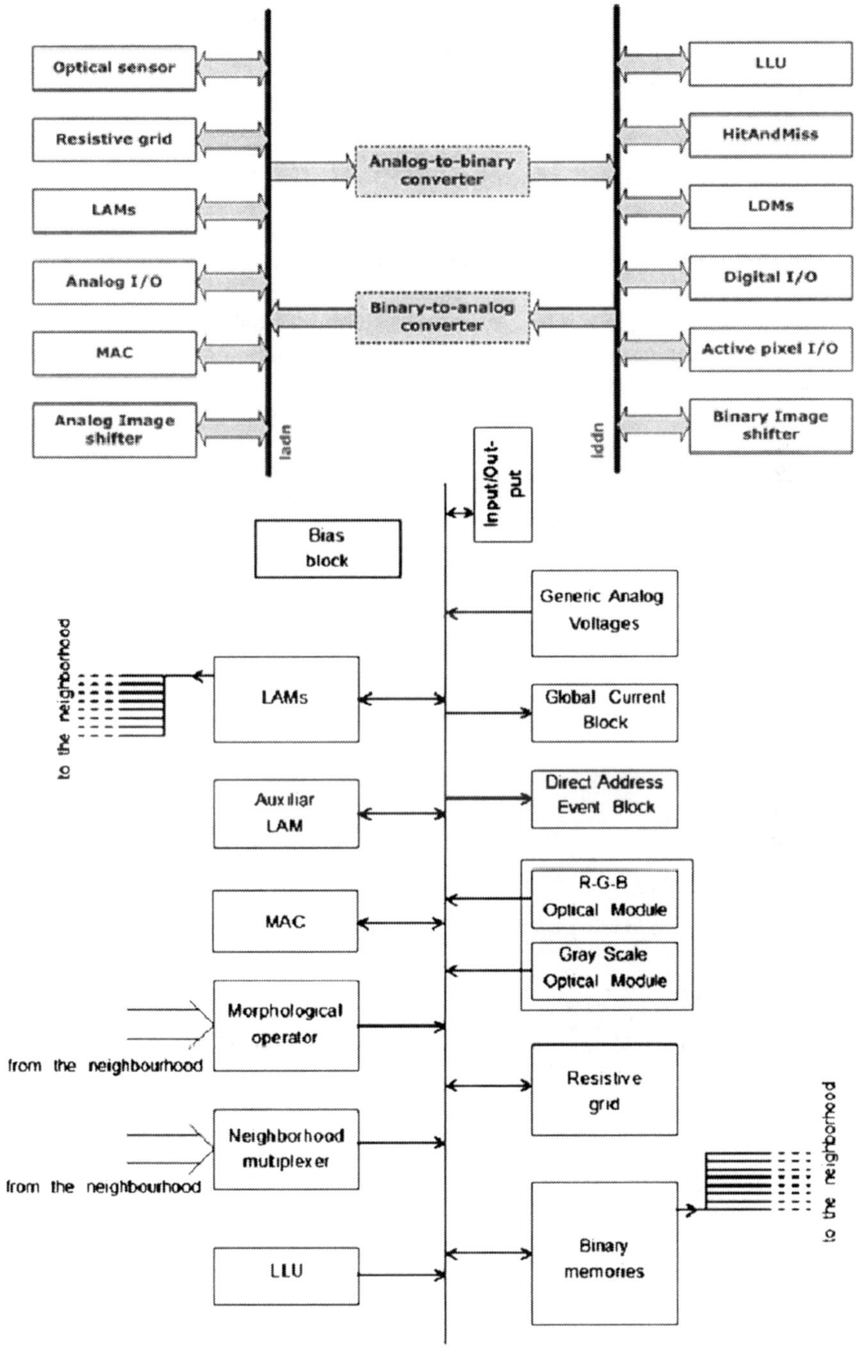

Fig. 6.6 Two alternative conceptual representations of the block diagram of the Q-Eye sensory-processing pixel

- Improved photo-sensors with a sensitivity of 1 V/(lux s) at 550 nm and an average value of 3.2 V/(lux s) within the visible spectrum.
- High-speed nonrolling electronic shutter with programmable exposure time (down to 20-ns step).
- Multi-mode image sensing: linear and high dynamic range (HDR).
- 300-mW typical power consumption with 1 W peak during gray-scale image I/Os.

Figure 6.7 illustrates image arithmetic operations realized by the Eye-RIS_v2.1 front-end. The images at the left are inputs, while those at the right are outputs. Pictures themselves highlight the subjective, perceptual quality of the operation outcome. Quantitative analysis shows that errors in the operation remain below 3% – low enough for vision tasks.

Besides allowing to reduce the dimensionality of data, the tight coupling between sensors and processors eases signal acquisition enhancement. Particularly, sensor gains can be controled pixel-by-pixel to allow HDR image acquisition. It is illustrated in Fig. 6.8 where pictures at the top are acquired by suing linear integration and those at the bottom are acquired by using a well-capacity adjustment algorithm (Decker et al. 1998). DR enhancements up to 75 dB are achieved. Actual enhancement can be programmed in real-time.

Figure 6.9 shows inputs and outcomes for different linear and nonlinear diffusion processes realized by using the *resistive grid* embedded in the Q-Eye. Both the filter

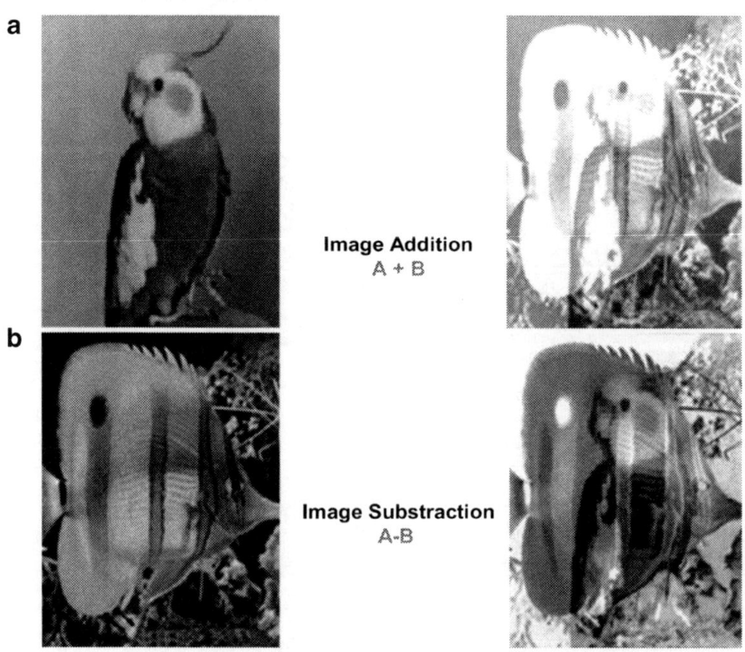

Fig. 6.7 Illustrating the realization at arithmetic operations at the Eye-RIS_v2.1 front-end

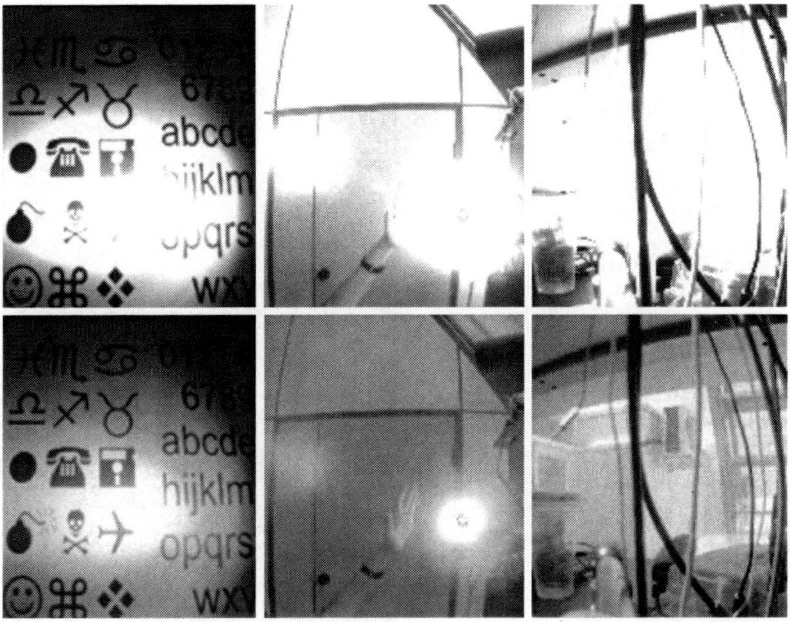

Fig. 6.8 Acquisition of HDR images using the Eye-RIS_v2.1. At each case, the bottom picture is acquired in linear integration mode and HDR mode, respectively. HDR acquisition is achieved by processing right at the pixel level utilizing an algorithm based on the *well capacity adjustment* technique (Decker et al. 1998)

type and the spatial band-width of the diffusion process can be controlled by the user. The results of performing low-pass, high-pass, and band-pass spatial filtering on the input image of Fig. 6.9a are shown in Fig. 6.9b–d, respectively. Different values of sigma are employed. Figure 6.9e shows the output of a masked diffusion process (bottom figure), the mask being the binary figure at the top.

Figure 6.10 shows major steps of the processing chain implemented within the Eye-RIS_v2.1 to find deffective parts in a production line. Eye-RIS_v2.1 identifies deffective parts based on feature analysis instead of brute force pattern matching. It enables speed improvements (number of pieces per second) of several orders of magnitude as compared to conventional systems (Cognex Ltd.). The figure illustrates the progressive reduction of data along the processing chain. Out of the some 26 kB raw data acquired by the sensors at the Q-Eye front-end, only some 100 bytes remain after pre-processing and are actually coded in digital domain and sent to the post-processing stage. Such reduction, together with the intensive parallel processing performed at the front-end is instrumental to achieve this data reduction and hence the overall efficiency enhancement in the completion of the part finding task.

The Eye-RIS_v2.1 can be software-programmed for a large variety of applications such as distributed video surveillance networks, industrial inspection, factory automation, automotive, military, toy industry, etc. A complete set of programming

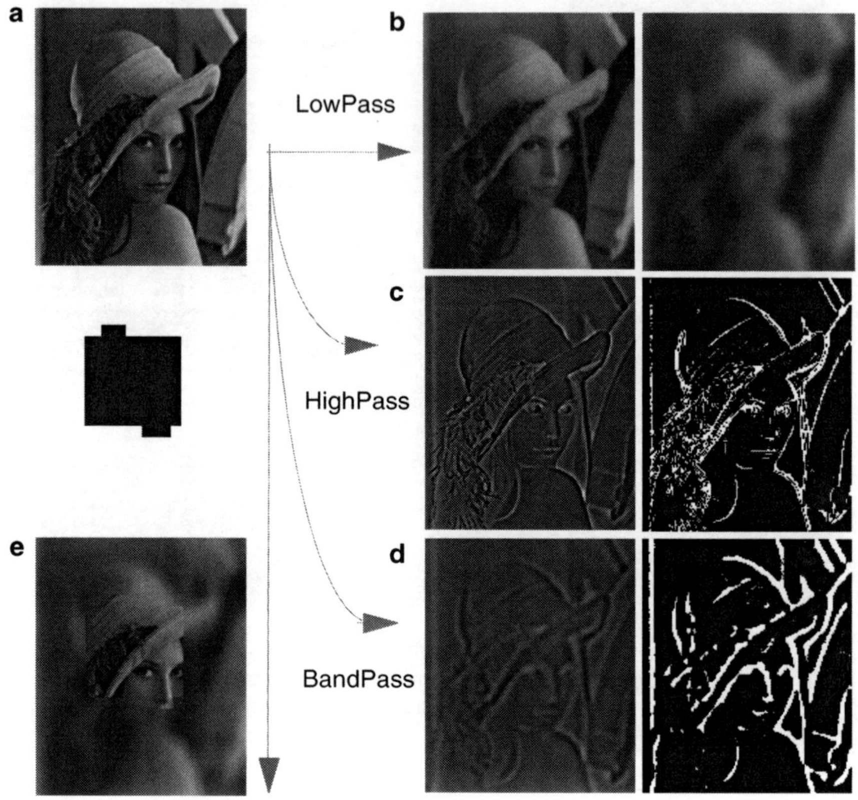

Fig. 6.9 Illustrating the realization of different kind of spatial filterings at the Eye-RIS front-end. Low-pass: $\sigma = 1$ (*left*); $\sigma = 4$ (*right*). High-pass: $\sigma = 1.4$ (*left*); BW-hp (*right*). Band-pass: $\sigma = 1.4/2.4$ (*left*); BW-bp (*right*). Masked diffusion: $\sigma = 4$

tools, embedded into an application development kit (AnaFocus Ltd.) is available for system-level users to develop and debug specifics algorithms and programs for each application. Although programming languages are standard, applications engineers must use resources at both the pre-processing and the post-processing stages in order to take full advantage of the system capabilities.

Figure 6.11 illustrates a typical processing flow. It corresponds to the sequence of operations needed to guide an unmanned vehicle across a road. The splitting of tasks between pre- and post-processor illustrates the capabilities featured by the processing circuitry close to the sensors.

6.6 Discussion

Full integration of the vision functional features into single chips represents an important stage of machine-vision system evolution. Actually, the design of imaging

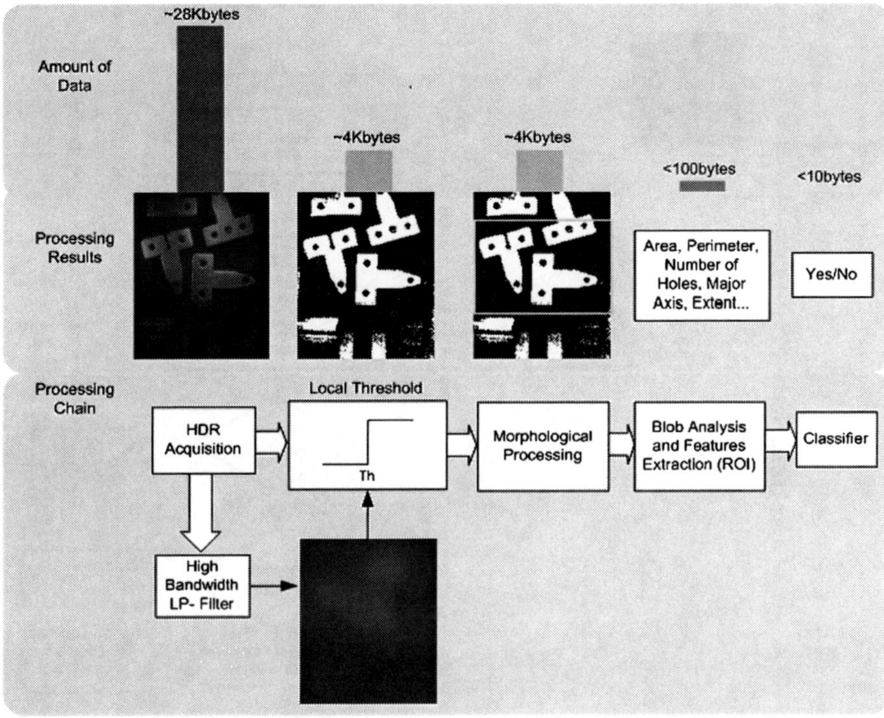

Fig. 6.10 Illustration of the progressive data reduction of data along the vision processing chain as it actually happens in the Eye-RIS_v2.1. All steps of the processing chain above but the last one are completed in the Q-Eye sensory-processing front-end. Thus, the dataset delivered for processing by the host digital processor is quite small

systems (sensors + readout + data conversion + controller + drivers) on CMOS chips has been making good progress during the last decade. In the same direction, the design of systems-on-chip has been also progressing during the last years. However, the design of VSoC has not. There is hence a gap between the current art and the requirements of industry.

During the last decade, massive parallel approaches are receiving an important attention from industry and academia. Excluding dedicated chips (Komuro et al. 2003), typical architectures are based on the use of multi-core processors or DSPs, where each processing core may be capable of some data-parallel operation. Devices such as Nvidia's graphics processors (GPUs) (Morris 2008), Intel's Larrabee GPUs (Seiler et al. 2008), the Cell processor (Pham et al. 2005), SPI's stream processors (Kapasi et al. 2003), Tilera's Tile64 (Wentzlaff et al. 2007), and other many-core devices (Vangal et al. 2007; Duller et al. 2003; Yu et al. 2008; Keckler et al. 2003) define, in terms of performance, current state-of-the-art in general-purpose programmable parallel computing chips. These devices are offering coarse-grain parallelism with many relatively complex autonomous or semi-autonomous processors, and are suitable for a wide range of applications

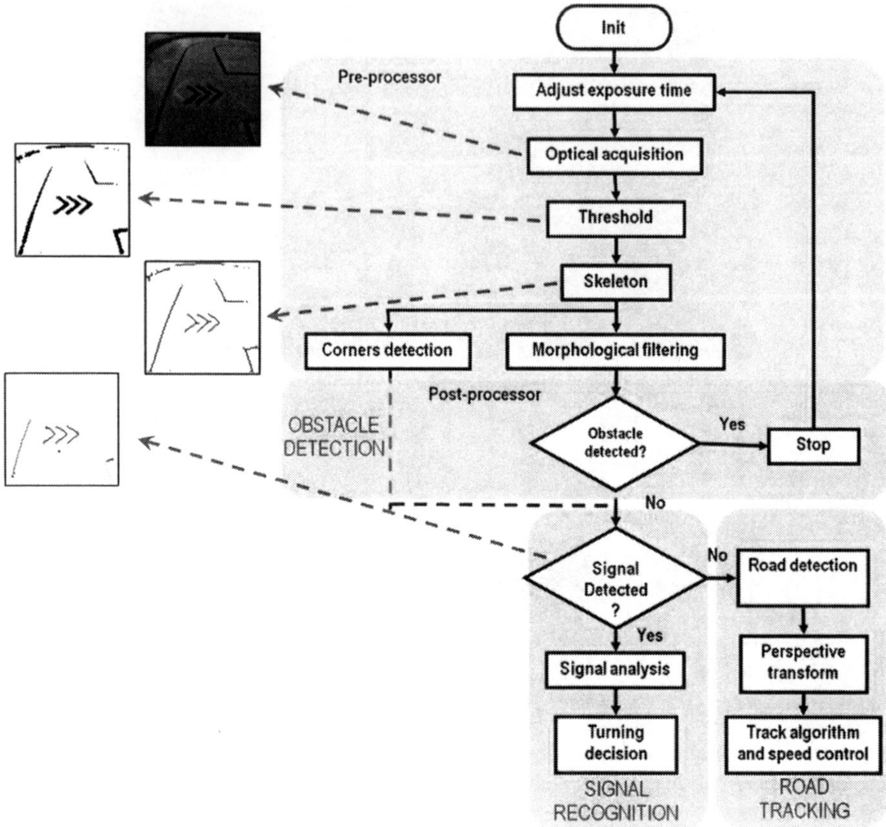

Fig. 6.11 Typical chip processing flow

requiring high computational demands. However they are sub-optimal in terms of efficiency. For example, the high-performance Nvidia's chip consumes over 100 W and features some 10 GOPS/W (Morris 2008). Alternative architectures able to extract the intrinsic parallelism of imaging applications are hence highly required.

One possible solution is resorting to *artificial retina* chips. Pioneering works on programmable artificial retina chips and focal-plane processing chips were completed already in the early 1990s (Bernard et al. 1993; Eklund et al. 1996). However, they are never reached the levels of performance and flexibility required for industrial applications. Another alternative is using digital parallel-processors with single instruction multiple data (SIMD) architecture. One dimensional (1D) SIMD processors chips have been reported elsewhere – for instance at Abbo et al. (2008) and Kyo et al (2005). However, they do not take the full advantage of a VSoC solution since they do not embed optical sensors. Other devices including sensors (Cheng et al. 2008; Lindgren et al. 2004) do not address the data reduction challenge at the border between the sensing and the processing sections.

The Eye-RIS_v2.1 combines artificial retina concepts and 2D SIMD processor concepts to obtain a *general-purpose*, true VSoC with a processing capability of 250 GOPS and a power consumption of 4 mW/GOP. It is very well suited for applications where *compactness*, *cost*, energy *consumption efficiency*, and *operation speed* define major targets.

Acknowledgments Authors would like to acknowledge fruitful discussions with Dr. Ricardo Carmona, Dr. Gustavo Liñán, Dr. Akos Zarandy, and Dr. Piotr Dudek.

The work of Prof. Rodríguez-Vázquez has been partially supported by the Spanish project 2006-TIC-2352 and the PIMA program of the CICE/JA.

References

Abbo AA, et al (2008) Xetal-II: A 107 GOPS, 600 mW massively parallel processor for video scene analysis. IEEE J Solid-State Circuits 43(1):192–201

AnaFocus Ltd., http://www.anafocus.com

Bernard T, et al (1993) A programmable artificial retina. IEEE J Solid State Circuits 28(7):789–797

Carmona R, et al (2003) A bio-inspired 2-layer mixed-signal mixed-signal flexible programmable chip for early vision. IEEE Transact Neural Networks 14(5):1313–1336

Chih-Chi Cheng et al (2008) iVisual: An intelligent visual sensor SoC with 2790fps CMOS image sensor and 205GOPS/W vision processor. IEEE Int Solid-State Circuits Conf Dig Tech Papers 306–307, San Francisco CA

Chua LO, Roska T (2002) Cellular neural networks and visual computing. Cambridge University Press, Cambridge, UK

Cognex Ltd., http://www.cognex.com/ProductsServices/InspectionSensors

Decker SJ, et al (1998) A 256 × 256 CMOS imaging array with wide dynamic range pixels and column-parallel digital output. IEEE J Solid State Circuits 33:2081–2091

Delbruck T, Lichsteiner P (2006) Freeing vision from frames. Neuromorphic Eng 3:3–4

Devaraj G, et al (2008) Applying algorithms. Vis Syst Design 13(11):17–20; 85–87

Duller A, et al (2003) Parallel processing -the picoChip way! Communicating process architectures – 2003. IOS Press, pp 125–138

Eklund J, et al (1996) VLSI implementation of a focal plane image processor — a realization of the near-sensor image processing concept. IEEE Transact VLSI Syst 4(3):322–335

El Gamal A, Eltoukhy H (2005) CMOS image sensors. IEEE Circuits Devices Mag 6–20

ENIAC working group (2007) Strategic research agenda, 2nd edn. European Technology Platform Initiative, November 2007*.

Green WE, et al (2004) Flying insect-inspired vision for autonomous aerial robot maneuvers in near-earth environment. Proc IEEE Int Conf Robotics Automat 2347–2352, New Orleans LA, April-May

International Technology Roadmap for Semiconductors (ITRS) (2007) Edition emerging research devices, http://www.itrs.net/Links/2007ITRS/Home2007.htm

Kapasi UJ, et al (2003) Programmable stream processors. IEEE Comput 36(8):54–62

Keckler S, et al (2003) A wire-delay scalable microprocessor architecture for high performance systems. IEEE Int Solid-State Circuits Conf Dig Tech Papers 1:168–169

Komuro T, et al (2003) A digital vision chip specialized for high-speed target tracking. IEEE Transact Electron Devices 50(1):191–199

Kyo S, et al (2005) An integrated memory array processor architecture for embedded image recognition systems. Proc 32nd Int Symp Comput Arch (ISCA'05) 134–145

Lindgren L, et al (2004) A multi-resolution 100-GOPS 4-Gpixels/s programmable smart vision sensor for multi-sense imaging. IEEE J Solid-State Circuits 40(6):1350–1359

Liñán G, et al (2004) A 1000FPS@128 × 128 vision processor with 8-bit digitized I/O. IEEE J Solid-State Circuits 39(7):1044–1055

Morris K (2008) A passel of processors: NVIDIA's Tesla T10P blurs some lines. FPGA Structured ASIC J, June 2008 (on-line: http://www.fpgajournal.com/articles_2008/20080617_nvidia.htm)

Pham D, et al (2005) The design and implementation of a first-generation CELL processor. IEEE Int Solid-State Circuits Conf Dig Tech Papers 184–185

Philipp RM, et al (2006) A 128 × 128 33 mW 30 frames/s single-chip stereo imager. IEEE Int Solid-State Circuits Conf (ISSCC 2005) Digest Tech Papers 2050–2059

Rodríguez-Vázquez A, et al (2004) ACE16k: The third generation of mixed-signal SIMD-CNN ACE chips toward VSoCs. IEEE Transact Circuits Syst-I 51(5):851–863

Roska T, Rodríguez-Vázquez A (2001) Towards the visual microprocessor. Wiley, Chichecter UK

Russ JC (1992) The image processing handbook. CRC Press, Boca Raton

Seiler L, et al (2008) Larrabee: A many-core x86 architecture for visual computing. ACM Transact Graphics 27(3)

Vangal S, et al (2007) An 80-tile 1.28TFLOPS network-on-chip in 65nm CMOS. Int Solid-State Circuits Conf Dig Tech Papers 98–99

Wentzlaff D, et al (2007) On chip interconnection architecture of the TILE processor. IEEE Micro 27(5):15–31

Yu Z, et al (2008) Architecture and evaluation of an asynchronous array of simple processors. J Signal Processing Syst 53(3):243–259

Chapter 7
Cellular Multi-core Processor Carrier Chip for Nanoantenna Integration and Experiments

Akos Zarandy, Peter Foldesy, Ricardo Carmona, Csaba Rekeczky, Jeffrey A. Bean, and Wolfgang Porod

Abstract A new generation of IR and sub-millimeter wave radiation detector imager array with integrated per channel high-gain capacitive amplifiers and digital signal processing/enhancement circuitry was designed. The multi-core processor carrier chip, with the analog interface and the digital processor array were implemented in standard 0.18-μm CMOS technology and verified within a compact measurement system. Characterization with external photosensors has been completed and the associated measurement results are presented. A concept for nano antenna type detector array integration to the processor carrier was also developed, and some preliminary experiments have been conducted with metal-oxide-metal (MOM) diodes.

7.1 Introduction

Micro- or *nanoantenna* coupled *metal-oxide-metal (MOM) diodes* (Sanchez et al. 1978; Hochstedler et al. 2006) for capturing images in exotic wavelengths, such as long wave infrared or, *sub-millimeter wave* or *THz* (wavelength of 0.3 mm to 10 μm), were intensively researched in the last years (Esfandiari et al. 2005). They have several advantages, such as small foot-print, high bandwidth selectivity, very

---

A. Zarandy (✉), C. Rekeczky, and P. Foldesy
Eutecus Inc, 1936 University Ave., Ste 360, CA94704
e-mail: zarandy@eutecus.com

R. Carmona, J.A. Bean, and W. Porod
University of Notre Dame

A. Zarandy and P. Foldesy
On leave from the Computer and Automation Research Institute of the Hungarian Academy of Sciences

R. Carmona
On leave from IMSE-CNM-CSIC, Seville (Spain) on a Postdoctoral Mobility Program from the Spanish Ministry of Education and Science (grant No. PR2006-0353)

C. Baatar et al. (eds.), *Cellular Nanoscale Sensory Wave Computing*,
DOI 10.1007/978-1-4419-1011-0_7,

low latency, and they do not require cooling. This makes them very good candidates to built high-speed hyper-spectral detectors in the IR region or even longer wavelength regions. However, as the initial experiments showed, their response signals is relatively small, hence, they require very strong amplification. *High-gain amplifier*s always increase noise; therefore, strong noise filtering is a must.

Due to theses special requirements (high-gain amplification and strong digital noise filtering), previously single detectors were built and various scanning techniques were applied to build images. These scanning methods make possible the capturing of stable or quasi-stable scenes or objects only (Kalkbrenner et al. 2005). To overcome this shortage, we introduce a *nanoantenna* array carrier chip, which contains per channel *high-gain amplifier*s and digital filters. The *nanoantenna* array carrier chip is a fully scalable smart read out integrated circuit (ROIC).

As a first demonstrational prototype, Eutecus, Inc. (involving analog chip design expertise from Notre Dame University also) designed, built and characterized an 8×8 *nanoantenna* carrier array (Fig. 7.1.) implemented on a standard 0.18 micron CMOS technology chip. Using operational *carrier chip*, *nanoantenna* integration experiments have been already started at Notre Dame University.

This chapter introduces the signal path (Sect. 7.2), the new *nanoantenna carrier chip* architecture (Sect. 7.3), the details of the integration of the *nanoantenna*s (Sect. 7.4), and the built measurement system (Sect. 7.5). At the end of the chapter, we conclude our results (Sect. 7.6).

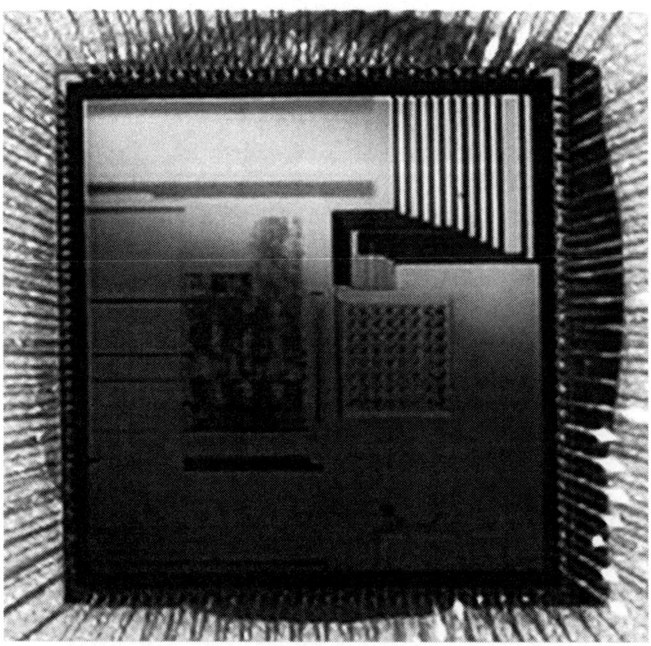

Fig. 7.1 Die photo of the 8 × 8 nanoantenna carrier chip (source: Eutecus, Inc.). The die size is $5 \times 5\ mm^2$

7.2 Algorithmic Considerations

In a noisy environment, the modulated or narrow-band frequency detection can solve suppression of the noise; hence increase the signal-to-noise ratio. This requirement is increasingly important in the *far-IR* and *THz* band detection, among other reasons, due to low power sources and absorption of the atmosphere. We can distinguish two basic approaches: active and passive methods. In active illumination case, the illumination source is modulated (e.g., by a mechanical chopper) and the detection uses synchronized phase sensitive detection at the sensor site. In passive detection mode, we have no way to modify the illumination; rather the detector's input is closed to create a reference measurement for the detection. Both active and passive methods are commonly adopted techniques for detection of *far-IR*, *THz* radiation.

Later on, the resulting images (that could be formed by, e.g., direct acquisition on arrays, or reconstructed from synthetic aperture imaging, scanned, etc.) may require further noise filtering on top of the phase sensitive detection as well by methods like the Wiener filtering. Meanwhile, if the detected phenomenon – incoming radiation distribution – is a complex event, it is most effectively detected by spatio-temporal filter configurations.

The *nanoantenna carrier chip* has been prepared for both active and passive detection. The analog signal path is composed of a low noise current integrating module, followed by a high-gain AC coupled amplifier. After digitalizing the analog signal, in the digital domain programmable subsampling and spectrum calculation are performed at different modulation frequencies. The resulting components are then combined and the signal strength is derived separating the modulated signal from the background noise. The implemented signal flow can be seen in Fig. 7.2.

To increase the time frame, in which samples can be processed with limited memory resources, we analyzed several spectral filtering algorithms. We considered discrete and fast *Fourier transformation*s (DFT, FFT), momentary

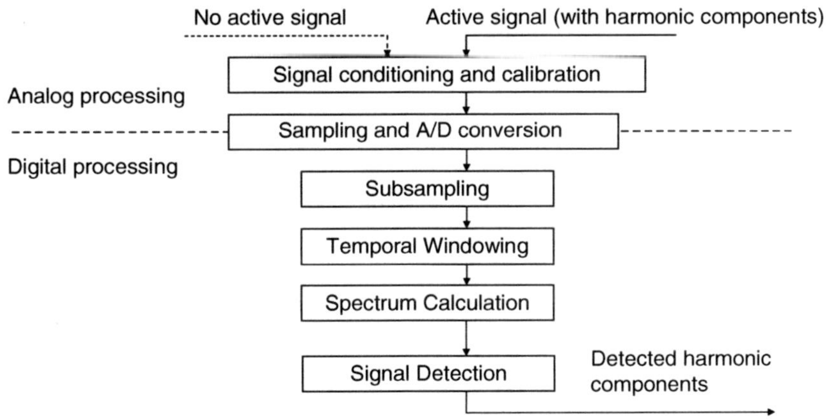

Fig. 7.2 Block diagram of the algorithm implemented on the nanoantenna carrier chip

Fourier transformation (MFT) (Papoulis 1977; Dudás 1986), and resonator-based transformation (RBFT) (Péceli 1989, Varkonyi-Koczy 1995). In all these cases, continuously or at specific timesteps all spectral lines have to be calculated, meanwhile memory requirement is proportional to the sample number.

7.2.1 Numeric Precision

The question of numeric precision affects the data flow differently at each step. For the *near sensor processing*, we have considered the analog backend signal quality and the analog to digital conversion quantization error. As core signal processing chain we supposed a few discrete complex spectral component calculations – what is required by the phase sensitive detection.

The floating-point arithmetic is although advantageous, costly in hardware realization. On the other hand, iterative and even more recursive operations keep the inherent risk of error accumulation that can be the source of result instability. This holds as well, when working with usual floating-point representations. The known problem is that in the floating-point form numbers are not regularly spread over their range.

Hence, we have chosen a fixed-point representation (scaled integer). Assuming small footprint 8–10 bit A/D converters, 10,000 sample size to be analyzed at once, it turns out that the intermediate precision should be at least 21 or 23 and the constant factor 17 or 21 fractional bit (plus sign bit). After numeric experiments, we have selected 24 bit signed integer and $20 + 4$ bit signed fixed-point representation as the internal precision of the *near sensor processing*.

7.3 Architecture of the Nanoantenna Carrier Chip

In this section, the first experimental implementation is introduced with 8×8 sensor array resolution. The block diagram of the architecture is shown in Fig. 7.3. It contains 64 (8×8) sensor interface circuits, each equipped with a physical *nanoantenna* interface. Sixteen of these interface blocks may alternatively receive input from external pins too. This makes possible the testing of the interface circuits before the *nanoantenna* integration.

The amplified output signal of the sensor interface units are stored by sample and hold (S&H) circuits. Hence, the data are sampled in the entire array simultaneously. Then, the sampled signals are digitized with the analog to digital converters (ADC). There is one ADC in each row, each is connected to the sensor interfaces through an 8:1 multiplexer. The output of the ADCs is read out by the digital processors through a frame buffer. Each of the processor handles one row of pixel values (8 pieces). The processed digital pixel values finally are read out through an output multiplexer.

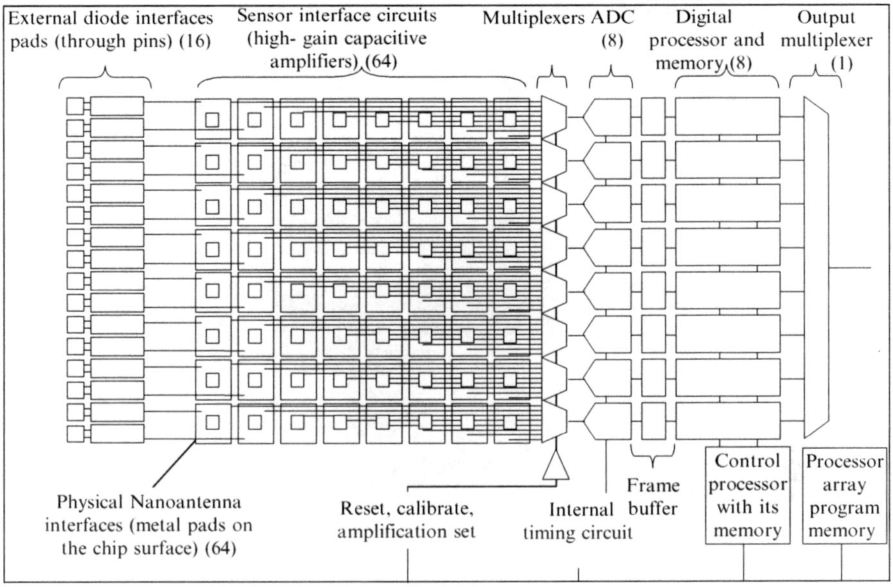

Fig. 7.3 The architecture of the 8 × 8 nanoantenna carrier chip

The analog and mixed signal part receives the control externally, except the ADCs, for which an internal timing control unit provides the control signals. The digital processors are *single instruction multiple data* (SIMD) type processors, hence one common instruction memory serves all. The program flow is controlled by the control processor. The SIMD instruction set is embedded into the assembly of the control processor, extending its processing power.

The layout of the chip is shown in Fig. 7.4. The functional blocks are marked. Besides the previously mentioned function blocks, the chip contains separate (off array) test interface circuit and ADC. The die size is 5 × 5 mm. As it can be seen, roughly one fourth of the available silicon area is used. The reason of the large unused area is that the post *nanoantenna* integration process cannot handle smaller dies. On the other hand we did not want to build a larger test implementation in the first round.

The sensor interface array with the internal *high-gain amplifier*s is well isolated to minimize the picked up digital noise coming from the other components of the chip.

7.3.1 High-Gain Sensor Readout Channel

The sensor interface in this chip is an array of 8 × 8 high-gain readout channels that match the overlaid 8 × 8 sensor array. These sensors can be either deposited

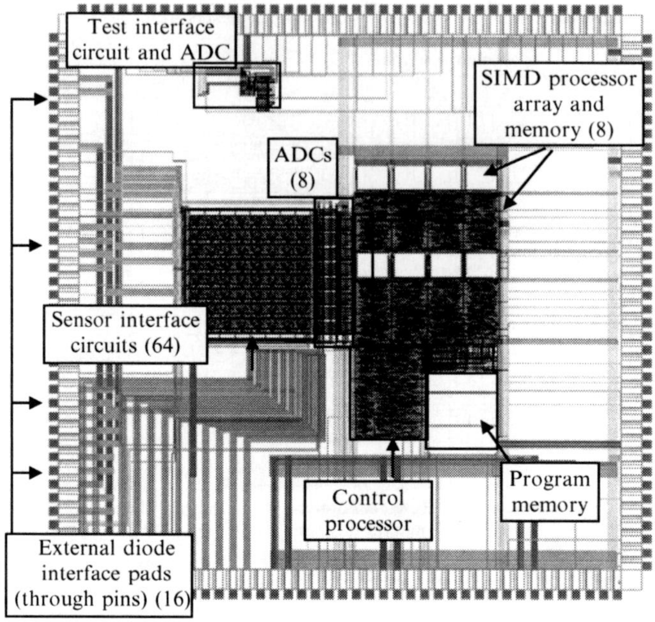

Fig. 7.4 The layout of the nanoantenna carrier chip (source: Eutecus, Inc.)

antenna-coupled *nanodiodes* for mixing and detecting signals of the required wavelengths or thin-film diodes of a semiconductor with the appropriate energy band distribution. The only requirement on them will be to have a current output within the specified range. In our design, the currents that will be sensed range from 0.1 to 10nA. The front-end of the sensor interface is a capacitive *transimpedance amplifier* (CTIA), as depicted in Fig. 7.4 with offset and noise sources and a simplified detector model. It has been employed traditionally to interface arrays of passive pixels (PPS) (Plummer and Meindl 1972). However, it is used nowadays in high-precision image sensors (Kleinfelder et al. 2004; Helou et al. 2006), because of its high charge-to-voltage conversion ratio and its low readout noise (Fowler et al. 2000). In a CTIA active pixel sensor (APS), the charge-to-voltage conversion factor is set by a well-controlled sensing capacitor, C_s, as compared to the loosely controlled value of the parasitic diode capacitance in the conventional 3T APS. In addition to this, it allows for properly biasing the sensor device, with the help of the virtual ground of the opamp. If changes in the input current are assumed to vary much slower than the sampling rate, $1/\Delta t$, and the sampling period is uniform, the CTIA output voltage is proportional to the input current, and hence to the power of the incident radiation following the sensor constitutive law:

$$V_{o_1}(t_o + \Delta t) = V_{bias} + V_{os} - \frac{I_{in}(t_o)}{C_s}\Delta t \tag{7.1}$$

where V_{os} is the offset voltage of the opamp. Both bias and offset are sampled at a previous step in the subsequent stages, therefore implementing *correlated double sampling* (CDS) (Enz and Temes 1996), and thus reducing the FPN introduced by variations on the offsets at each pixel. The input referred noise of the opamp determines the minimum detectable current signal (Lv et al. 2008):

$$\mathrm{MDS} = \frac{1}{R_{\mathrm{det}}} \int_{f_o}^{f_c} \frac{v_n^2}{\Delta f} \mathrm{d}f \tag{7.2}$$

If the equivalent input noise of the opamp is $10\,\mathrm{nV}/\sqrt{\mathrm{Hz}}$, and the useful signal is limited to 100 Hz, a total input noise of $v_{\mathrm{nT}} = 0.1\,\mu\mathrm{V}_{\mathrm{rms}}$ results. For a detector output resistance of $R_{\mathrm{det}} = 1\,\mathrm{k}\Omega$, it means that the CTIA will not resolve for currents below $0.1\,\mathrm{nA}_{\mathrm{rms}}$. For $R_{\mathrm{det}} = 100\,\Omega$, then $\mathrm{MDS} = 1\,\mathrm{nA}_{\mathrm{rms}}$.

On the other side of the range, the largest sensing capacitor implemented is 1 pF – a control signal allows switching to a $C_s = 0.5\,\mathrm{pF}$. In addition, V_{o_1} will not excurse more than 1 V in order to avoid distortion. For a maximum refresh rate of 10 kHz, there is a maximum sensor current of 10 nA beyond which the pixel saturates in the prescribed integration time.

The second stage of the analog front-end is an ac-coupled capacitive amplifier (Fig. 7.5). The fact that we are more interested in changes in the sensor reading from frame to frame, rather than the exact dc value, made us select this structure reducing the number of switches to be employed. Signals ϕ_{cal} and ϕ_{amp} are employed to set the calibration and amplification phases. These phases must not overlap. The output of the second stage is then given by:

$$V_{o_2}(n+1) = V_{\mathrm{REF}} - \frac{c_{\mathrm{in}}}{c_{\mathrm{F}}} \left[V_{o_1}(n+1) - V_{o_1}(n) \right] \tag{7.3}$$

during the amplification phase. $V_{o_1}(n)$, i.e., the input voltage of the second stage during the calibration phase, contains the reference level from which the changes in the output are measured. This implements in-pixel CDS, eliminating FPN due to offset. Gain errors from pixel to pixel, that are also a source of FPN, are less

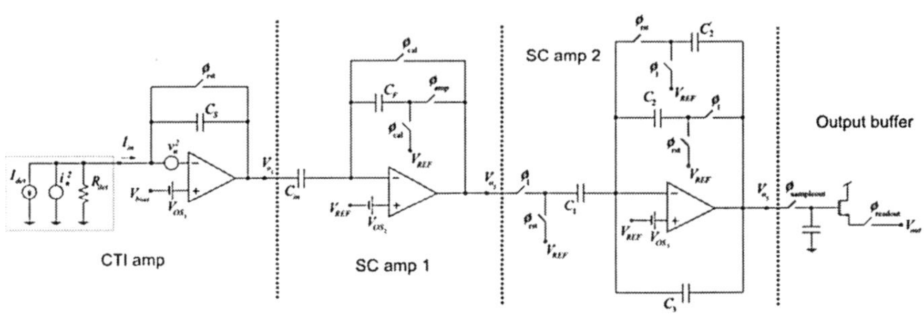

Fig. 7.5 Capacitive transimpedance amplifier for the analog front-end

noticeable as the pixel gain is determined by a ratio of capacitors. A control signal permits switching between 4× and 8× amplification at this stage.

The last stage of the readout channel is another SC amplifier (Johns and Martin 1997). This one is a resettable-gain circuit that cancels the opamp offset and reduces $1/f$ noise. It has a switchable gain between 10× and 20×. In this amplifier, a capacitive reset, via the extra capacitor c_2', avoids large excursions of the output voltage – this prevents the opamp from slewing to V_{REF} at each clock cycle. Also a deglitching capacitor, C_3, has been inserted (Matsumoto and Watanabe 1987). This will render a cleaner input for the A/D converters outside the array. Clock phases here do not correspond to the calibration and amplification phases at the second stage, intended to clamp the reference offset. Instead, they are aligned with the integration reset of the first stage, thus fixed at the sampling rate. ϕ_1, that is the phase in which this stage is amplifying, is a nonoverlapping counter-phase of ϕ_{rst}.

An output sample and hold and a voltage follower complete the high-gain sensor readout channel as depicted in Fig. 7.5. The transimpedance of the first stage goes from 100 MΩ at 10 kHz, with a 1 pF sensing capacitor, to 2 GΩ at 1 kHz, with $C_s = 0.5$ pF. The gain of the following two SC gain stages can be 40, 80, or 160.

7.3.2 Digital Processor Architecture

The development of the digital signal processor architecture is motivated by different requirements: sufficient computational power to perform filtering, low power consumption, small area footprint, and scalability. The absorbed electric power yields self-heating and heat radiation that can directly affect the background noise of the long-wave infrared detectors. Hot spots near the sensor area also could distort the spatial uniformity of the detection. Scalability and small footprint requirements are related to speed up the development, by later design reuse and efficient area utilization.

The functional architecture of the *carrier chip* can be seen in the Fig. 7.6. The major blocks of the chip besides the sensor interfacing are the frame buffer of a complete image, channel-wise processing array, and a control processor. This section describes the implementation considerations and details of these units.

7.3.3 Partitioning

From algorithm point of view, we faced two main challenges, namely the algorithmic partitioning and hardware-software mapping and the numeric representation. The sensor to image signal flow chain can be mapped efficiently to hardware due to its well-defined and deterministic nature. The higher-level operations require more complex processors and higher flexibility, possibly nonlinear program flow.

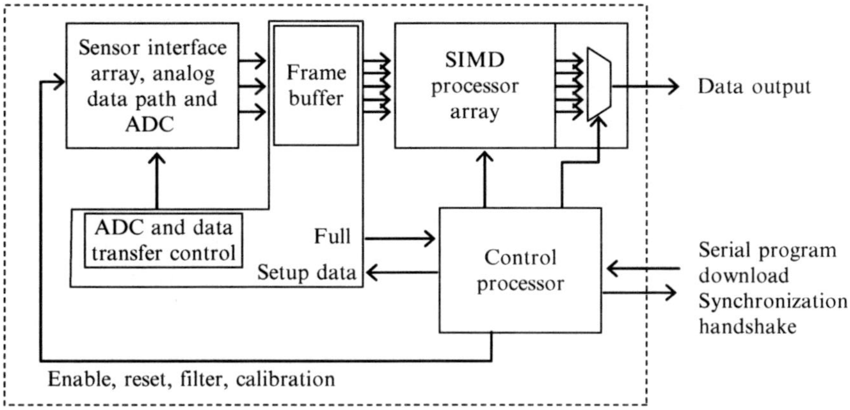

Fig. 7.6 The functional architecture of the carrier chip

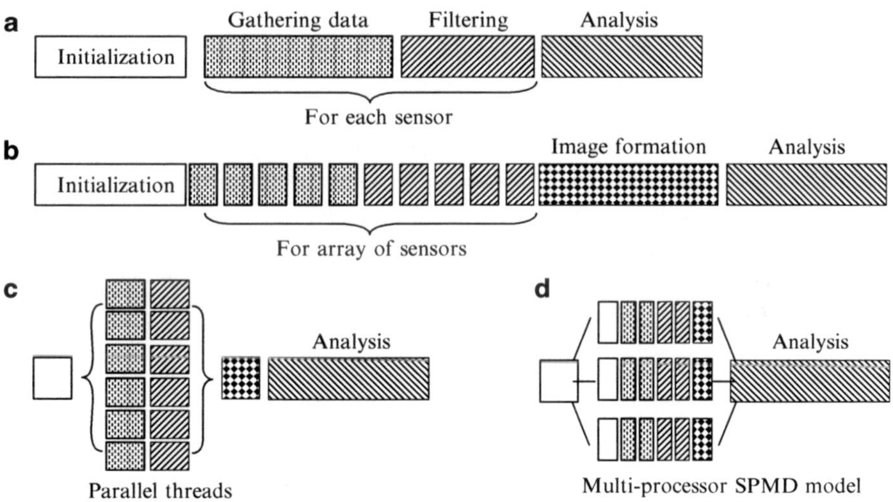

Fig. 7.7 (**a**) Mapping the general algorithm flow, (**b**) mapping from single signal to many signals, (**c**) than to parallel tasks, and (**d**) to multi processor architecture

The low-level operations are typically simpler, run in real-time, and demand higher computational effort. Moving from one signal to array of signals, we might have to deal with computational model. The array signal processing including the image formation is easily described by the *single program multiple data* (*SPMD*) (Darema 2001) computational model. Since the data amount, rate, and per pixel task is known in advance, the *SPMD* parallelization is straightforward. Furthermore, the *near sensor processing* has linear program flow; hence the sensor processors can work in the same way, forming a data parallel *SIMD* array (see Fig. 7.7).

The partitioning also needs to deal with data transfer requirements. The near sensor operations – like phase sensitive detection – uses high bandwidth, while the image post-processing steps, due to the relatively small image format, requires significantly lower data rate. The ratio of the near sensor and processed image data amount is in the range of 100:1 to 1,000:1 due to the phase sensitive detection scheme.

As a conclusion, we have implemented the *near sensor processing* in the *carrier chip* shared to a control processor and an *SIMD* data path unit, putting the data and image analysis tasks to an external host processor. In practice, these tasks are further separated for the chip hosting Bi-i (www.eutecus.com) system and a desktop PC.

7.3.4 Control Processor

The control processor manages the algorithm flow. This autonomous unit has the typical capabilities of a commodity microcontroller, such as to communicate with external components through general-purpose ports, interrupts, call stack, ports for analog interface management (set operational conditions like filter cut-off frequency, amplification), and to feed special instructions for the *SIMD* processor array. One of the main tasks of the controller is the internal synchronization with the embedded ADCs (start operation, setup operation, and wait for full frame ready). This synchronization is achieved by barrier software construct.

To free up the control processor from the continuous monitoring of the *SIMD* array, self-timed sensor-interfacing modules, data buffers with full/empty type interrupt capabilities are embedded. The architecture of the control processor can be seen in Fig. 7.8.

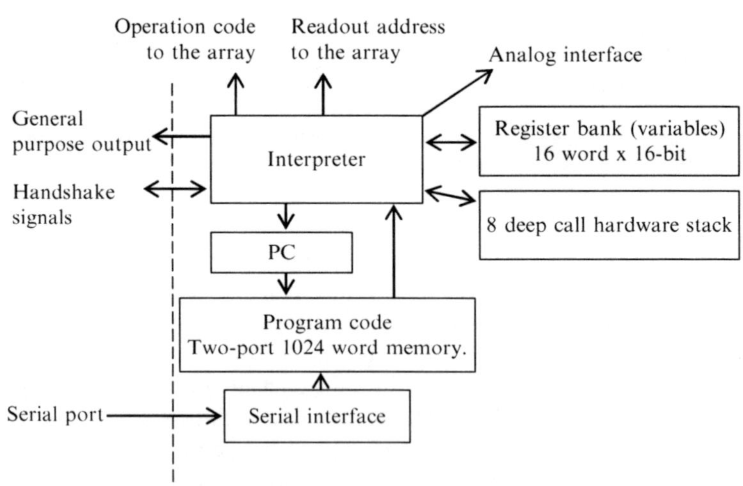

Fig. 7.8 The architecture of the control processor

To achieve compactness, the program of the *SIMD* processor array is mapped into the program space of the control processor. The *SIMD* array's memory is mapped into the memory as well. Finally, to lower the required external components, standard serial communication serves for programming the chip before operation. The controller has a compact instruction set, and can be programmed by macro assembler.

7.3.5 SIMD Processor Array

The processor array is composed of identical arithmetical slices, which are composed of a datapath ALU, memory, and working registers. Each unit has the same structure, and executes the same code in an *SIMD* model. The instruction coding is compact containing source, target, and operation codes like the VLIW machines do (very long instruction word). Each instruction is executed within a single clock cycle. The datapath ALUs are arranged functionally in a vector (one-dimension line), instead of an array. Their connectivity is restricted to their associated ADC data vector, the program controller, and the data output bus (no inter-ALU connection defined). The architecture can be seen in Fig. 7.9.

The ALU is capable to add, subtract, multiply, shift, and saturate data coming from work registers, immediate constant from the code (in case of common value),

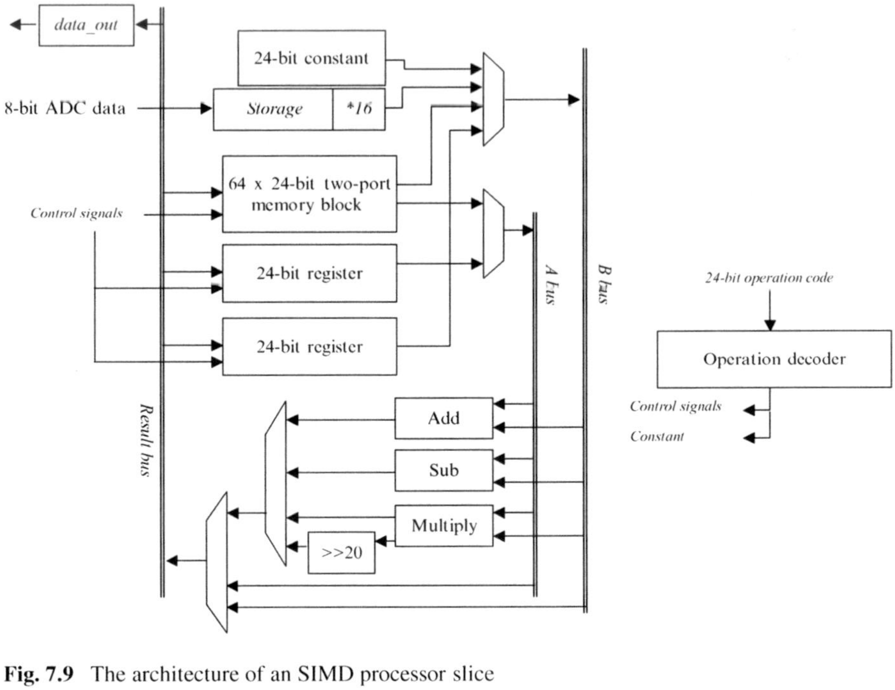

Fig. 7.9 The architecture of an SIMD processor slice

the register file, and the connected ADC frame buffer. The output of the arithmetic can be stored at various targets at the same time. The register file is two-port type that enables read–modify–write operations in one cycle. The amount of memory – 64 words per slice – is defined based on the recursive algorithm evaluation.

7.4 Nanoantenna Integration

As explained in Sect. 7.2, each of the readout channels of the sensor array interface contains a signal processing chain that allows for current sensing, filtering, and amplification. The first stage, employed to integrate the current of the photodiodes, is a charge amplifier (CTIA in Fig. 7.5). This permits to bias the diodes while, at the same time, the current is sensed and integrated. In the case of the antenna-coupled MOM *nanodiode*s, the operation is based on a different principle.

7.4.1 Antenna Coupled Nanodiode Interfacing

Figure 7.10 depicts the structure built to implement the antenna-coupled *nanodiode*. Two rotated L-shaped metal stripes slightly overlap in the elbow area thus creating a vertical MOM diode. Because of the construction method, oxides in the nanometer scale are obtained. The reduced area, below 0.01 μm^2, results in the capability for rectification and mixing of signals up to 30 THz (Fumeaux et al. 1998).

A lumped-circuit model for this structure (Matyi 2004) is shown in Fig. 7.11. The labels for the circuit nodes establish the correspondence with the physical structure. Given the polynomial approximation of the I–V characteristic of the MOM diode in (Sanchez et al. 1978) – that has been verified for the *nanodiode*s built at Notre Dame (Hochstedler et al. 2006), the application of an ac signal of amplitude V_{ac}, due to IR radiation, and a bias voltage V_b, results in a rectified voltage of:

$$V_{rect} = \frac{R_s}{2R_D}(m + 3nV_b)\,V_{ac}^2 \tag{7.4}$$

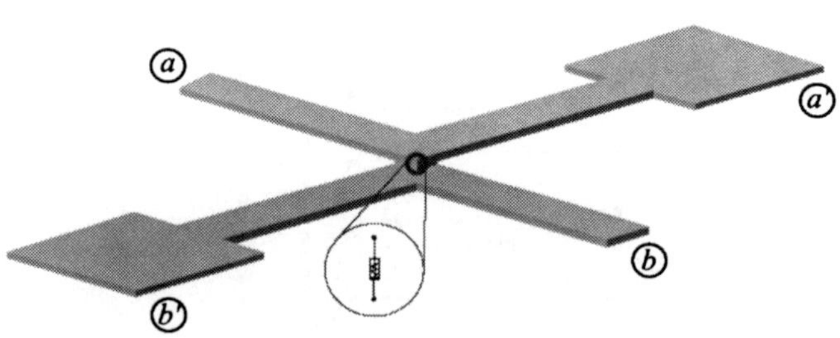

Fig. 7.10 Conceptual diagram of the antenna-coupled nanodiode

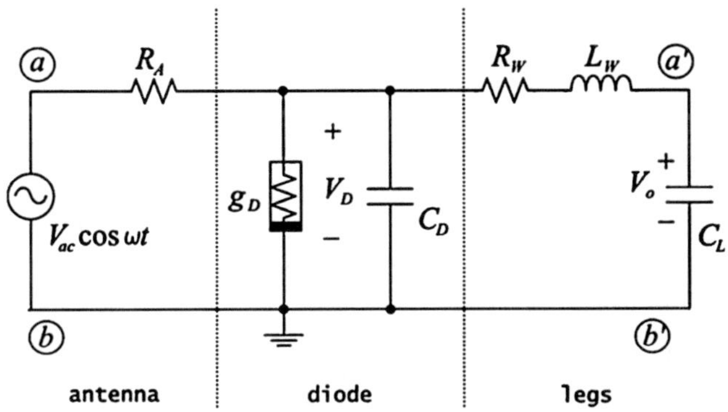

Fig. 7.11 Model for the antenna-coupled nanodiode

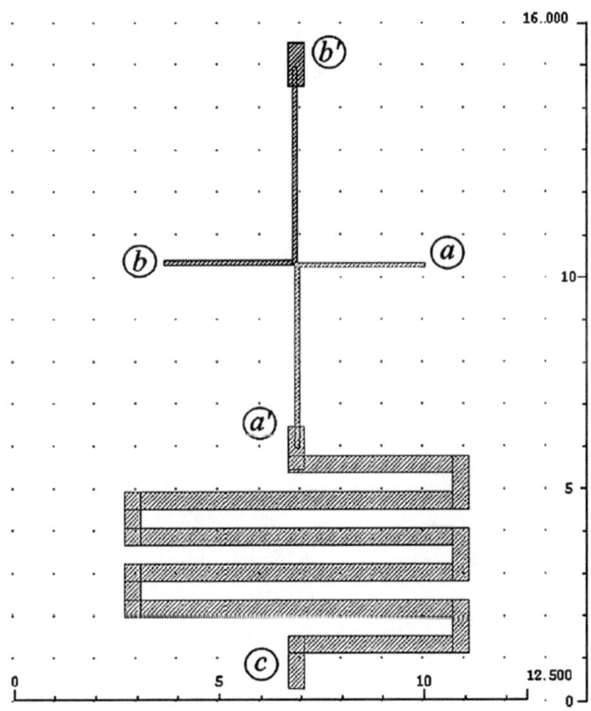

Fig. 7.12 L-shaped antenna legs and serpentine resistor

added to the dc voltage measured at V_O when in absence of radiation. In order to be detected by the CTIA at the first stage of the readout channel, this microvolt range voltage needs to be converted to a current. For that, a resistor can be added to the *nanoantenna* with a serpentine structure (Fig. 7.12). For a typical 50 mOhm/sq. sheet resistance, this resistor will have a few ohms, therefore currents will be high and the refresh rate must be increased to avoid saturation of the sensing capacitor.

7.4.2 Physical Integration of the Nanoantenna Array

There are several points at the integration, which had to be addressed with regards to the design of the *carrier chip*. These items deal with the placement and electrical connection of the *nanoantenna* devices. The Notre Dame group has developed a fabrication procedure for dipole antenna-coupled MOM diodes with ultrasmall contact areas (around 50 nm × 50 nm) suited for the detection of 10.6 μm infrared radiation. Both symmetrical and asymmetrical diodes were fabricated using a one-step electron beam lithography followed by a *double-angle evaporation*.

The Fig. 7.13 demonstrates experimental microphotos of the implemented structures.

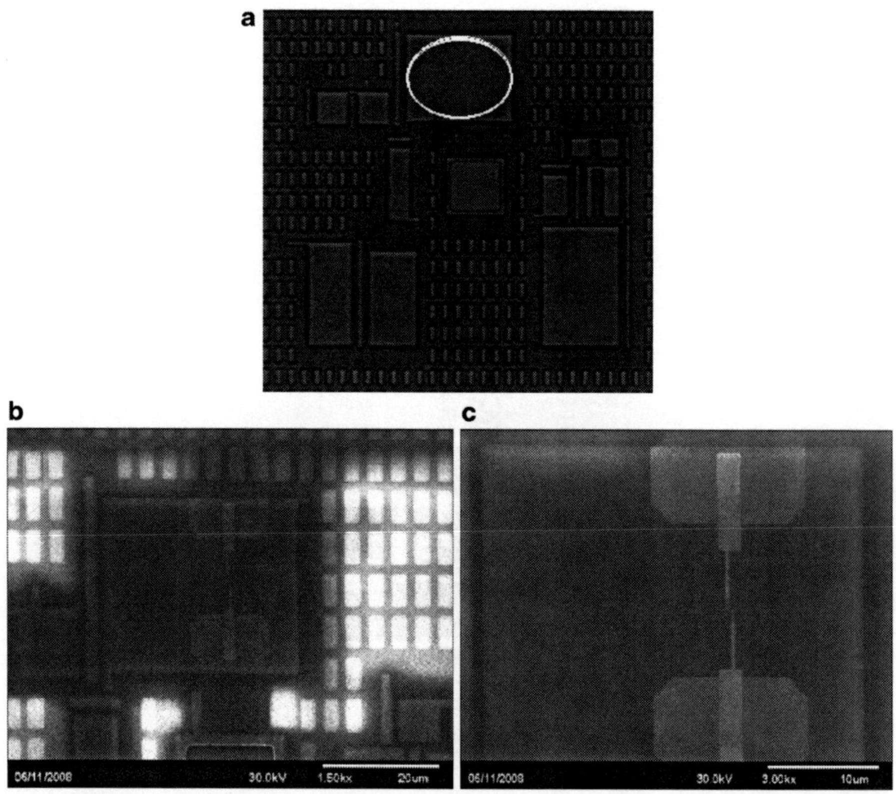

Fig. 7.13 The three images show the nanoantenna integration. (**a**) The visible microscope image shows a single readout circuit and the location of the nanoantenna to be implemented. (**b**,**c**) Closer views of nanoantenna structures

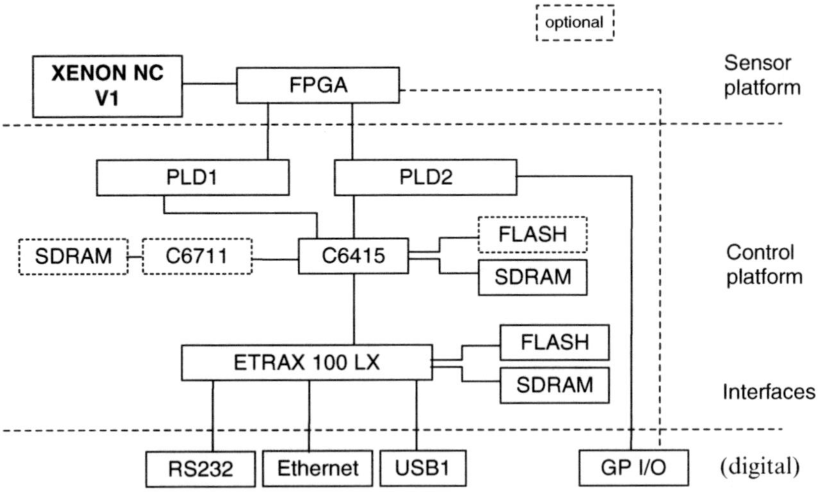

Fig. 7.14 Block diagram of the measurement system for the nanoantenna carrier chip

7.5 Measurement Environment

To be able to test and characterize the *nanoantenna carrier chip* a measurement environment is needed to be built. This system should be able to communicate with the *nanoantenna carrier chip*, provide power for it, and should be able to communicate with remote computers, at which users can control the measurements. Moreover, the system should be too large, to be able to fit on an optical table.

We have selected our Bi-i V3 smart camera (Zarándy and Rekeczky 2005; Bi-i V301) as the host unit. It is a standalone camera system with modular sensor platform (Fig. 7.14), which can be redesigned and replaced to host a new sensor type. Its main board, called control platform contains two processors. One is a high-performance DSP, the other is a communication processor running an embedded LINUX. As to its interfaces, it is equipped with an Ethernet, two RS232 lines, a USB1, and a GPIO. Its cubature is relatively small (7 × 5 × 2.5 in. without optics).

7.6 Concluding Remarks

Nanoantenna carrier chip architecture and prototype was introduced. The chip is prepared to carry an array of 8×8 nanoantennas sensitive in the IR or sub-millimeter waveband. Each nanoantenna is interfaced with a high-gain capacitive amplifier. The chip is equipped with 8 A/D converters and 8 digital processors for signal extraction and noise filtering. The carrier prototype is characterized. Initial nanoantenna integration experiences are also reported.

Acknowledgments This work was carried out by Eutecus, Inc., Berkeley, California, and the University of Notre Dame, South Bend, Indiana. The work was supported by the Office of Naval Research, ONR (STTR contr. # N00014–05C-0370), which is a MURI complementary program.

References

Bi-i V301 High-Speed Smart Camera (2009) http://www.analogic-computers.com/ProdServ/Bi-i/Bi-iv301.html, Accessed 27 February 2009

Darema F (2001) *SPMD* model: past, present and future. Recent advances in parallel virtual machine and message passing interface: 8th European PVM/MPI Users' Group Meeting, Santorini/Thera, Greece, September 23–26, 2001. Lect Notes Comput Sci 2131:1

Dudás J (1986) The momentary fourier transform. Ph.D. thesis, Technical University of Buda-pest

Enz CC Temes GC (1996) Circuit techniques for reducing the effects of op-amp imperfections: autozeroing, correlated double sampling, and chopper stabilization, Proc IEEE 84(11): 1584–1614

Esfandiari P, Bernstein G, Fay P, Porod W, Rakos B, Zarandy A, Berland B, Boloni L, Boreman G, Lail B, Monacelli B, Weeks A (2005) Tunable antenna-coupled metal-oxide-metal (MOM) uncooled IR detector. Proc SPIE 5783:470

Fowler BA, Godfrey M, Balicki J, Canfield J (2000) Low-noise readout using active reset for CMOS APS. Proc SPIE's Sensors Camera Syst Scientific, Industrial, Digital Photography Appl 3965:126–135

Fumeaux C, Herrmann W, Kneubühl FK, Rothuizen H (1998) Nanometer thin-film Ni–NiO–Ni diodes for detection and mixing of 30*THz* radiation. Infrared Phys Technol 39(3):123–183

Helou JN, Garcia J, Sarmiento M, Kiamilev F, Lawler W (2006) 0.18um CMOS fully differential CTIA for a 32 × 16 ROIC for 3D ladar imaging systems. Proc SPIE's Infrared Photoelectronic Imagers Detector Devices II 6294: 9–13

Hochstedler J, Stroube B, Bean J, Porod W (2006) Antenna-coupled *metal-oxide-metal diodes*. University of Notre Dame, Dept. of Electrical Engineering, Technical Report

Johns D, Martin K (1997) Analog integrated circuit design. Wiley, New York

Kalkbrenner T, Håkanson U, Schädle A, Burger S, Henkel C, Sandoghdar V (2005) Optical microscopy via spectral modifications of a *nanoantenna*. Phys Rev Lett 95(20):200801

Kleinfelder S, Yandong C, Kwiatkowski K, Shah A (2004) High-speed CMOS image sensor circuits with in situ frame storage. IEEE Trans Nuclear Sci 51:1648–1656

Lv J, Jiang YD, Zhang DL (2008) Ultra-low-noise readout integrated circuit for uncooled microbolometers, Electron Lett 44(12):733–735

Matsumoto H, Watanabe K (1987) Spike-free switched-capacitor circuits. IEE Electron Lett 23(8):428–429

Matyi G (2004) *Nanoantennas* for uncooled, double-band, CMOS compatible, high-speed infrared sensors. Int J Circuit Theor Appl 32(5):425–430

Papoulis A (1977) Signal analysis. McGraw-Hill, New York

Péceli G (1989) Resonator-Based Digital Filters. IEEE Transactions on Circuits and Systems Cas-36(1):156–159

Plummer JD, Meindl JD (1972) MOS electronics for a portable reading aid for the blind. IEEE J Solid-State Circuits 7(2):111–119

Sanchez A, Davis CF, Liu KC, Javan A (1978) The MOM tunneling diode: theoretical estimate of its performance at microwave and infrared frequencies. J Appl Phys 49(10):5720–5277

Varkonyi-Koczy A (1995) A recursive fast *Fourier transformation* algorithm. IEEE Transact Circuits Syst II: Analog Digital Signal Processing 42(9):614–616

Zarándy A, Rekeczky Cs (2005) Bi-i: a standalone ultra high speed cellular vision system. IEEE Circuits Syst Mag second quarter:36–45

Chapter 8
Circuitry Underlying Visual Processing in the Retina

Frank S. Werblin

Abstract Early retinal processing is involved with managing the set point for retinal neurons, taking care to keep all neural activity of each cell at a neutral set point, about midway between its maximal and minimal activity levels. Most retinal neurons are active at their midpoint, receiving and transmitting even under ambient conditions. Light input alters the patterns of activity among these neurons.

There is a general organizational plan for the inner retina whereby vertically oriented inhibition is carried by a population of many different amacrine cell types, defined by morphology. Many of these amacrine cells are narrowly diffuse glycinergic amacrine cells. For the most part these vertical cells carry information from the ON to the OFF systems, and provide "crossover inhibition" that serves to correct for the rectification inherent in all synapses. Wide field inhibition is carried by laterally oriented GABAergic amacrine cells. This inhibition forms a second tier of antagonistic interaction. Wide field inhibition is mediated by at least five different antagonistic surround possibilities: Horizontal cell feedback, horizontal cell GABA and electrical feedforward, GABAergic wide amacrine cell feedback, GABAergic wide amacrine cell feedforward, and glycinergic amacrine cell crossover inhibition. In addition to the general plan, there are specific circuitries that account for the unique behavior of individual ganglion cell types. A few examples of this specific circuitry are now available, and have been described above near the end of this chapter.

8.1 Introduction

The last decade has seen a burgeoning of research uncovering many of the physiological and morphological features of retinal processing. The wealth of information is so vast that it is often difficult to organize into a comprehensive view of retinal

F.S. Werblin (✉)
Vision Research Laboratory, Department of Molecular and Cell Biology, University of California, Berkeley, CA 9472, USA
e-mail: werblin@berkeley.edu

C. Baatar et al. (eds.), *Cellular Nanoscale Sensory Wave Computing*,
DOI 10.1007/978-1-4419-1011-0_8, © Springer Science+Business Media, LLC 2010

circuitry. We've learned for example, that the retina generates about 12 different output streams, each carrying a different space–time representation of the visual world (Roska et al. 2006; Roska and Werblin 2001). Each of these representations is formed by a specific retinal circuitry, and each has its unique functional characteristics. We also know that there is a bewildering array of more than 30 different types of amacrine cells the main retinal inhibitory interneuron, and it has been difficult to assign roles of specific interneurons and their circuitry interactions to the 12 different visual streams. But stepping back from the details it's possible to view most of what we've learned about retinal circuitry in general terms that can greatly simplify our understanding of the apparent complexity of the retina. This chapter is composed of two parts: first I try to show that most of what we know about retinal circuitry can be described as a canonical background circuit, an organization that applies to most of the 12 major retinal output streams. Second I try to show that each of the specific visual streams incorporates this canonical circuitry but is enhanced by the addition or modification of specific components in the basic canonical circuit.

8.1.1 Background Circuit Organization

It is generally agreed that information passes from cones to bipolar cells to ganglion cells via glutamatergic synaptic transmission. The most significant part of this pathway is the division, at the bipolar cell dendrites into ON and OFF activity. This difference is mediated by distinct receptor types at the dendrites of the bipolar cells: dendrites of the OFF bipolar cells express ionotropic receptors so these cells respond in phase with the photoreceptors; those of the ON bipolars express metabotropic receptors, so these cells respond out of phase with the photoreceptors, inverting the response to light. This division into ON and OFF visual streams is carried through to ganglion cells and continued at each level of processing at higher visual centers including the LGN and primary visual cortex where ON and OFF activity continues to be expressed. The glutamate pathways through the retina are illustrated in Fig. 8.1.

These basic glutamatergic synaptic pathways are intersected at the cone to bipolar level and at the bipolar to ganglion cell level by laterally oriented interneurons that introduce spatiotemporal components into the neural interactions, most important for visual processing. At the outer retina, the horizontal cells feed back to the cones and feed forward to the bipolar cells as shown in Fig. 8.2. The mammalian retina expresses only a few different types of horizontal cells, and these are strongly interconnected via electrical coupling. Feedforward synaptic activity to both ON and OFF bipolar cells is mediated by GABA, but the mechanism mediating feedback to photoreceptors remains controversial.

Different chloride concentrations in the bipolar dendrites allow horizontal cell feedforward to antagonize the ON and OFF pathways, polarizing the dendrites in opposite directions (Miller and Dacheux 1983; Vardi et al. 2000). These horizontal cell interactions mediate several essential visual functions: The strongly interconnected

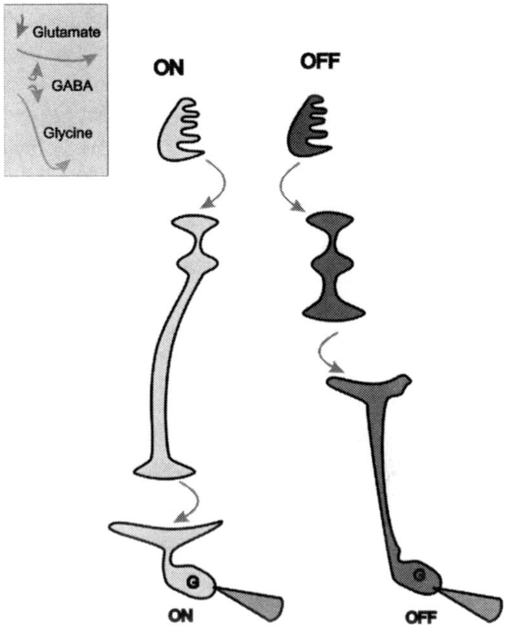

Fig. 8.1 Glutamate pathways through the retina. Photoreceptors drive the ON and OFF bipolar cells that initiate the ON and OFF pathways that can be found throughout all levels of visual processing, well into the visual cortex. In this and following figures, these arrows will represent glutamate synapses

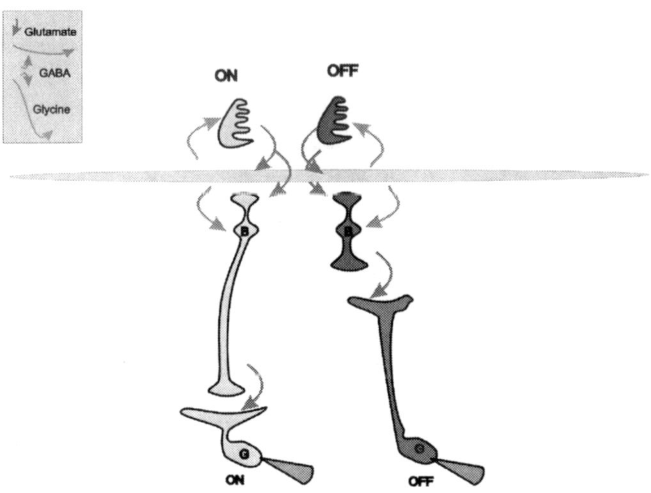

Fig. 8.2 Photoreceptor-to-bipolar pathway is modulated by horizontal cell activity that feeds back to photoreceptors and forward to bipolar cells. In this and following figures, the additional outside arrows indicate these inhibitory pathways

horizontal form a highly blurred "neural image" of the visual world. This blurred image interacts with the sharper image carried by the cone array in two important modes to accomplish different visual functions: on the one hand, the blurred image is subtracted from the sharper image to generate a neural image that is the difference of Gaussians. This difference is thought to enhance or accentuate the neural representation of edges in the visual scene. On the other hand, the blurred neural image carried by the horizontal cells feeds back to cones to modulate cone to bipolar cell gain as a function of cone activity. It thereby serves to normalize the representation of intensity within the visual scene, a form of local gain control. This resolves, for example, the situation where one is attempting to take a photograph through a bright window. For camera with a global aperture or shutter speed, the room is either well lit and the view through the window is saturated or the room is dark and one sees more clearly through the window. The neural image, adjusted by horizontal cell feedback and normalized, is "read out" by the two bipolar cell types, now corrected for local intensity variations. There is an additional gain control mechanism at the bipolar terminals that accommodates for changes in contrast, termed contrast gain control. Contrast gain control assures that the neural image brought to the ganglion cells falls within the dynamic range of the bipolar cell synaptic release and the limits of the ganglion cell voltage and spiking response (Demb 2008).

8.1.2 Extreme Complexity of Amacrine Cell Interactions

Moving to the inner retina there is an additional broad lateral population of interneurons, the wide and narrow field GABAergic amacrine cells. These interneurons feed back to the bipolar cells and feed forward to the ganglion cells and modulate transmission between the bipolar and ganglion cells. They affect both the spatial and temporal properties of bipolar to ganglion cell transmission. The processes for the cells tend to be confined to single strata and they can extend of an up to millimeter laterally (Volgyi et al. 2001). The apparent neatness of the bipolar-to-ganglion cell pathways is interrupted by interactions with a bewildering array of about 30 different amacrine cell types (MacNeil and Masland 1998). There appear to be two general morphological classes of amacrine cell: (1) narrowly ramifying, diffuse amacrine cells that run vertically through the IPL and span the ON–OFF boundary, shown to be glycinergic (Hsueh et al. 2008). (2) Amacrine cells that run horizontally through the IPL and are often confined to a single or limited number of IPL layers, shown to be GABAergic. Some of these extend broadly; others more narrowly. These two geometrically orthogonal classes of amacrine cell play specifically different roles in organizing the visual message. Including these interneurons in the scheme leads to the following circuitry.

8.1.3 A Dozen Different Representations

The basic circuit motif shown in Fig. 8.3 is repeated at least 12 times in an elaborate layering of the IPL first described by Cajal and more recently shown by Masland's lab (Euler and Masland 2000; MacNeil and Masland 1998; Rockhill et al. 2002). There appear to be about ten discrete layers, each subserved by a different class of

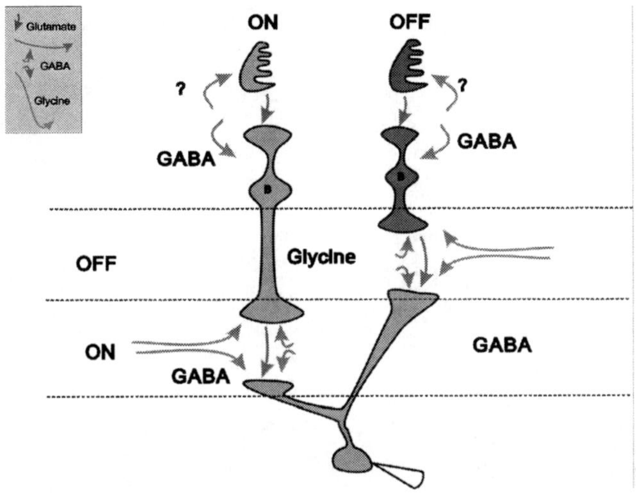

Fig. 8.3 Adding the wide field GABAergic amacrine lateral interneurons to the retinal circuitry, these GABA pathways shown as lateral arrows. These amacrine cells are of the wide and narrow variety, and they feed forward to ganglion cells and (possibly) back to bipolar cells, forming two additional antagonistic surrounds

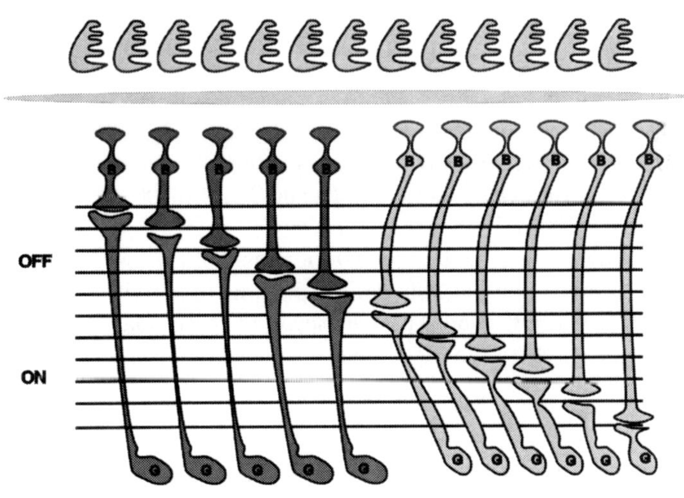

Fig. 8.4 Sketch of the layer by layer connectivity between bipolar cells and ganglion cells. The synaptic terminals of each of the ten bipolar cell types lie at a separate and distinct stratum (Some axon terminals are a bit more diffuse). The dendrites of each ganglion cell type ramify at a distinct stratum or strata

about ten different bipolar cell types, distinguished here by the layered location of their synaptic terminals. The activity formed at each of these layers is "read out" the specific set of ganglion cell dendrites that ramify at that layer as shown schematically in Fig. 8.4. Roughly speaking, at each layer of the IPL, a population of a single

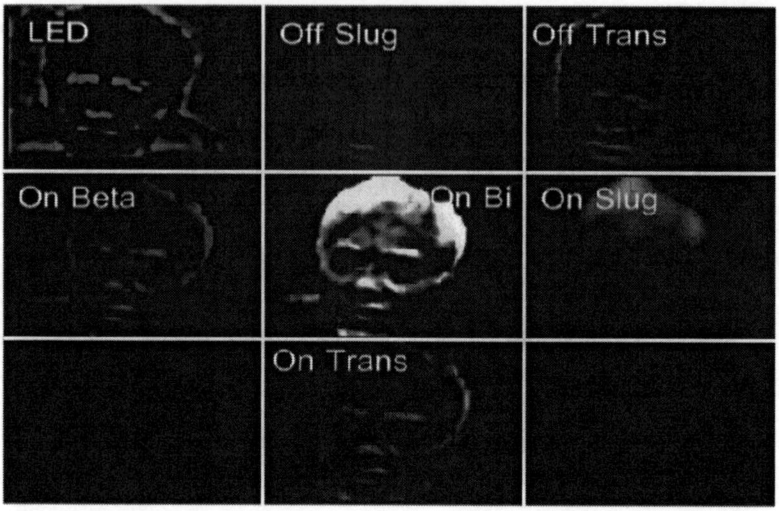

Fig. 8.5 Seven of the 12 different representations of a face, each generated by a different population of ganglion cell types. These images were generated using a CNN model of retina that included many of the synaptic interactions described in this and other papers (Roska et al. 1998, 2000)

bipolar cell type drives a population of a single ganglion cell type (MacNeil and Masland 1998). (There are some exceptions where ganglion cell dendrites are bistratified.) The layering at the IPL has important functional significance: Each layer in the IPL has been shown recently to generate a specific space–time representation of the visual world (Roska et al. 2006; Roska and Werblin 2001; Werblin et al. 2001; Werblin and Roska 2004).

Each layer generates a separate space–time movie of the visual world.

In a modeling study, we have shown the patterns of activity generated by each layer looking at a natural scene – a face. Seven of these images, shown as single frames of a movie are shown in Fig. 8.5.

8.1.4 Each of the Ganglion Cell Outputs Extends over a Specific and Different Space–Time Domain

Figure 8.6 shows the different domains over for each of the retinal outputs. These blobs were constructed by generating a linear model, and tuning it in space and time so that the patterns generated by the model in response to a flashed square approximated those of the individual retinal outputs.

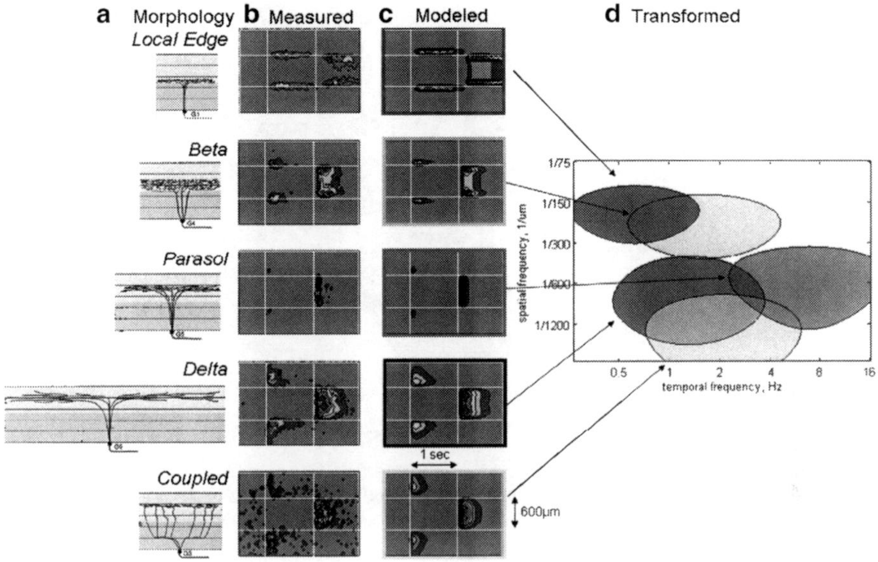

Fig. 8.6 Comparison of measured and modeled outputs of the retina. The left column shows the morphologies of five different retinal cell types. The "measured" column shows the measured patterns of activity, in space and time, for the retinal cells. The "modeled" column shows the modeled linear approximations to the measured results. The right column shows the "blobs" of activity in space and time, derived from the model

8.1.5 Crossover Circuitry of Vertical Amacrine Cells Affects Bipolar Amacrine and Ganglion Cells

The vertical amacrine cells carry information across the ON–OFF boundary of the IPL. Their interactions occur at the synaptic terminals of the bipolar cells and are shown in the illustration below. This crossover inhibitory interaction between amacrine cells appears to be a fundamental motif, governing many of the interactions at the inner plexiform layer. The majority of bipolar amacrine and ganglion cells receive a glycinergic inhibitory input of opposite phase from excitation: ON cells receive OFF glycinergic inhibition and OFF cells receive ON glycinergic inhibition (Hsueh et al. 2008) as shown in Fig. 8.7.

The circuitry at the bipolar terminal is entirely consistent with the patterns found in electron micrographs of the mammalian and salamander retinas (Dowling and Boycott 1966; Dowling and Werblin 1969). A typical circuitry is represented in Fig. 8.8. By this scheme, ON glycinergic amacrine cells inhibit OFF bipolar, amacrine and ganglion cells. A similar but complementary interaction is generated by the OFF glycinergic amacrine cells to the ON cells. All of the synaptic contacts necessary for these interactions exist in the synaptic pathways defined through electron microscopy at the bipolar cell terminal at the "diad" synapse at the bipolar cell terminal as sketched in Fig. 8.9. The circuitry suggests that the on glycinergic

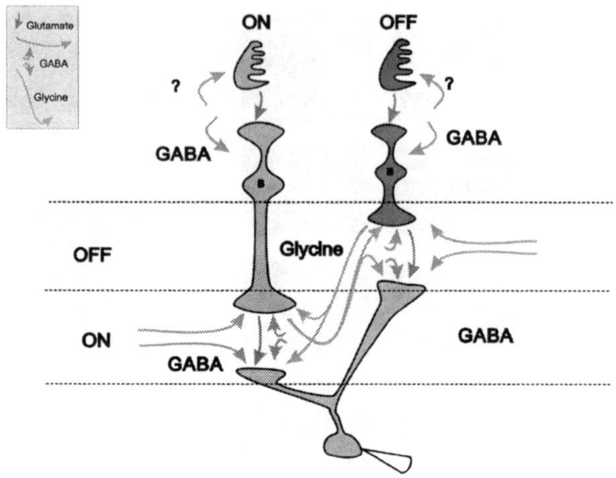

Fig. 8.7 Crossover inhibition in ganglion cells. ON ganglion cells receive OFF inhibition and OFF ganglion cells receive ON inhibition. This diagram indicates that bipolar, amacrine and ganglion cells all receive crossover inhibition, a circuitry that is verified by measurements of excitation and inhibition in each cell type

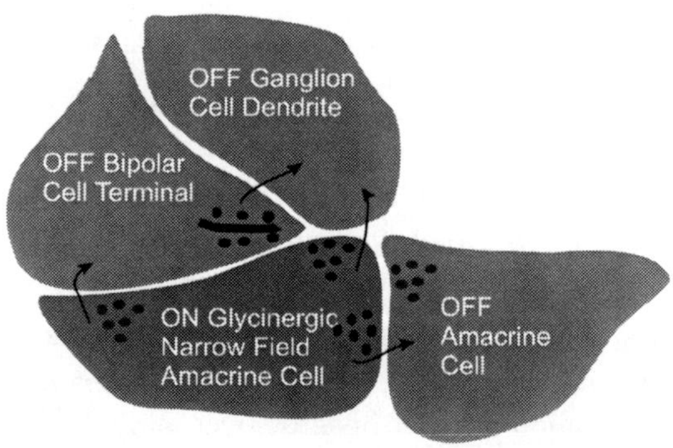

Fig. 8.8 Sketch of electron micrograph of a synaptic terminal of a bipolar cell terminal diad showing the synaptic pathways typically found in these images. Bipolar cell drives a ganglion cell G and an amacrine cell, A. Amacrine cell feeds back to bipolar cell and forward to ganglion cell. The amacrine cell also inhibits a neighboring amacrine cell

amacrine cell shown here can provide inhibitory feedback to three different cell types: the OFF bipolar terminal, to the OFF ganglion cell, and to the OFF amacrine cells as well. The existence of these very inhibitory pathways has been borne out through experiment. The majority of OFF bipolar and ganglion cells receive ON inhibition, and about half of the ON bipolar and ganglion cells receive OFF inhibition (Molnar and Werblin 2007a). These crossover pathways in the general retinal scheme are represented by the following circuitry.

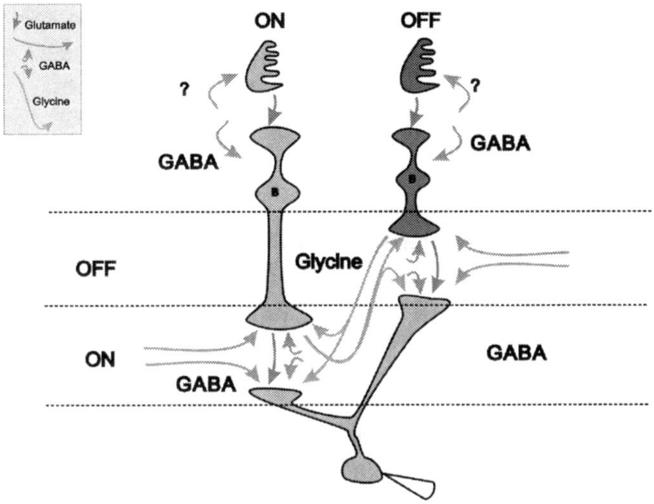

Fig. 8.9 Schematic showing the full interactive circuitry of GABA and glycine pathways in the mammalian retina

8.1.6 The Visual Functional Roles of Crossover Circuitry

Some of our recent work suggests that these vertically oriented amacrine cells perform an essential function in the retina, compensating for nonlinear distortions that occur at most synapses. Synaptic transmission throughout the nervous system is, by its very nature, outwardly rectify, distorting the signals carried along the neural stream of activity. Transmitter release depends upon calcium entry at the synaptic terminals. Release is related to calcium entry mediated by voltage-gated calcium channels, and the activation of calcium channels is nonlinear. As a consequence of this nonlinearity, the postsynaptic currents generated by presynaptic depolarizations are larger than currents generated by presynaptic hyperpolarizations. This is a particularly difficult problem in the retina where most transmission is mediated by graded, spikeless activity. So in order to maintain a linear processing stream in the retina it's necessary to compensate for these nonlinearities at every synapse. The vertically oriented amacrine cells appear to correct for this nonlinear transmission through a circuitry motif defined above as "crossover inhibition" whereby ON excitation is combined with OFF inhibition and OFF excitation is combined with ON inhibition at each stage of retinal processing, including bipolar, amacrine and ganglion cell levels.

Crossover inhibition carried by the vertically oriented glycinergic amacrine cells serves at least four different visual functions when it linearizes visual streams of activity that have been distorted by synaptic transmission: (1) it improves the ability of retinal circuitry to enhance edges by creating an active feedforward inhibitory surround at bipolar and ganglion cells, (2) it allows retinal activity to distinguish between brightness and contrast, (3) it allows neurons to average photon count across their receptive fields, and (4) it maintains a relatively constant input impedance. Each of these functions would be compromised by the nonlinearities inherent in synaptic transmission. The next section summarizes the circuitry that mediates this crossover effect, describes how the nonlinearity is corrected, and outlines how crossover inhibition enhances the integrity of the visual signal.

8.1.6.1 Active Surround Mediated by Crossover Inhibition in Ganglion Cells

Crossover inhibition, carried by narrow field diffuse glycinergic amacrine cells, underlies a significant form of lateral interaction that acts to enhance edges. How can a population of narrow field amacrine cells be involved in generating a broad field surround? As an example, each neuron in the OFF bipolar cell population carries a broad antagonistic surround, initiated by horizontal cell activity, and represented as a reduction of the hyperpolarizing response to an increase in center intensity. The surround signal is therefore an incremental depolarization of the OFF bipolar cells that serves as an incremental excitation to OFF amacrine cells. The OFF amacrine cells are therefore excited by surround illumination. When these amacrine cells "crossover" to inhibit ON ganglion cells, they provide a direct inhibitory input to the ON ganglion cells in response to surround illumination. This is combined with a decrease in excitation from the ON bipolar cells (Fig. 8.10).

8.1.7 Crossover Inhibition Helps to Distinguish Brightness from Contrast (Molnar et al. 2008)

An example of how crossover inhibition corrects for about nonlinearities aboriginal synapses is shown in Fig. 8.11. The stimulus here covering the center of the receptive field is a slow sine wave modulated faster sine wave. Excitation and inhibition to the postsynaptic cell, shown in red and blue, are shown in the center of the figure. Because of the distortion to the rectification, the brightness level is now confused with contrast. By subtracting one of the signals from the other it's possible to reestablish a steady brightness in a modulated contrast shown on the right. This is a good example how signal reconstruction, mediated by crossover inhibition, can eliminate signal distortion.

8.1.7.1 Crossover Inhibition Allows Neurons to Linearly Add Intensities Distributed Across the Receptive Field Center for Ganglion Cells (Molnar and Werblin 2007b)

Many ganglion cells respond to small changes within the receptive fields, a consequence of nonlinear summation across the many bipolar cells that provide synaptic input. But other ganglion cells simply integrated intensity across the receptive field

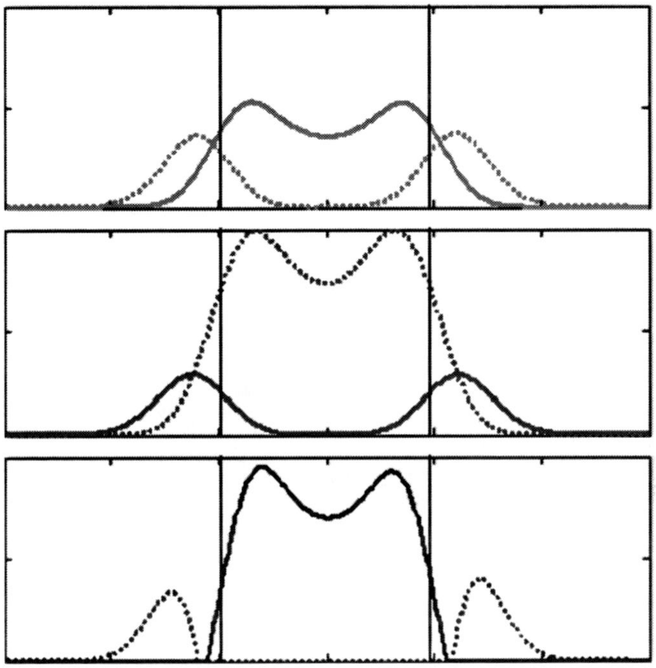

Fig. 8.10 Crossover inhibition brings an active inhibition to the ganglion cells, enhancing the antagonistic surround that was formed at the outer retina. The *top row* this figure shows the profiles for center in surround activity and there is no interaction between them. The *bottom row* shows how excitatory activity can be shaped and compressed within the boundaries of the original stimulus as a consequence of crossover inhibition. Here the outer wings served to actively suppress responses the edges of the center excitatory activity

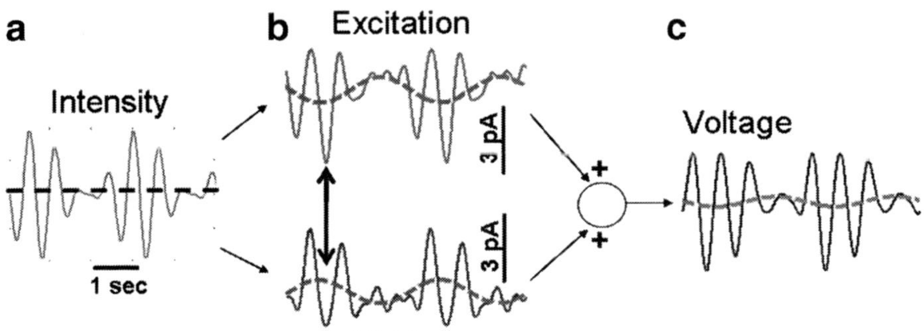

Fig. 8.11 Crossover inhibition corrects for the nonlinearity that confuses brightness with contrast. The waveform on the *left* is the input: a fast sign wave that is itself slow sine wave modulated. This can be interpreted as a constant brightness in the presence of the modulated contrast. The *center* waveforms, showing excitation above and inhibition below, show how signal rectification distorts this image. The waveforms on the *right* show how the image has been reconstructed through crossover inhibition

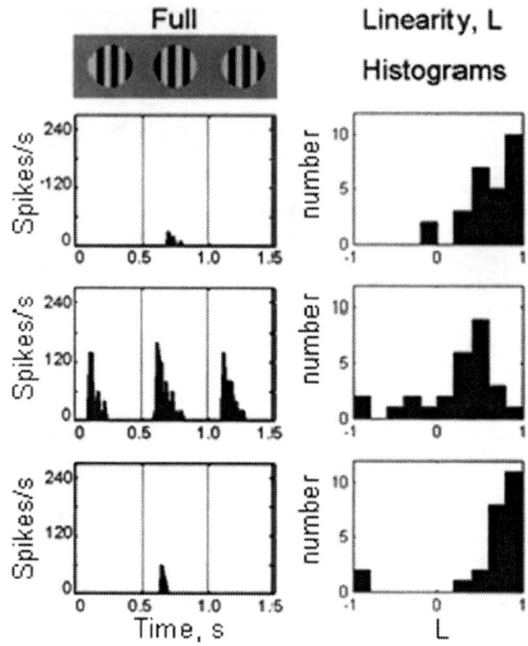

Fig. 8.12 Crossover inhibition allows ganglion cells to maintain linear properties, responding to average luminance across the receptive field in the presence of a flipping grating. *Top row*: no response in this so to something grating at constant luminance across the receptive field. *Middle row*: response at each flip a crossover inhibition has been pharmacologically blocked. *Bottom row*: null responses restored when the pharmacological blocker is removed. This example shows how crossover inhibition linearize the center receptive field response of the cell

so the signals that the integrated must be linear. This linearization is mediated by crossover inhibition, as supported by the experiments shown below. The center of the receptive field stimulated by an inverting great for each light stripe is replaced by a dark stripe and vice versa every half second. Because there is no intensity change across the receptive field, the response is null. But one crossover inhibition is interrupted pharmacologically as shown in the center row, the cell responds to each flip up the grating. This is a consequence of the nonlinear input that has not been compensated because crossover inhibition has been blocked (Fig 8.12).

Crossover inhibition maintains a more constant input impedance. Because retina neurons receive many synaptic inputs, it's important that the dominant excitatory input not act as a shunt on the other excitatory and inhibitory inputs. To control for this crossover inhibition moves memory conductance in an equal and opposite direction to excitation thereby maintaining a more constant input conductance (Fig. 8.13).

8.1.7.2 In-layer Interactions are Mediated by GABAergic Pathways

There are some additional pathways from ON cone bipolar cells to other ON cone bipolar cells and to rod bipolar cells mediated by GABA. We found no inhibition from OFF to OFF bipolar cells. The circuitry for glycinergic inhibition between amacrine cells is also remarkably simple. Again, most amacrine cells receive crossover inhibition. But OFF to ON inhibition is more prevalent than ON to

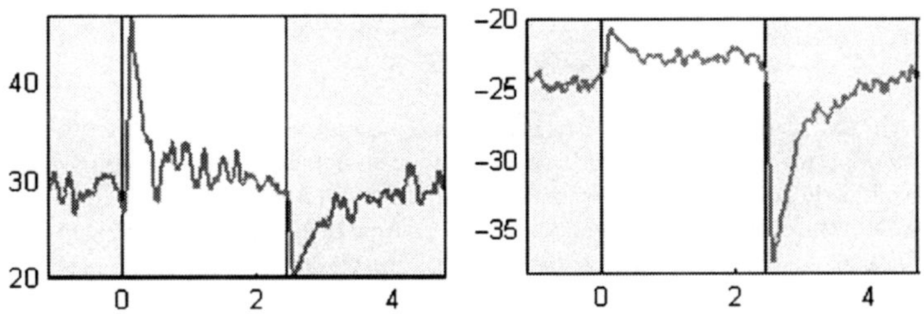

Fig. 8.13 Crossover inhibition maintains a constant input impedance in retinal neurons. *Left*: excitatory input showing how would current transient ON an inward current transient at OFF. *Right*: inhibitory input showing output current transient at ON an inward current transient and OFF. These currents are generated by equal and opposite conductances at ON and OFF. This combination of conductances keeps the input conductance of the neuron relatively constant

OFF inhibition. There is very little ON to ON or OFF to OFF amacrine cell inhibition. We found no inhibition whatever impinging upon the ON–OFF wide field amacrine cells. Therefore, the wide field amacrine cells may be the only retinal neurons that receive no inhibition whatsoever (Bloomfield and Volgyi 2007; Volgyi et al. 2005).

Furthermore, we found no evidence for GABAergic ON–OFF inhibition to either bipolar cells or amacrine cells. This suggests that the GABAergic ON–OFF amacrine cells only feed forward and only to some ganglion cells.

8.1.8 Specific Ganglion Cell Circuitries

8.1.8.1 Directionally Selective Ganglion Cells

There are a few amacrine cells that have now been identified with very specific personalities. For example, starburst amacrine cells named for the characteristic starburst pattern of their processes, span about 200 μm. Starburst cells contain a release of both GABA and acetylcholine. They are the key elements in the organization of directional selectivity in the retina. A large population of starburst amacrine cells is associated with each directionally selective (DS) ganglion cell, and neighboring DS ganglion cells likely share many starburst amacrine cells. Starburst cells are inherently directionally selective, generating more release for centrifugal movement. One likely mechanism involves calcium-initiated calcium release, but this remains an area of intense exploration. Release occurs along the outer 1/3 of the starburst processes. These processes not only release GABA, but they are also GABA sensitive. This creates a mutual inhibition between starburst cells that acts to amplify directional motion sensitivity as shown in Fig. 8.14 (Lee and Zhou 2006). Starburst cells inhibit the DS cells asymmetrically, with stronger inhibition arriving from the null side than from the preferred side (Fried et al. 2002, 2005). These three mechanisms, inherent directional selectivity in the starburst cells themselves, mutual antagonistic

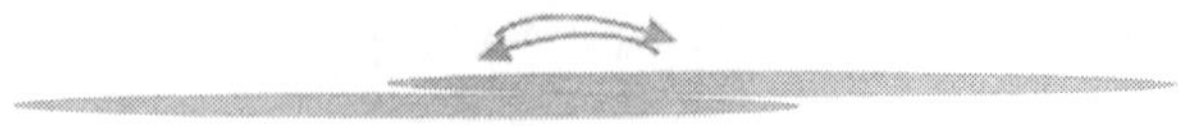

Fig. 8.14 Mutual inhibition between starburst amacrine cells amplifies the directional properties of the starburst network. Here, two starburst amacrine cells, themselves directionally selective, are mutually inhibitory. Individual starburst cells are directional for movement away from their centers. They release GABA at their circumference. This GABA release inhibits neighboring starburst cells and starburst cells on the "null" side of the DS ganglion cell also inhibit

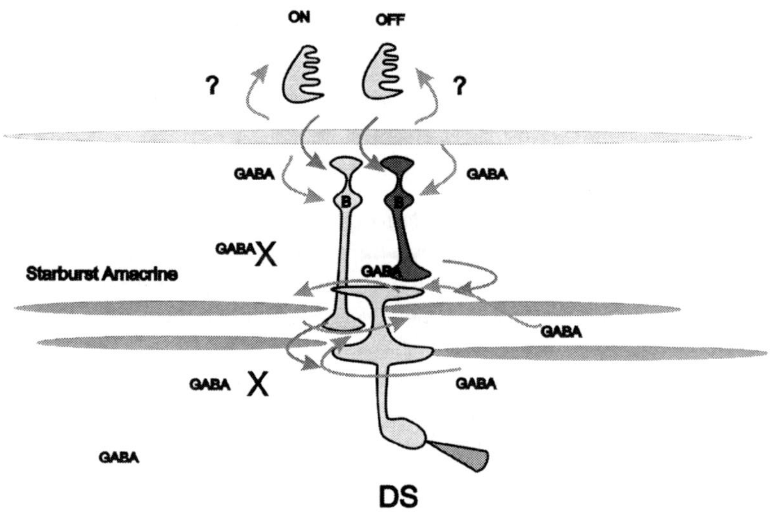

Fig. 8.15 Pathways underlying the behavior of the directionally selective ganglion cell. Starburst amacrine cells, themselves, directionally selective, are mutually inhibitory. They inhibit by feeding back to bipolar cells and forward to ganglion cells. This inhibition is asymmetric: it is stronger on the NULL side than on the PREFERRED side, thereby endowing the DS ganglion cell with directional properties

interaction between neighboring starburst cells, and asymmetrical inhibition acting both pre- and postsynaptically at the ganglion and bipolar cells, endow the DS cell with some of its directional properties as shown in Fig. 8.15.

The circuitry puzzle regarding the DS cells is far from solved, but the general organizational rules listed above still apply. The lateral inhibitory interneuron is GABAergic, following the GABA rule for laterally oriented cells.

8.1.8.2 Alpha Ganglion Cells

Alpha cells have the largest dendritic diameter in the retina ranging up to almost a millimeter in diameter. These cells have recently been shown to have an unusual form of crossover inhibition. OFF cells receive an electrical synapse from AII amacrine cells as well as a chemical crossover input mediated by glycine from the ON pathway shown in Fig. 8.16. It appears that the AII amacrine cells serve

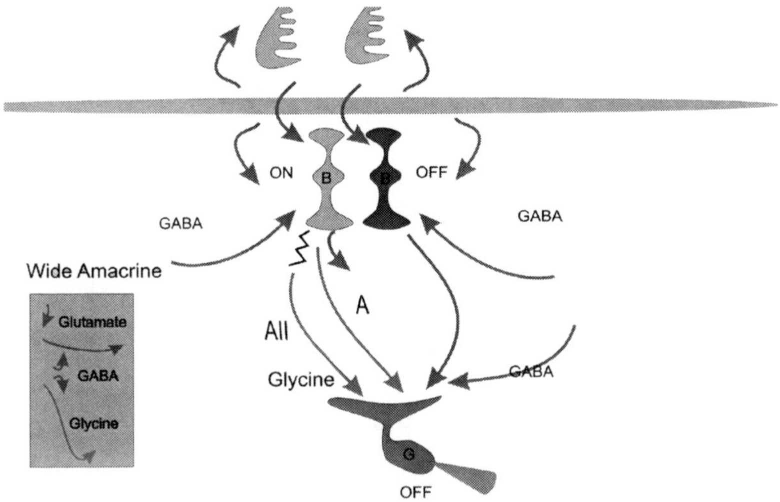

Fig. 8.16 Pathways underlying the behavior of the alpha cell/looming detector

roles other than simply coupling the rod to the cone system. Because the diameter of the cell some large and because retinal ganglion cells tend to tiled retina the alpha cells have been hard to study physiologically.

There are numerous other examples of special purpose circuitry that utilize laterally oriented amacrine cell interneurons. The polyaxonal amacrine cells, studied in detail (Volgyi et al. 2005) are thought to mediate saccadic suppression (Roska and Werblin 2003). In other cases, the same cell type has been implicated in mediating object motion sensitivity (Olveczky et al. 2007). It is likely that other amacrine cell types also serve specific functions, but their properties have not yet been identified.

8.1.8.3 Local Edge Detectors

At about the time that Levick (Levick 1965) was characterizing the directionally selective ganglion cell, he also described another ganglion cell that he turned the local edge detector (LED). This cell appears to be unique in that it was activated by local edge in the form of moving gratings at the center of its receptive field, and that activity was suppressed by moving edges in the surround. Later studies (van Wyk et al. 2006) have recently gone on to characterize some of the special temporal properties of this neuron. It receives both excitation and inhibition at both ON and OFF, and is inhibited by edge stimuli presented at the surround. A more recent study has shown that inhibition at both ON and OFF at the receptive field center are mediated by glycine, and that broader lateral inhibitory feedback input is mediated by GABA. Both inhibitory components follow the general rule of vertical glycinergic and lateral GABAergic activity as shown in Fig. 8.17. The role of this neuron in the overall scheme of vision remains obscure, but it likely is involved in high resolution, slow temporal response activity.

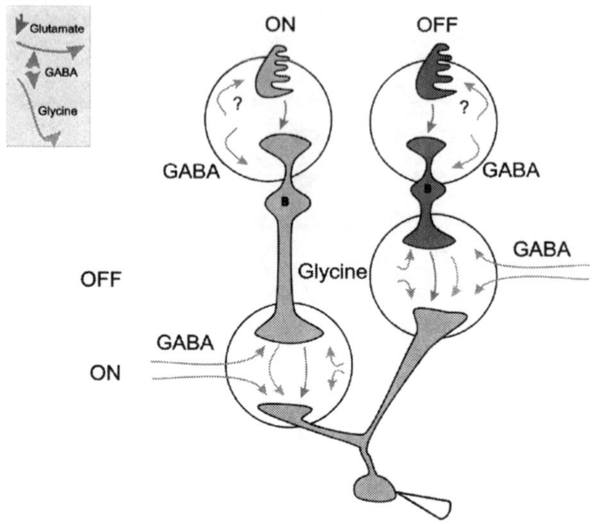

Fig. 8.17 Pathways underlying the response properties of the local edge detector. This cell receives local ON and OFF inhibition that is glycinergic, broad field inhibition that is GABAergic. But GABA is only fed back, not forward in this cell type

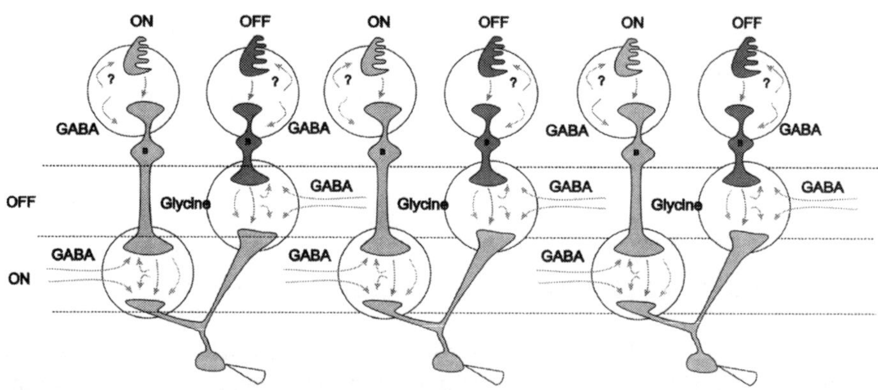

Fig. 8.18 Center surround circuitry for the LED showing that small detail in the surround inhibits the response to small detail in the center. The inhibitory signals are carried by wide field horizontal cells carrying information from surround to center

8.1.8.4 ON Beta Cells

ON beta cells encompass most of the circuitry described for the general retina. The cells receive local glycinergic inhibition and also wide and narrow GABAergic inhibition that is fed both forward and back. Surprisingly, these cells also receive an OFF excitatory input but is only visible when all inhibition is blocked. The excitatory input is modulated by GABA feedback to bipolar cells, and there appears to be both ON and OFF glycinergic narrow field input as well (Figs. 8.18 and 8.19).

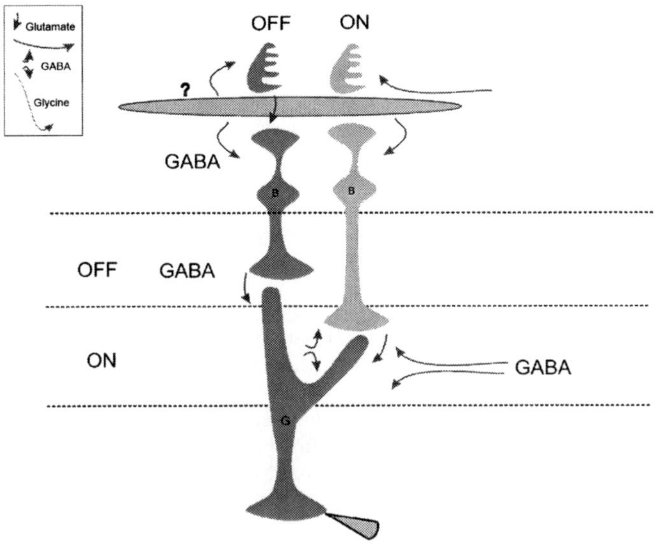

Fig. 8.19 ON Beta cell circuitry. These cells receive a full complement of inhibitory inputs from wide and narrow GABAergic amacrine cells. They also receive input from glycinergic ON and OFF cells (not shown here)

References

Bloomfield SA, Volgyi B (2007) Response properties of a unique subtype of wide-field amacrine cell in the rabbit retina. Vis Neurosci 24:459–469

Demb JB (1966) Functional circuitry of visual adaptation in the retina. J Physiol

Dowling JE, Boycott BB (1966) Organization of the primate retina: electron microscopy. Proc R Soc Lond B Biol Sci 166:80–111

Dowling JE, Werblin FS (1969) Organization of retina of the mudpuppy, Necturus maculosus. I. Synaptic structure. J Neurophysiol 32:315–338

Euler T, Masland RH (2000) Light-evoked responses of bipolar cells in a mammalian retina. J Neurophysiol 83:1817–1829

Hsueh HA, Molnar A, Werblin FS (2008) Amacrine-to-amacrine cell inhibition in the rabbit retina. J Neurophysiol 100:2077–2088

Levick WR (1965) Receptive fields of rabbit retinal ganglion cells. Am J Optom Arch Am Acad Optom 42:337–343

MacNeil MA, Masland RH (1998) Extreme diversity among amacrine cells: implications for function. Neuron 20:971–982

Miller RF, Dacheux RF (1983) Intracellular chloride in retinal neurons: measurement and meaning. Vision Res 23:399–411

Molnar A, Werblin F (2007a) Inhibitory feedback shapes bipolar cell responses in the rabbit retina. J Neurophysiol 98:3423–3435

Molnar A, Werblin FS (2007b) Inhibitory feedback shapes bipolar cell responses in the rabbit retina. J Neurophysiol 98:3423–3435

Olveczky BP, Baccus SA, Meister M (2007) Retinal adaptation to object motion. Neuron 56: 689–700

Rockhill RL, Daly FJ, MacNeil MA, Brown SP, Masland RH (2002) The diversity of ganglion cells in a mammalian retina. J Neurosci 22:3831–3843

Roska B, Werblin F (2001) Vertical interactions across ten parallel, stacked representations in the mammalian retina. Nature 410:583–587

Roska B, Werblin F (2003) Rapid global shifts in natural scenes block spiking in specific ganglion cell types. Nat Neurosci 6:600–608

Roska B, Nemeth E, Werblin FS (1998) Response to change is facilitated by a three-neuron disinhibitory pathway in the tiger salamander retina. J Neurosci 18:3451–3459

Roska B, Nemeth E, Orzo L, Werblin FS (2000) Three levels of lateral inhibition: A space-time study of the retina of the tiger salamander. J Neurosci 20:1941–1951

Roska B, Molnar A, Werblin F (2006) Parallel processing in retinal ganglion cells: how integration of space-time patterns of excitation and inhibition form the spiking output. J Neurophysiol 95(6):3810–22

van Wyk M, Taylor WR, Vaney DI (2006) Local edge detectors: a substrate for fine spatial vision at low temporal frequencies in rabbit retina. J Neurosci 26:13250–13263

Vardi N, Zhang LL, Payne JA, Sterling P (2000) Evidence that different cation chloride cotransporters in retinal neurons allow opposite responses to GABA. J Neurosci 20:7657–7663

Volgyi B, Xin D, Amarillo Y, Bloomfield SA (2001) Morphology and physiology of the polyaxonal amacrine cells in the rabbit retina. J Comp Neurol 440:109–125

Volgyi B, Abrams J, Paul DL, Bloomfield SA (2005) Morphology and tracer coupling pattern of alpha ganglion cells in the mouse retina. J Comp Neurol 492:66–77

Werblin FS, Roska, B (2004) Parallel visual processing: a tutorial of retinal function. Int J Bifurcation and Chaos 14:843–852

Werblin F, Roska B, Balya D (2001) Parallel processing in the mammalian retina: lateral and vertical interactions across stacked representations. Prog Brain Res 131:229–238

Chapter 9
Elastic Grid-Based Multi-Fovea Algorithm for Real-Time Object-Motion Detection in Airborne Surveillance

Balazs Gergely Soos, Vilmos Szabo, and Csaba Rekeczky

Abstract In this chapter, a generic multi-fovea video processing architecture is presented, which supports a broad class of algorithms designed for real-time motion detection in moving platform surveillance. The various processing stages of these algorithms can be decomposed into three classes: computationally expensive calculations can be focused onto multiple foveal regions that are selected by a preprocessing step running on a highly parallel topological array and leaving only the nontopological (typically vector-matrix) computations to be executed on serial processing elements. The multi-fovea framework used in this chapter is a generalized hardware architecture enabling an efficient partitioning and mapping of different algorithms with enough flexibility to achieve good compromise in the design tradeoff between computational complexity versus output quality. We introduce and compare several variants of four different classes of state-of-the-art algorithms in the field of independent motion analysis and detection. On the basis of the analysis, we propose a new algorithm called the Elastic Grid Multi-Fovea Detector characterized by moderate hardware complexity while maintaining competitive detection quality.

9.1 Introduction

9.1.1 Unmanned Aerial Vehicles

Unmanned aerial vehicles offer economic solutions for vegetation classification, in flood and fire defense and for large area surveillance. Today unmanned planes are capable of flying over the operation zone following a predefined path using intelligent navigation system based on GPS and motion sensors. During the flight,

B.G. Soos (✉) and V. Szabo
Pázmány Peter Catholic University, Budapest, Hungary
e-mail: soos@digitus.itk.ppke.hu

C. Rekeczky
Eutecus Inc., Berkeley, California, USA

C. Baatar et al. (eds.), *Cellular Nanoscale Sensory Wave Computing*,
DOI 10.1007/978-1-4419-1011-0_9,

they can gather information and transmit to a ground station via radio connections. Recorded video shots can be analyzed after landing in offline mode; consequently, through analysis is feasible either by human experts or using machine intelligence. The flight path can be modified when interesting events are detected in order to collect more detailed information. The aim of this research was to devise an optimal architecture for an onboard visual system capable of making these decisions. The proposed framework is designed to be universal for any visual surveillance task. It is reviewed and analyzed focusing on the specific application area of independent motion detection.

9.1.2 Multi-Fovea Approach

Processing the entire data captured by an image sensor at full resolution is computationally expensive, and in most cases, unnecessary. Even in the human visual system, data convergence could be observed: the amount of data processed and transferred from photoreceptors in the retina to cortical structures via the optic nerve significantly decreases, whereas the abstraction of the information extracted increases. Light intensity is captured by roughly 130 million sensory cells and is transferred by 1 million ganglion cells only.

In the *input video flow*, frames have fix resolution and are discretized in time at a constant *frame rate*. In our artificial visual system, a decision can be made at an early stage of the *image processing algorithm* to locate interesting regions. Thus, the computational effort can be focused on critical areas, and an efficient processing scheme can be formulated with moderate data transfer between modules. Hardware realization can be designed to solve parallel tasks in each region, or existing vision processors can be utilized. Selected regions are called *foveal windows* analog to the fovea of the mammalian retina. These regions are rectangular regions covering a part of the original input frame depending on the *scale factor*. This model was first described in Rekeczky et al. (2004). The aim of this chapter is to present algorithms utilizing this concept.

High-level elements of the motion detection algorithms are as follows: first interesting regions are selected by using mainly *topological 2D operators* (Class 1), then the regions are processed using local adaptation in each region (Class 2) and some numerical descriptors are extracted. Finally, depending on the topology of the windows and the extracted values, global decision is made (Class 3) for aligning consecutive frames.

These three steps are highly different in terms of the required operator set. We propose an *abstract architecture* for optimal computation with three different types of processors, the *frontend processor array (FPA)*, the *foveal processor array (FVA)*, and the *backend processor (BP)*. They communicate via an intelligent *memory manager unit*. The abstract architecture can be realized on various hardware components. We also propose some feasible variations of topological array processors and pipeline architectures.

To describe a general *video processing algorithm*, a *flowchart diagram* will be used (*modeling*). Then, all *processing blocks* will be mapped to an *abstract processor* architecture depending on the required operator set (*partitioning*). For a given underlying hardware platform, the individual blocks will be implemented, and code segments and parameters could also be optimized (*implementation*).

9.1.3 Airborne Motion Detection

In large field airborne surveillance applications (Hu et al. 2004), the detection of moving ground objects is a key issue. After detection of these objects, they can be followed by the plane, and with enough information, they can be identified as well. A good review for tracking can be found in Yilmaz et al. (2006). Besides military applications, another application field is traffic monitoring (Molinier et al. 2005).

For medium-altitude video flows (100–300 m), main streams in detection are optical flow (Adiv 1985; Argyros et al. 1996; Black and Jepson 1994) and registration-based methods using background subtraction. For low-altitude videos, real 3D analysis of the scene is required (Sawhney et al. 2000; Irani and Anandan 1998; Zhu et al. 2005; Manolis et al. 1998; Fejes and Davis 1999). However, in surveillance tasks, medium altitude is more common.

For a good review on general optical flow methods and registration methods, refer to Barron et al. (1992) and Zitova and Flusser (2003), respectively.

In this chapter, feature-based registration methods for background subtraction are reviewed and compared to highlight the capability of our framework. This approach for independent motion detection is popular among researchers (Kumar et al. 2001; Morse et al. 2008; Ali and Shah 2006; Pless et al. 2000). Creating panoramic images from frames captured by a rotating camera is also an active research field. This problem covers similar registration tasks but may use offline algorithms with much larger computational need (Hsieh 2004; Brown and Lowe 2003; Szeliski 2006; Sawhney and Kumar 1999; Kaaniche et al. 2005).

Mikolajczyk and Schmid (2005) recently compared local descriptors. They highlighted the efficiency of the popular *scale invariant feature transform* (SIFT, Lowe 2004). We will compare the SIFT-based algorithm and the Lucas-Kanade-Tracker (Lucas and Kanade 1981; Shi and Tomasi 1994) with traditional block matching (Zhu and Ma 2000) and Harris corner (Harris and Stephens 1988)-based corner pairing algorithm (CPA). On the basis of the overall analysis, we propose a new algorithm called the Elastic Grid Multi-Fovea Detector (ELG), which is characterized by moderate hardware complexity while maintaining competitive detection quality.

More detailed description of the framework and the algorithms are published in Soos et al. (2009).

9.2 Independent Motion Analysis

9.2.1 Images and Video Frames

Let us assume that the airplane flying over the inspection area faces to the ground. The camera captures frames on regular time instances. Frames $\{\mathbf{I}_t(\mathbf{x})\}$ ($t \in \{1, 2, \ldots, K\}$) are sampled light intensities that are projected to the image plane (sensor array) collected into a list for all time instances.

Homogenous representation of points on the image plane is a column vector $\mathbf{x}_{3\mathrm{H}}(x1, x2, x3) = [x1, x2, x3]^T$, $x1, x2, x3 \in \mathbb{R}$, where the corresponding point in Cartesian coordinates is $\mathbf{x}_2(x1', x2')$, $x1' = x1/x3; x2' = x2/x3$. Scene points (points in 3D world) are represented by Cartesian coordinates in most cases $\mathbf{x}_3(x1, x2, x3)$, $x1, x2, x3 \in \mathbb{R}$. Homogenous representation will be denoted by the symbol "H" in subscript over the dimension.

Images are described by functions, and defined and stored using matrices. In practice, video sensors have finite resolution; therefore, intensity values in frames are defined at integer coordinated pixels only – m rows × n columns by the image matrix, $\mathbf{I}_k$, in horizontal and vertical order, respectively ($u{=}1, \ldots, n, v{=}1, \ldots, m$), $\mathbf{I}_k(u, v) := [\mathbf{I}_k]_{v,u}$. For noninteger points, it can be interpolated – $\mathbf{I}_k(\mathbf{x}_2)$, $\mathbf{I}_k(\mathbf{x}_{3\mathrm{H}})$.

The camera projects scene points to image points:

$$\mathbf{x}_{3\mathrm{H}} = P(\mathbf{x}_{4\mathrm{H}}) \tag{9.1}$$

P is defined more precisely in the Sect. 9.2.2. It assigns a ray of 3D points to an image point.

In a simplified capturing model, we have light sources and reflecting surfaces. Pixel value in a frame is the total intensity coming from the specific ray; therefore, we are interested in the point $\mathbf{x}_3$ where the ray intersects a surface element of the scene. We consider surfaces with diffuse reflection. It means intensity for an image point depends on the incoming intensity and emission at the corresponding 3D location but not on the relative orientation of the surface element and the camera since the surface causes omnidirectional reflection.

$$\mathbf{I}_k(\mathbf{x}_2) = \mathbf{I}(\mathbf{x}_3) \tag{9.2}$$

Detailed description of epipolar geometry and camera models can be found in Hartley and Zisserman (2000) and Zhang (1998).

9.2.2 Background and Objects

The scenes considered, namely large open-field areas, or highways with region of interest constraints may be regarded as flat surfaces, since the variation in height of the ground is small compared with the distance to the camera. Thus, we can model the ground as a plane with a texture map $\mathbf{B}(\mathbf{x}_2)$. This texture is the *background* image, describing the intensity values of the static empty screen. In some cases, a small part of the sky is also visible in the frames. The bounded volumes of the 3D scene having nonnegligible height or changing their position are objects. Objects in frames can be described by their *shapes* and *appearances*. Silhouette of an object is the region where it covers the background. Shape is the description of the silhouette, and the appearance is the model how it alters the background. All properties are time-dependent because of the camera motion. By definition, areas where shadow is cast also belong to the specific object.

9.2.3 Global Image Motion Model

Using a homogenous vector representation of image coordinates $\mathbf{x}_{3\mathrm{H}}$ and world points $\mathbf{x}_{4\mathrm{H}}$, camera mapping (9.1) may be directly described as a 3×4 linear projection:

$$\mathbf{x}_{3\mathrm{H}:k}\left(x', y', 1\right) = P(\mathbf{x}_{4\mathrm{H}}) = H^{k}_{3\times 4}\,[x, y, z, 1]^{T} \tag{9.3}$$

This representation may be used for pinhole or orthographic camera models representing camera pose-dependent external parameters and internal parameters as well. The world coordinate system may be defined as the ground plane lying in the "x–y" plane. The camera at time instant k is located at $\mathbf{c}_3$ and has a specific orientation. During the frame-by-frame time, the camera center is moved and its orientation is changed. Points from the surface are projected to image planes, forming video frames $\mathbf{I}_k(\mathbf{x})$ and $\mathbf{I}_{k+1}(\mathbf{x})$. Since for all background points z coordinate component is zero, mapping can be simplified. The plane to plane transformation for the actual image can be described by a 3×3 linear assignment.

$$\begin{aligned}\mathbf{x}_{3\mathrm{H}:k} &= H^{k}_{3\times 3}\,[x, y, 1]^{T}\\ \mathbf{x}_{3\mathrm{H}:k+1} &= H^{k+1}_{3\times 3}\,[x, y, 1]^{T}\end{aligned} \tag{9.4}$$

Or direct relation may be expressed between points in images k and $k+1$:

$$\mathbf{x}_{3\mathrm{H}:k+1} = H^{k+1}_{3\times 3}\left[H^{k}_{3\times 3}\right]^{-1}\mathbf{x}_{3\mathrm{H}:k} \tag{9.5}$$

$$\mathbf{x}_{3\mathrm{H}:k+1} = H^{k+1,k}_{3\times 3}\,\mathbf{x}_{3\mathrm{H}:k} \tag{9.6}$$

This transformation maps points from the coordinate system of kth frame to representation as in $k+1$. The geometrical transformation may be calculated for all image points of $\mathbf{I}_k$:

$$\begin{aligned}&\mathbf{I}_k \rightarrow \mathbf{J}_k\left(H^{k+1,k}_{3\times 3}\,[u, v, 1]^{T}\right) = \mathbf{I}_k\left([u, v, 1]^{T}\right)\\ &u \in \{1, 2, \ldots, n\};\, v \in \{1, 2, \ldots, m\}\end{aligned} \tag{9.7}$$

This means that frames containing common parts from the background can be aligned by a linear transformation matrix by using homogenous representation. In the most general case, this can be a projective transformation. This is our global model for image motion (*global motion model*) describing the effect of the camera motion in consecutive frames.

To calculate a smooth transformation, integer coordinates are used in the target coordinate frame, and interpolation is applied in the source frame (inverse mapping):

$$\begin{aligned}&\mathbf{J}_k\left(\left[u',v',1\right]^T\right)=\mathbf{I}_k\left(\left[H_{3\times 3}^{k+1,k}\right]^{-1}\left[u',v',1\right]^T\right)\\&u'\in\{1,2,\ldots,n\};v'\in\{1,2,\ldots,m\}\end{aligned}\tag{9.8}$$

9.2.4 Motion Detection, Object Extraction, and Global Background Mosaic

Using a global motion model, more frames can be aligned to a common coordinate frame. A large *mosaic* image can be created from aligned images combining image matrices where they overlap (*blending*), and fill uncovered regions with a default value.

In most cases, the plane flies above an unknown field, which means the background image is unknown. On the contrary, if it is known, then the pose of the plane is unknown. Indeed for the InputFrame ($\mathbf{I}_{k+1}(\mathbf{x})$), the previous image BaseFrame may be used as reference after estimating the proper global motion and AlignedFrame can be calculated ($\mathbf{J}_k(\mathbf{x})$) from $\mathbf{I}_k(\mathbf{x})$. They both cover parts of the background and different snapshots of the moving objects. The detection is the process of creating DetectionMask with "1" elements for locations that are recognized to be part of an object silhouette in the frame of $\mathbf{I}_{k+1}(\mathbf{x})$. The clusters in DetectionMask are listed in separate masks $\{\mathbf{O}_j(\mathbf{x})\}$ (ObjectMasks).

The first task is to calculate frame-to-frame alignment. If it is reliable for a sequence of consecutive frames, a local background mosaic can be constructed from them. It is a robust estimate for a part of the background image, more reliable then using only a single frame from the past. For slowly moving objects or objects with special motion vectors, a small projected motion vector arises, resulting in small changes for shapes in consecutive frames. For a steady camera, the solution is to decrement the frame rate, but for a moving observer, large overlap is also needed for efficient frame-to-frame registration. Small errors in frame-to-frame registration do not limit the detection capability. However, the time span for reliable *local background mosaics* is limited, since the error accumulates.

Building a reliable global mosaic for estimating the background image and to track the full path of the plane (Simultaneous Localization and Mapping) is a difficult problem and it is not covered in this chapter. Our main objective was to solve the detection task.

9.3 Multi-Fovea Framework: Abstract Hardware Model

To describe a *video flow processing algorithm*, a possible option is to create a flowchart diagram. This is the *modeling step* of the algorithm design.

In Soos et al. (2009), an abstract hardware architecture called *multi-fovea framework* is proposed comprising three different types of processors for ideal computation of each image processing step, which communicate via a complex *memory manager unit* (Fig. 9.1).

The first processing unit is called frontend processor array (FPA) for preprocessing, and also containing the sensor for image capturing. Usually preprocessing (noise reduction, spatial filtering for feature extraction), is highly parallel on pixel level. Operators are either defined on a small neighborhood of pixels (typically 3×3), for example, convolution, or combining two images point by point, for example, image subtraction. *Topological 2D operators* are also referred to as *templates*. The input and most of the intermediate images are gray scale, they are called *maps*. Some operators result in binary images or *masks*. Image arrays are extended with some virtual pixels defining neighborhood values for pixels around boundaries.

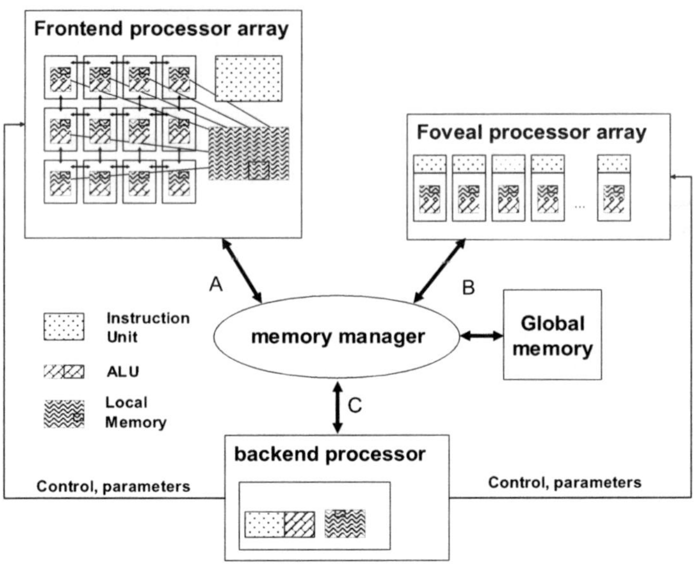

Fig. 9.1 Main processing elements of the abstract hardware architecture: a frontend processor array for data-parallel steps with processing elements in 2D topology, a foveal processor array for task parallel steps and a backend processor responsible for control, organization, and clasification. A processing element (*PE*) consists of some registers and a logic and arithmetic unit (*ALU*) and optionally some local memory. An instruction unit can support multiple PEs. Images can be stored in a distributed way in the frontend processor array to grant fast access to mapped image parts if communication link for neighbors are present for sharing overlapping data. Processors interact via an intelligent memory manager and some direct control lines. At communication channel "**a**" scalars and images with As^s size are transferred. Channel "**b**" is for images with Aw size and scalars. At channel "**c**" images with arbitrary size and scalars are transferred

The data-parallel structure of the problem allows the usage of a large number of independent *threads*, each processing small, possibly overlapping partitions of the image maps. Since the data and operators rely on 2D pixel topology, it is practical to identify the threads with 2D ID-s. Since the threads are branchless, *processing elements* may share a common *instruction unit*.

The definition is abstract, but the underlying implementation of the FPA can be a single threaded processor or a pixel-pipeline. Alternatively, a real array of cores may be designed with distributed local memory and communication links to neighbors for sharing overlapping data either in a course grain or fine grain configuration.

As a result of preprocessing, the fixed sequence of operators produces some filtered versions of the input frame combined with some images from the past. Combination of gray scale maps should produce at least one feature map indicating interesting locations.

Preprocessing should run in real time keeping up with the frame rate of the input source. This unit must have enough local memory to store all intermediate data in the processing step of a given input frame – *short-term local memory (STLM)*, and even some extra memory to store results from a previous time instant – *long-term local memory (LTLM)*.

The resolution of the sensor array is a_0 (m rows and n columns). In some cases, smaller resolution is enough for describing the scene. Support for downsampling to create images with $As^s = (1/4)^s\, a_0$ pixels is desirable.

After preprocessing, foveal regions with a resolution of $mw \times nw$ (Aw size) are selected and stored in a list. Individual windows are referred to as w_i, whereas the coordinate of the corresponding center is referred to as $\mathbf{w}_i$. Foveal processors (cores inside FVA) are fed by the Memory Manager Unit. This unit maps the corresponding windows of the filtered images – foveal image list (the same region from each) to the memory space of a processing unit. Improved analysis needs more sophisticated algorithms with branching; therefore, these steps are task-parallel rather than data-parallel. Furthermore, the foveal windows can be distributed in various configurations and their overlapping is small, thus topological thread-processing element mapping is no longer reasonable. Operations can use large neighborhood ($mt \times nt$).

To describe a feature by a support region, some fixed number of pixels is required to have enough variance. In most cases, this means that window size and template size in pixels do not depend on the scale of the given map. Instead the size of the window is fixed; therefore, the coordinates of the centers are scaled.

Since the number of the foveas may be much larger than the number of the processing elements, LTLM is not available at this level. All results needed for the next iteration need to be saved. The frame is processed when all foveas are ready. Foveal processors may have more sophisticated programs with branches and limited iterations as well, optionally supported with high level data-parallel instructions implemented in hardware. In this case, templates may have large radii and being executed only at given locations not for all possible placements inside the foveal windows.

Output of a fovea may be an image part backprojected by the memory manager unit to a global image using the position of the fovea in the original frame or some scalars collected to a list.

The BP is a serial processor that can access any global memory space and does all the serial calculation. This is capable of setting up the window configuration for foveas and the program for both foveas and frontend units.

Algorithmic steps should be analyzed, and depending on their properties, different mappings could be applied. Considering data transfers computational steps should be assigned to the appropriate abstract hardware module. This is the *partitioning* step of the algorithm design.

9.4 Algorithms

As it was described in Sect. 9.2, the series of input image frames are considered as the main input to the system. They are projections of the scene at different camera locations and orientations since the plane is moving.

In most cases, objects alter the background image in a special way thus separate images can be analyzed for spatial features (e.g., colorful cars on the gray street). If the size of the object is known, even a filter tuned for a certain spatial frequency can be used. Since the background may also be textured and it is difficult to link features to form contour, it is more tempting to extract primitive spatial features and evaluate the change of their position in time. This means spatio-temporal analysis of the flow.

First, feature pairs are (1) extracted and (2) matched. Using this point-to-point correspondence, (3) a *global motion model* can be estimated. Finally, (4) this transformation can be calculated for all pixels' points in a frame using interpolation. The first four steps (Fig. 9.2a–d) of the process are called registration (Zitova and Flusser 2003). Since numerous feature pairs can be part of an object, a robust technique is necessary.

An *error measure* can be defined on the intersecting frame regions, and outstanding regions can be detected. Since background regions must fit with small error, extracted regions are objects. This concept works only if the objects cover a small portion of the frame. For a basic solution, necessary steps are summarized in Fig. 9.2.

The first step (a) is FeatureSelection. Feature points are selected from the new frame captured by the sensor (called InputFrame, or $\mathbf{I}_{k+1}$). For the extraction, either $l1$ numbers of foveas are used or the full image is processed. The result is a list of point locations $\{\mathbf{fp}\}$, containing $l2$ elements. Some feature locations are robust so they are selected for tracking: BasePoints $\{\mathbf{bp}\}$. BasePoints used at a given step are derived from $\mathbf{I}_k$.

The second step (b) is FeatureMatching. On the basis of image parts extracted from $\mathbf{I}_k$ from the vicinity of BasePoint locations and on FeaturePoints, a list of vectors is created, called InputPoints. For all elements in BasePoint, a location is

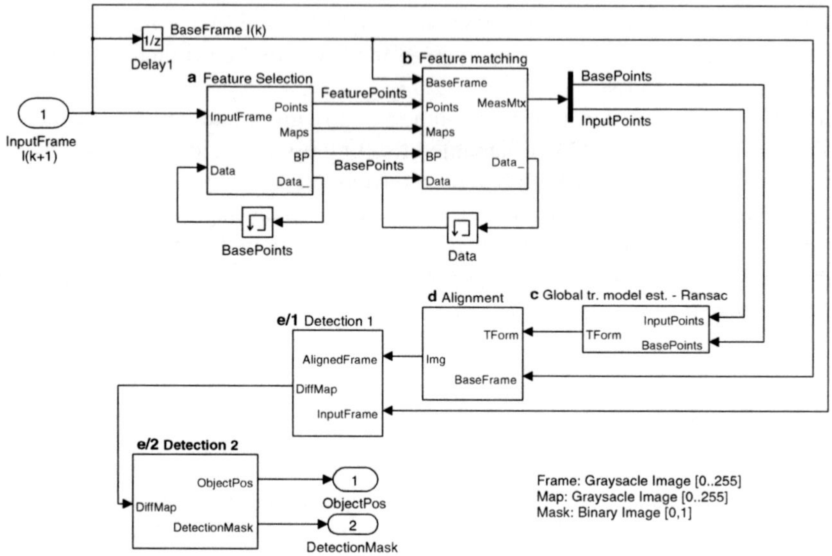

Fig. 9.2 Global registration-based algorithm family

(a) Feature/template selection: locates robust feature point locations on the incoming frame, InputFrame $\mathbf{I}_{k+1}(\mathbf{x})$.

- Some gray scale maps are extracted along with the vector of robust feature point locations, FeaturePoints $\{\mathbf{fp}_j\}$.
- BasePoints or $\{\mathbf{bp}_i\}$ is a list of feature points selected for tracking in frame $\mathbf{I}_k(\mathbf{x})$.

(b) Feature/template matching: matches feature pairs, finds corresponding InputPoints $\{\mathbf{ip}_i\}$ on $\mathbf{I}_{k+1}(\mathbf{x})$ (or select form $\{\mathbf{fp}_j\}$) for all BasePoints.

- $\{\mathbf{ip}_i\} = \text{NULL}$ if $\{\mathbf{bp}_i\}$ is lost.
- $\{mu_i\}$ is also defined holding similarity measure for matching pairs.
- Certain maps are stored for the next frame to support localization.

(c) Global transformation model estimation: estimates transformation on point correspondences.

- A robust transformation matrix $\mathbf{H}_{3\times 3}^{k,k+1}$ is calculated to map points in $\mathbf{bp}\{i\}$ to $\mathbf{ip}\{i\}$

(d) Alignment: calculates transformation for $\mathbf{I}_k(\mathbf{x})$ and interpolates it.

- The full image $\mathbf{I}_k(\mathbf{x})$ is transformed to the coordinate system of $\mathbf{I}_{k+1}(\mathbf{x})$
- The resulting image, AlignedFrame $\mathbf{J}_k(\mathbf{x})$ should be defined for all pixel coordinates, thus inverse mapping is applied with interpolation in the frame of $\mathbf{I}_k(\mathbf{x})$

$$\mathbf{J}_k\left([u,v,1]^T\right) = \mathbf{I}_k\left(\left[\mathbf{H}_{3\times 3}^{k,k+1}\right]^{-1}[u,v,1]^T\right)$$

(e) Detection: e/1; calculates error map – DiffMap(AlignMap).

- DiffMap $(\mathbf{E}(\mathbf{x}))$ is a gray scale image highlighting possible objects. Global registration-based algorithms use AlignMap $(\mathbf{E}_A(\mathbf{x}))$.

$$\mathbf{E}_A(\mathbf{x}) = \text{absdiff}\left(\mathbf{I}_{k+1}(\mathbf{x}) - \mathbf{J}_k(\mathbf{x})\right)$$

e/2; performs segmentation to create DetectionMask.

- The result of the segmentation is a binary mask, ObjectMask.

assigned with a *similarity measure* value (*mu*). If a point is lost, $mu\{i\}$ will be zero; if matching is robust, then $mu\{i\}$ will be equal to one. Matching is done by using $l2$ numbers of foveal windows. Typically, this is the length of $\{\mathbf{bp}\}$ list. The signed difference between $\mathbf{ip}\{i\}$ and $\mathbf{bp}\{i\}$ is the ith displacement vector, $\mathbf{h}\{i\}$. The number of point pairs is $l3$.

Steps (a) and (b) can be done simultaneously (block matching algorithms). The regions around point pairs can be matched. There exists a transformation that maps one region to its corresponding pair in the consecutive frame regarding the chosen error measure. For short-time intervals, even pure displacement can be used as a *local motion model.*

After extracting point features and forming pairs, based on (9.6), a transformation matrix can be linearly estimated using four point-to-point pairs. This is the third step of the algorithm (c). Since points are located with moderate precision in frames, some error arises even for background pairs. If the matrix is used for registering the full image afterward, it is crucial to use more correspondences with some robust fitting technique, for example, RANSAC or Least Median Square. Outliers after the fitting indicate moving objects with high probability.

The BaseFrame can be aligned using the estimated transformation (d). DiffMap is a gray scale description with high pixel vales for suspected object regions. Global registration-based methods calculate an error measure, AlignMap taking the absolute-difference of the Inputframe and the aligned version of the previous frame. For this group of algorithms, DiffMap is defined to be equal to AlignMap. Some methods, however, use an alternative solution for highlighting moving objects.

Since frames have finite resolution, fine features – textures and region boundaries – are mapped to discrete pixels, the exact location depending on the interpolation strategy. This one pixel ambiguity can lead to high registration error around edges. Another reason for possibly high error values is when the underlying assumption on the flat world model is violated. In those cases when an object changes its position between frames, high error values also arise around present and previous silhouette locations. Thus, the analysis of the error map can highlight objects, especially moving ones. This method can identify object boundaries and non-overlapping object parts but not the exact object shape. Therefore, this process is called *moving object detection* as opposed to *object extraction* where the goal is to recover the exact object shape. However, this detection framework is considered to give a focusing mechanism for shape extraction. Foveas can be directed to these regions and further analysis is required to extract the object shape in a more computationally effective way.

If an object is detected in more frames, a tracker can be initialized to describe the motion of the object and possibly to build up a better object shape. Later on, the track can be classified as belonging to a moving or a static object.

In the next four subsections, four different methods will be briefly described. All of them utilize the basic algorithmic concept but focus different amount of computational effort on specific stages of the estimation–detection procedure.

9.5 Corner Pairing Algorithm

One of the most widely used point feature extractors is the Harris Corner Detector (Harris and Stephens 1988). It uses autocorrelation-function to extract locations with a small support region that robustly differ from their neighborhood, that is, have large intensity change in both x and y directions inside their surrounding regions. These feature points are likely to be present in the next frame as well. Corners are extracted from the incoming frame and stored for matching in the next time step. If the support region of a corner in the BaseFrame is similar to a support region in the InputFrame, they are considered as projections form the same 3D region and paired. Feature extraction and matching routines were taken from Torr's toolbox (Torr 2002), which uses the sum of the absolute differences (SAD) as similarity measure for matching.

For constructing correspondence, there exist more sophisticated methods, for example, graph cut (Kolmogorov and Zabih 2002). As an alternative, simple exhaustive search may also be applied with gating based on Manhattan distance to keep complexity low (e.g., three closest corners in $k + 1$th frame are considered for each BasePoint). The exhaustive approach is used.

Since the feature extraction can be done with small neighborhood, it is tempting to do this step on the FPA. Then, for each location in frame k, the support window is extracted and matched with three windows from frame $k + 1$. This step is within the capabilities of a foveal processor. If one matching is stronger than the others and also larger than a predefined constant, the pairing is considered to be successful. Since there is no search (possible locations are predefined), the window size can be equal to the template size. In the comparisons, this algorithm will be referred to as FP.

9.5.1 Block Matching Algorithms

If there is no hardware to support efficient array calculation to estimate autocorrelation for all pixels, larger regions can be handled together. One possibility is to define BasePoints statically as points of a sparse grid without locating feature points and find displacement for support regions centered at grid points. These techniques are called *block matching algorithms* (*BMAs*) or pattern matching algorithms.

A rectangular pattern, that is the *template*, is extracted from $\mathbf{I}_k(\mathbf{x})$ around BasePoint locations and matched against displaced image parts of the same size in $\mathbf{I}_{k+1}(\mathbf{x})$. Since there are no previously determined possible locations, a search is performed in a given range. These possible search locations are displacements with integer values. They can be represented by a similarity map centered around zero displacement. The basic operator of the search is the calculation of the similarity measure between the template and the corresponding image part of a given displacement at every try. In most cases, this measure is the SAD or the sum of the squared differences. If the search radius is large, the brute-force or full search method (*BMA-FS*) with exhaustive search can be outperformed by suboptimal or adaptive methods and solutions such as the Spiral Search, which focuses on smaller displacements at the beginning. They make an effort to keep count of already processed locations when selecting the next one. On the contrary, they calculate less elements of the similarity map than the Brute-force search. BMAs are widely used in video encoding for motion-compensation [MPEG1, MPEG2].

Diamond Search (*BMA-DS*) is one of the preferred adaptive methods. Diamond Search uses two diamond shaped *search patterns*: a *large diamond search pattern* (LDSP; 5×5) and a *small diamond search pattern* (SDSP; 3×3). The similarity measure is calculated at every displacement grid point masked by the actual pattern and registered, thus overlapping possibilities are calculated only once. However, all of them are considered when optimum is chosen for the current step. Search starts with LDSP step which is repeated until the actual optimum is at the center of the mask, when a final SDSP is applied to find the exact solution.

The search needs large template thus the computation cannot be solved by the frontend processor array, it is mapped to the foveal processor.

9.5.2 KLT Algorithm

The *KLT algorithm* is a well-known solution for tracking feature points in a video flow. The basic concept of Lucas-Kanade optical flow calculation was presented in 1981 (Lucas and Kanade 1981) and later extended to track feature points (Shi and Tomasi 1994).

Point features are extracted exploiting the properties of the selected local matching model. In the basic realization, a pure displacement model is used for consecutive frames, although an extension for affine changes also exists.

The template is extracted from the BaseFrame and matched in the new InputFrame. The similarity measure is the (weighted) sum of squared differences for all the pixels of the template. The matching is done with subpixel accuracy; therefore, interpolation is needed.

$$E = \sum_{T} \left[\mathbf{I}_{k+1}(\mathbf{x}+\mathbf{h}) - \mathbf{I}_{k}(\mathbf{x})\right]^2 \tag{9.9}$$

The optimization for the minimal similarity measure is done using zero constraint for the gradient.

$$\begin{aligned} 0 &= \frac{\partial E}{\partial \mathbf{h}} \\ 0 &= \frac{\partial \sum_T \left[\mathbf{I}_{k+1}(\mathbf{x}+\mathbf{h}) - \mathbf{I}_{k}(\mathbf{x})\right]^2}{\partial \mathbf{h}} \end{aligned} \tag{9.10}$$

If h is small, $\mathbf{I}_{k+1}(\mathbf{x}+\mathbf{h})$ may be estimated by its Taylor polynomial.

$$0 = \frac{\partial}{\partial \mathbf{h}} \sum_T \left[\mathbf{I}_{k+1}(\mathbf{x}_i) + \left[\frac{\partial \mathbf{I}_{k+1}}{\partial x}(\mathbf{x}_i) \;\; \frac{\partial \mathbf{I}_{k+1}}{\partial y}(\mathbf{x}_i) \right] \mathbf{h} - \mathbf{I}_k(\mathbf{x}_i) \right]^2 \tag{9.11}$$

$$0 = \sum_T 2 \begin{bmatrix} \frac{\partial I_{k+1}}{\partial x}(\mathbf{x}_i) \\ \frac{\partial I_{k+1}}{\partial y}(\mathbf{x}_i) \end{bmatrix} \left[I_{k+1}(\mathbf{x}_i) - I_k(\mathbf{x}_i) + \left[\frac{\partial I_{k+1}}{\partial x}(\mathbf{x}_i) \;\; \frac{\partial I_{k+1}}{\partial y}(\mathbf{x}_i) \right] \mathbf{h} \right] \tag{9.12}$$

$\mathbf{x}_i$ elements are taken from a rectangular area; therefore, $\mathbf{I}_k(\mathbf{x}_i)$ and $\mathbf{I}_{k+1}(\mathbf{x})$ values can be collected after interpolation to $\mathbf{F}$ and $\mathbf{G}$ matrices, respectively. Using subscript notations "x" and "y" for spatial derivatives and $\bullet$ for element wise product, Eq. (9.12) translates to:

$$\begin{bmatrix} \sum \mathbf{G}_x \bullet \mathbf{G}_x & \sum \mathbf{G}_x \bullet \mathbf{G}_y \\ \sum \mathbf{G}_y \bullet \mathbf{G}_x & \sum \mathbf{G}_y \bullet \mathbf{G}_y \end{bmatrix} \begin{bmatrix} [\mathbf{h}]_1 \\ [\mathbf{h}]_2 \end{bmatrix} = \begin{bmatrix} \sum (\mathbf{F} - \mathbf{G}) \bullet \mathbf{G}_x \\ \sum (\mathbf{F} - \mathbf{G}) \bullet \mathbf{G}_y \end{bmatrix} \tag{9.13}$$

$$\mathbf{Z}_{2\times 2}\mathbf{h}_2 = \mathbf{e}_2 \tag{9.14}$$

This linear equation system can be solved, thus the local optimum can be found for the displacement vector. In order to calculate **h**, **Z** matrix must be invertible. This holds true if both eigen values are large positive numbers. This property is used for selecting good features to track. This feature selection is analogue to Harris corner extraction. The linearization error is moderate only for small displacements; therefore, an image pyramid is created to support coarse-to-fine processing. Furthermore, an iterative search is applied on all levels to handle large displacements.

The pyramid creation can be supported by the FPA, whereas the displacement estimation fits to the FVA.

9.5.3 SIFT Algorithm

SIFT (Lowe 2004) is a state-of-the-art solution for key point matching with two algorithmic steps. It extends the local displacement model with rotation and scale. The first phase extracts a scale invariant point set from Gaussian scale-space, whereas the second phase creates a distinctive descriptor vector that enables highly reliable feature point correspondence matching. This description is quasi invariant to affine transformations and illumination changes. The major drawback of the method is the numerical complexity; thus, it cannot be realized exclusively on serial processors.

First, Gaussian *scale-space pyramid* is generated using a series of convolution of the input image and a Gaussian kernel $\mathbf{G}(x, y, \sigma)$. Parameter σ describes scaling. For consecutive *octaves*, σ of the Gaussian convolution kernel doubles, whereas the effective resolution of the image decreases by half. By resampling every second pixel, a starting image for the next octave is generated. The σ values are selected to span O octaves, with ns subdivisions in each. When the pyramid is ready, filtered images with consecutive scales are subtracted from each other to produce the *difference of Gaussian scale-space* (approximation of the Laplacian of Gaussian operator).

The feature points (key points) are selected from this three-dimensional image stack. A point is selected if it is a local maximum or minimum – depending on whether the luminance of the object was light or dark – of the neighboring ($3 \times 3 \times 3 = 27$) pixel values.

The size of the objects will shrink according to octaves, and due to the subdivisions in scale-space, small zooming effects may be cancelled. The SIFT descriptor is extracted from the vicinity of the key point (template region) in the corresponding scale-map. First, the gradient vectors for all pixels indexed by their magnitude and orientation are calculated, and an orientation histogram with 36 bins is created. To achieve rotation invariance, a transformed template is calculated for all regions by rotating the templates. The amount of rotation is determined by the maximum peak of the weighted histogram to align most edges in vertical direction. Multiple descriptors are created if several significant peaks exist, which increases the robustness. Second, the updated templates are divided into 4×4 subregions, and a 8-bin histogram is calculated from the gradient vectors for each subregion in the same fashion as in the first step resulting in a 128-long vector descriptor for all key points.

Descriptor vectors can be matched with gating on proximity using scalar product as a similarity measure.

9.5.4 Global Registration-Based Detection

InputPoints and BasePoints can be filtered to remove unreliable elements $\{\mathbf{bp}\} \rightarrow \{\mathbf{bpf}\}, \{\mathbf{ip}\} \rightarrow \{\mathbf{ipf}\}$. $l3$ denotes the number of point pairs. After point correspondences are extracted, alignment can be done by searching for optimal transformation. Our global motion model is a projection, which is estimated by the *direct linear transform* (*DLT*) method (Hartley and Zisserman 2000). To make this review self-contained, a brief summary is given.

Equation (9.6) can be rewritten using filtered points:

$$\forall k : \mathbf{ipf}_{3\mathrm{H}}\{k\} \overset{=}{\longleftrightarrow} H^{k+1,k}{}_{3\times 3}\mathbf{bpf}_{3\mathrm{H}}\{k\} \tag{9.15}$$

This mapping is defined on homogenous coordinates, which means that the vectors are not equal but parallel differing in a nonzero scale factor. It is better to emphasize that they are collinear by using cross product.

$$\mathbf{ipf}_{3\mathrm{H}}\{k\} \times H_{3\times 3}^{k+1,k}\mathbf{bpf}_{3\mathrm{H}}\{k\} = 0_3 \tag{9.16}$$

Or using $\mathbf{h}_3^{1T}, \mathbf{h}_3^{1T}$, and $\mathbf{h}_3^{1T}$ notation for rows in $H_{3\times 3}^{k+1,k}$, $[\mathbf{ipf}]_i$ for component of $\mathbf{ipf}_{3\mathrm{H}}\{k\}$, and $\mathbf{bpf}_3$ for $\mathbf{bpf}_{3\mathrm{H}}$

$$\begin{bmatrix} [\mathbf{ipf}]_2 \mathbf{h}_3^{3T} \mathbf{bpf}_3 - [\mathbf{ipf}]_3 \mathbf{h}_3^{2T} \mathbf{bpf}_3 \\ [\mathbf{ipf}]_3 \mathbf{h}_3^{1T} \mathbf{bpf}_3 - [\mathbf{ipf}]_1 \mathbf{h}_3^{3T} \mathbf{bpf}_3 \\ [\mathbf{ipf}]_1 \mathbf{h}_3^{2T} \mathbf{bpf}_3 - [\mathbf{ipf}]_2 \mathbf{h}_3^{1T} \mathbf{bpf}_3 \end{bmatrix} = \begin{bmatrix} 0 \\ 0 \\ 0 \end{bmatrix} \tag{9.17}$$

Furthermore, $\mathbf{h}_3^{1T} \mathbf{bpf}_3 = \mathbf{bpf}_3^T \mathbf{h}_3^1$

$$\begin{bmatrix} \mathbf{0}_3^T & -[\mathbf{ipf}]_3 \mathbf{bpf}_3^T & [\mathbf{ipf}]_2 \mathbf{bpf}_3^T \\ [\mathbf{ipf}]_3 \mathbf{bpf}_3^T & \mathbf{0}_3^T & -[\mathbf{ipf}]_1 \mathbf{bpf}_3^T \\ -[\mathbf{ipf}]_2 \mathbf{bpf}_3^T & [\mathbf{ipf}]_1 \mathbf{bpf}_3^T & \mathbf{0}_3^T \end{bmatrix} \begin{bmatrix} \mathbf{h}_3^1 \\ \mathbf{h}_3^2 \\ \mathbf{h}_3^3 \end{bmatrix} = \begin{bmatrix} 0 \\ 0 \\ 0 \end{bmatrix} \tag{9.18}$$

This gives equations for all corresponding feature pairs. Since the equations are corresponding to homogenous vectors, they are not independent. To solve the system, at least four point pairs are needed. The resulting overdetermined linear system can be solved by using SVD. The singular values comprise **H**.

To make the optimization robust against outliers, *RANdom SAmple Consensus* (*RANSAC*) method (Fischler and Bolles 1981) can be applied. Its concept is to use a minimal set of points selected randomly to determine a transformation and then calculate a score for this selection. The score depends on the number of *inlier*s consistent with the model of this transformation, that is the *symmetric distance measure* is smaller then a threshold limit. In this case, four point pairs are selected. Degenerate point sets with collinear points should be avoided: before running SVD, a test should be performed. The transformation with the largest number of inliers ($l4$) is selected among many tries.

If the probability of belonging to the background for any point pairs is q, the probability that any of the four selected points is part of the foreground can be estimated as:

$$1 - q^4 \tag{9.19}$$

since compared with the number of points, 4 is small.

To be sure to have selected only inliers at least once with, for example, 99% probability, more trials should be evaluated (N).

$$1 - \left(1 - q^4\right)^N > 0.99 \tag{9.20}$$

After estimating the transformation and having lin number of point pairs consistent with the actual best try, we can estimate q using the relative frequency:

$$\tilde{q} = \frac{l3}{\text{lin}} \tag{9.21}$$

Then, it is possible to evaluate (9.20) using the estimate $\tilde{q}$ and decide whether to generate further random sets. In addition, a hard limit for N can be defined to limit the number of iterations.

The best transformation candidate defines the final inlier set. As a last step, a DLT routine can be applied to all of the reliable pairs using first two independent lines of (9.18) to yield the final estimate.

The complexity of the small SVD for all tries is:

$$o\left(9^{*}12^{2}+12^{3}\right)=o\left(3,024\right) \tag{9.22}$$

whereas for the final DLT, step complexity is:

$$o\left(9^{*}\left(2*l4\right)^{2}+\left(2^{*}l4\right)^{3}\right) \tag{9.23}$$

Since this is cubic in the number of used pairs, $l4$, it is limited to 20.

For the implementation, the toolbox by Kovesi was used (Kovesi).

9.5.5 Elastic Grid Multi-Fovea Detector

The calculation of the projective transformation of the global motion model is rather time consuming since a global spatial transformation with interpolation is required. The algorithm described in this section gives an alternative solution by estimating the global transformation with tiles and local displacements. It performs a joint optimization process through coupling of the local displacement estimations utilizing the multi-fovea concept and the possibility of using foveal windows for efficient calculation.

Even projective transformation conserves collinearity: if a point lies on a line defined by others, the points will still be collinear after the transformation. This property can be used to define an adaptive iterative search mechanism.

Elastic contours are popular tools for image processing applications, for example, segmenting noisy images. The contour is built up from segments defined by control points. These points are iteratively moved in the image by a task-specific external force toward an exact segmentation result, whereas internal force balances this effect to keep contours pleasant (e.g., having low curvature). The elastic contour concept may be extended to an *elastic grid*, which could also be viewed as an extension and generalization of the block matching family. In this case, $\{\mathbf{bp}\}$ points are not located feature points but fix points placed along a regular sparse grid. Since they are placed in a 2D topology, they can be naturally indexed with {row, column} indices, $\mathbf{bp}\{k,l\}$.

The algorithm starts with calculating the similarity measure for the template and corresponding region with integer displacements in a given range using normalized SAD. SAD values may be collected into a potential map for all $\{i,j\}$ locations. Searching starts with $[0,0]^T$ displacement. During the search, a 3×3 box search pattern is used.

In all iterations of the elastic grid evolution, for all windows, the missing values are computed from the potential maps selected by the 3×3 search mask centered at current $\mathbf{ip}\{k,l\}$ locations, and the smallest among them is selected to compute

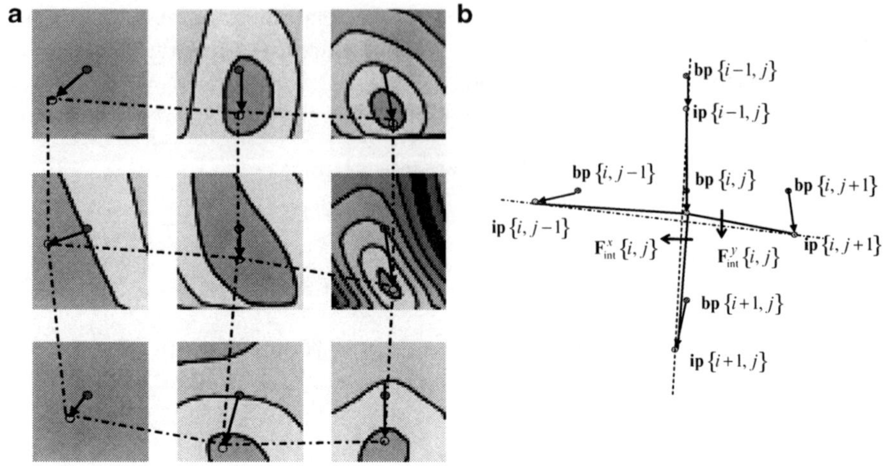

Fig. 9.3 BasePoints are not located but placed on a predefined 2D topology. They are indexed with 2D indices. Templates are extracted around BPs from the BaseFrame and matched against image parts from the InputFrame using the sum of absolute differences as a similarity measure. An elastic grid is defined on InputPoints. Grid starts from zero displacements and converges toward optimal displacement values. SAD values are arranged to form potential maps for external force calculation (**a**). Internal forces are calculated using 2 + 2 connectivity for x and y components (**b**)

the corresponding external force ($\mathbf{F}_{\text{ext}}$). The amplitude is the difference between potential values of the current center and that of the selected location pointing in its direction.

By construction, all $\mathbf{bp}\{k, l\}$ form collinear points with their neighbors and the same must stand for corresponding $\mathbf{ip}\{k, l\}$ points. An elastic grid can be defined in InputPoints as control points, with the internal forces having (2 + 2)-neighbor connectivity (Fig. 9.3). The collinearity constraint translates to the grid being pleasant, if connecting line segments are almost parallel or displacement vectors are close to the average of their neighbors. For calculating x and y components of internal forces ($[\mathbf{F}_{\text{int}}]_x$, $[\mathbf{F}_{\text{int}}]_y$), only data from neighbors in the West and East or in the North and South are used. Components of internal forces are defined as the difference from the sum of the corresponding displacement vector components weighted with their similarity measure:

$$[\mathbf{F}_{\text{int}}]_x\{i,j\} = mu\{i,j\}[\mathbf{h}\{i,j\}]_x - \left(\sum_{k=-1,1} mu\{i+k,j\}[\mathbf{h}\{i+k,j\}]_x\right) \tag{9.24}$$

$$[\mathbf{F}_{\text{int}}]_y\{i,j\} = mu\{i,j\}[\mathbf{h}\{i,j\}]_y - \left(\sum_{k=-1,1} mu\{i,j+k\}[\mathbf{h}\{i,j+k\}]_y\right) \tag{9.25}$$

Depending on the sum of internal and external forces, one neighboring element of the displacement grid is selected for all locations.

The search moves all control points toward smaller error values, but when the distortion of the grid is growing, it is lowered by climbing to a slightly worse location of its potential field.

This joint optimization method can find a good solution for untextured windows with flat potential maps and can find global optima without the need for exhaustive search.

In the elastic grid algorithm, global motion model is not calculated, and no global image alignment is done. Instead, the calculated displacements are applied to all corresponding regions.

A window containing an object with independent motion component would deform the grid that is mainly formed by the background features. It means that after a few iterations, locations with high-amplitude internal force highlight possible object regions.

The multiple displacement model gives a tiled-alignment used for DiffMap calculation that can be analyzed in the same way as for the first four algorithms. Alternatively, only the highlighted regions can be selected for analysis. More details on the algorithm are given in Soos and Rekeczky (2007).

9.6 Performance of Methods

9.6.1 Metrics for Quality

The quality of the algorithmic output can be assessed and compared both at the registration and at the detection level. The overall metrics is defined to take into consideration both aspects.

Registration is described by:

- the edge coverage defined on high-pass-filtered versions of images
$$\left(e_{\text{edge}} = 1 - \frac{||I_{\text{edge}} \cap J_{\text{edge}}||}{\min(||I_{\text{edge}}||, ||J_{\text{edge}}||)}\right)$$
- inlier ratio
- symmetric distance measure

If the global transformation estimation is successful, homogenous regions are perfectly overlapping and a high percent of the boundaries (edges) are covered. A large percent of feature points should be part of the background; thus during optimization, they should turn out to be inliers leading to a small global symmetric distance.

Ground-truth reference was created manually for all frames marking all objects with an independent blob (R_i). DetectionMap is labeled to result in a set of detection blobs (O_i). An object is detected if any detection blob intersects the corresponding reference. The set H contains objects that are detected. P_1 is the set of blobs that overlaps with any reference markings, whereas P_2 is the set of false positive detection patches.

$$e_{Hm1} = \frac{||P_1 \backslash R||}{||P_1||}; \quad 0 \le e_{Hm1} \le 1$$
$$e_{Hm2} = \frac{||P_2||}{||O||}; \quad 0 \le e_{Hm2} \le 1$$
$$e_{Hs} = H \leftrightarrow P_1; \quad 0 \le e_{Hs} \le 1$$

Normalized nonlinear Hausdorff distance

Time complexity of each algorithm is calculated using larger units. Detailed analysis can be found in Soos et al. (2009).

The steps defined in Sect. 9.4 (Fig. 9.2) are refined to functions and functions to elementary building blocks. The flowchart of an algorithm represents the elementary blocks and their connections. Complexity is given for all functions in the corresponding tables. Blocks within a given function are mapped to a common processor. The necessary data is fetched from the global memory and the results are written back in case they are needed for a function that is mapped to another processor or in case they do not fit to local memory.

Topological steps assigned to the frontend processor array or to the foveal processor array can be realized in serial, pipelined, or array hardware components. For more details on efficient implementation of topological operators see Chap. 10.

In case of fully serial solution for any calculations, all operands should be read to registers from local memory and subsequently, all results should be written back. Transfers and operators are considered to consume 1 unit time per pixel. Enough registers should exist to hold intermediate data and constant values during an elementary operation. Indirect memory addressing may be used for processing a full matrix pixel-by-pixel. For this, at least three pointers are needed. Incrementing address does not give extra time overhead. A core with a small number of registers can process all pixels and all the blocks of the flowchart in a serialized order. To store intermediate matrices, local short-term memory is needed.

It is tempting to use overlapping read, calculation, and write, since data can be processed in a well-defined serialized order. For overlapping neighborhoods, it is inefficient to fetch data multiple times. Instead, it is better to use an internal buffer from registers and pump data through it. For each time tick, one element from all input matrices is pushed in and after some delay, one element of the output is produced. The computation is not characterized by execution time but by pixel-delay. If all blocks are realized with independent cores and connected with extra smoothing buffers to equalize uneven delays, the full function can be realized with a pipeline.

In case of fully parallel array processors, images are stored in a distributed manner. All cores have a small portion of multiple images in their registers. Point-by-point arithmetic can be done in one step, whereas it takes a few extra communication steps to calculate a neighborhood operator. To evaluate a function, intermediate images must be stored locally and building blocks are processed in a serialized order.

During comparison, array processor implementation for FPA with one-to-one pixel processor mapping was considered, with enough local memory, supporting point–point arithmetic, gauss filtering, shift, downscaling, and logic operations. The complexity of foveal calculation was multiplied by the number of foveas. Since these parts are fully parallel, the number of physical execution units, and scale factor for execution time of these functions is roughly inversely proportional.

9.7 Comparison

To evaluate the capabilities and performance of the algorithms, output results for four video recordings have been compared. All videos had 240 × 320 pixel resolution. The first sequence is a rendered artificial 3D model, the artificial sequence. Three sequences – Godollo_1, Godollo_2, and Godollo_3 – were captured as part of the ALFA project by a mini-UAV above the airport of Gödöllõ (a city in Hungary). The most robust full-search method is capable of giving reliable frame-to-frame registrations for long time spans. In Fig. 9.4, the table contains some representative frames from the Godollo_2 sequence. To present the registration capability of the algorithm, and the correctness of the global motion model, borders of the aligned frames together with the first and the last edge images are overlaid and shown in Fig. 9.5. For results of the other sequences, the reader is referred to Soos et al. (2009).

After analyzing a large number of frame pairs, we can state that image pairs with e_{edge} – the error measure for full image alignment in high spatial frequency – less than 0.55 can be used to build local background mosaics and to track objects in the ground-based coordinate frame. In the case of larger error, a new mosaic should be started. Local mosaics can be used to detect larger parts of object silhouettes and for object extraction.

If e_{edge} is smaller than 0.7, AlignedFrame can be used for detection without yielding large false positive error.

t=200 t=230 t=250

Fig. 9.4 Representative frames from Godollo_2 sequence

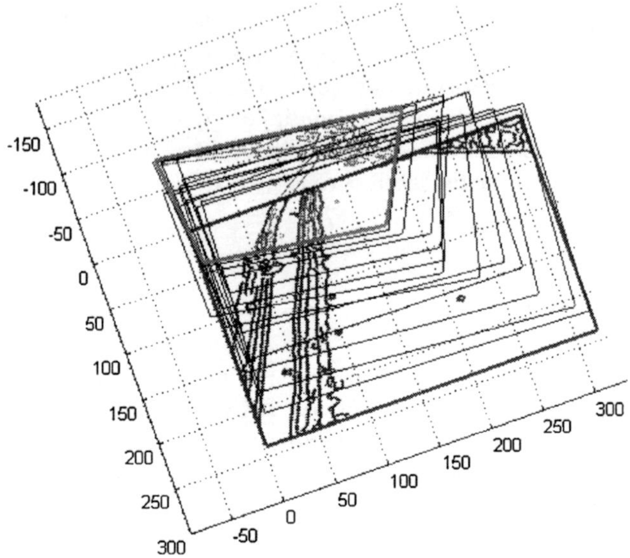

Fig. 9.5 Aligned frames for Godollo_2 sequence. 50 frames are aligned and displayed on the overlay image. The accumulated frame-to-frame registration error is apparently small

Table 9.1 True positive detections of the algorithms

	Artificial 135/130	Godollo_1 120/79	Godollo_2 300/230	Godollo_3 35/31
SIFT	130	52	200	29
Full search	130	58	194	29
Diamond search	128	61	208	29
KLT	130	52	217	29
Feature pairing	125	55	183	27
ELG	92	52	194	28

Total number of frames and frame number on which target '1' is visible are given for the sequences in the header

In the following measurements, quality and computational complexity is analyzed and presented for the Godollo_2 sequence with different parameters. In this case, maximal displacement between consecutive frames was measured as 12 pixels.

For hardware (computational) complexity, the analyzed parameters are the template width and the number of feature detection windows for region-based methods, whereas for the SIFT, it is the number of octaves (O) and intermediate scales (ns).

For detection quality outputs with template width equals to 4 and 80, windows are compared to the case when SHIFT was running on 2 octaves and 2 subscales (Table 9.1). The results show that ELG is characterized by moderate hardware complexity while maintaining competitive detection quality.

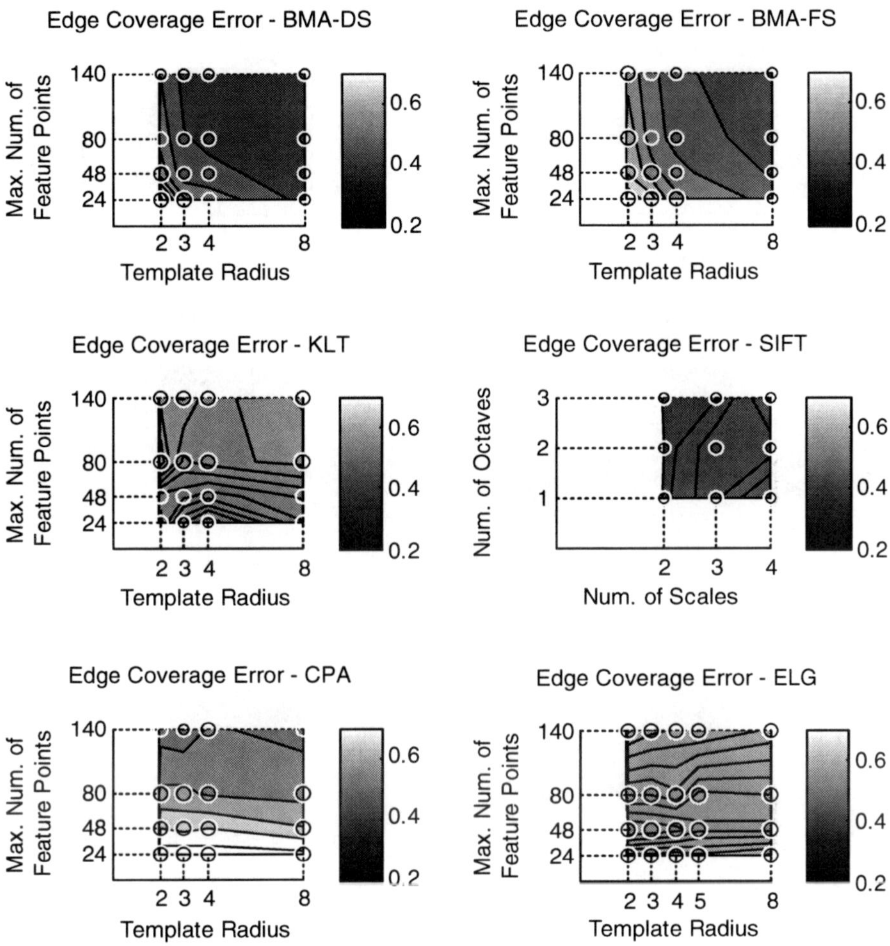

The following table shows comparison for all algorithms regarding the registration capability described by the mean e_{edge} error (e_{edge}: edge overlap ratio for a given frame and the registered pair of it) and the complexity of the calculation projected to operators needed by a serial processor and normalized to the input size.

9.8 Summary

A novel algorithm (Elastic Grid Multi-Fovea Detector) was proposed to utilize the advantages of the generic hardware architecture of the multi-fovea computational framework. This algorithm relies on topologically connected foveal processors (within the Elastic Grid Model) to create a "locally interacting" motion map of the observed field. It was experimentally shown that the multiple displacement

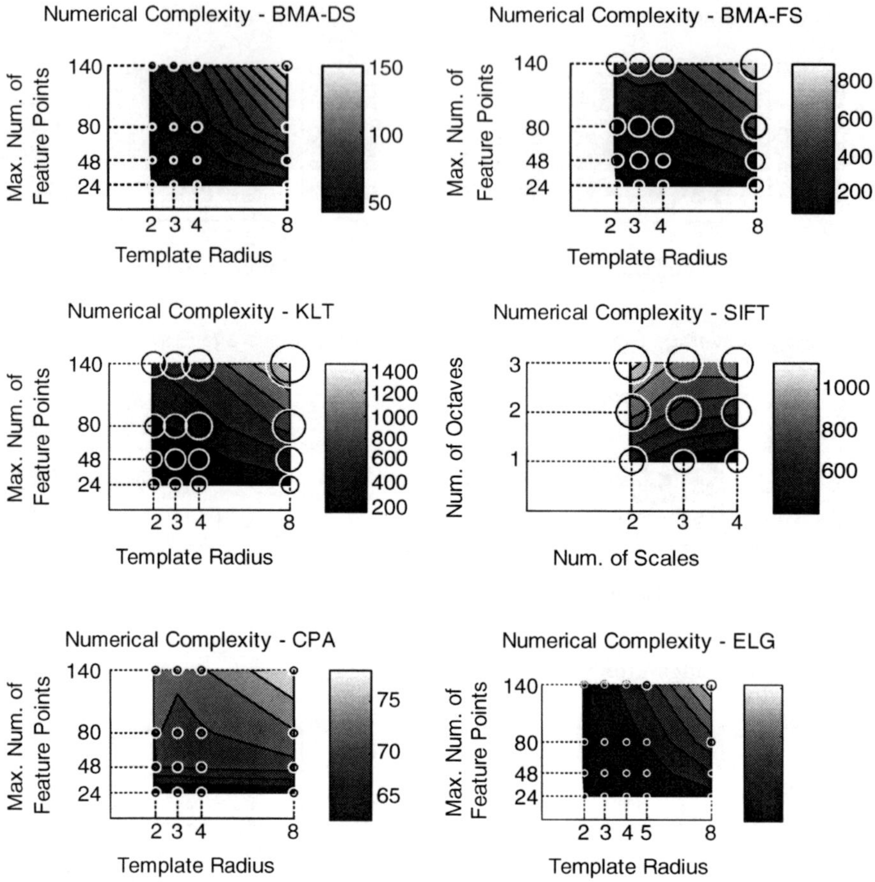

motion model used is appropriate for detecting objects moving on the ground from a mini-UAV. The proposed algorithm was compared with state-of-the art methods highlighting its good output quality and moderate computational complexity.

Appendix A

Complexities for the algorithms are briefly described in the following tables. Functions are described in rows. They can be optimally implemented on the frontend processor array (FPA) or on the foveal processor array (FVA) or on the serial backend processor (BP): one of them is marked. In case of foveal processing, the number of used foveal window is also displayed. The input/output is described using notation S for scalars, and p for points.

At first, complexity for the global registration-based detection part is given (Table 9.2), and then all the algorithms one by one (Tables 9.2–9.7).

Table 9.2 Complexity of the global registration-based detection part of the algorithms (steps c–e)

Detection part of algorithms	FPA	FVA	BP	Fovea	Read/Write	Algorithm Step	STLM	LTLM
c) Global tr. model est.								
RANSAC: for 13 feature pairs 14 of them will turn to be inlier in *N* iterations			X		R-13 × 2p W-14 × 2p	N* 15000		
Estimate: linear estimation for inliers			X		R-14 × 2p W-9S	$\begin{pmatrix} 9 \times (2 \times 14)^2 \\ + \\ (2 \times 14)^3 \end{pmatrix} \times 5$		
d) Alignment								
Transform: transform previous frame			X		R-1As R-9S W-1As	30As		
e) Detection								
Detect: AbsDiff + threshold + morphology	X				R-1As W-15 × p W-1As	ser*: 5As + 10As arr*: 3As	1As	1As (Frame)

Value is depending on the architecture, ser for serial and arr* for full-grain array implementation

Table 9.3 Complexity of feature pairing algorithm, together with global transformation registration-based detection

Feature pairing algorithm	FPA	FVA	BP	Foveas	Read/Write	Algorithm Step	STLM	LTLM
a) Feature/template selection								
ReadCamera: input frame from sensor	X				W-1As	ser*: 89As + 20As		1As (Frame)
Extract: Harris corner extraction	X				W-l1 × p	arr*: 12As + 16As	3As	
b) Feature/template matching								
SelectA: possible pairs with gating			X		R-l1 × p W-l2 × p	$5 \times l1^2$		
Check: correlation check		X		l2	R-2At W-S	5At		
SelectB: keep good pairs			X		R-l2xS R-l2 × p W-l3 × 2p	3 × l2 × 10		
c) Global tr. model est.								
RANSAC: for l3 feature pairs, l4 of them will turn to be inlier in *N* iterations			X		R-l3 × 2p W-l4 × 2p	N * 15,000		
Estimate: linear estimation for inliers			X		R-l4 × 2p W-9S	$\begin{pmatrix} 9 \times (2 \times l4)^2 \\ + \\ (2 \times l4)^3 \end{pmatrix} \times 5$		
d) Alignment								
Transform: transform previous frame			X		R-1As R-9S W-1As	30As		
e) Detection								
Detect: AbsDiff + threshold + morphology	X				R-1As W-l5 × p W-1As	ser*: 5As + 10As arr*: 3As	1As	1As (Frame)

Value is depending on the architecture, ser for serial and arr* for full-grain array implementation

Table 9.4 Complexity of the BMA algorithms

Block matching algorithm	FPA	FVA	BP	Fovea	Read/Write	Algorithm Step	STLM	LTLM
a) Feature/template selection								
ReadCamera: input frame from sensor	X				W-1As			1As (Frame)
Prefilter: autocorrelation filtering in 11 windows on a fixed grid (11 = Mc × Nc)		X		11	R-1At W-2S	[2*4 + 3*4]At + 20	3At	
SelectA: select good locations (12 from 11)			X		R-11 × p W-12 × p	411		
b) Feature/template matching								
CalcDisplacement: AbsDiff for templates with [full,diamond] search, q steps		X		12	R-2Aw W-5S	$2At + q \times 6At$ $+2 \times q^2$	4At 1Aw	
SelectB: select reliable matches (13 from 12)			X		R-12 × 5S W-13 × 2p	412		

For full search, $q = (2r)^2$, for diamond search, $q = 9 + 5r$, where r is the maximal displacement of the video flow. The global transformation-based detection is used in the same way as in the previous algorithms (steps c–e also described in Table 9.2)

Table 9.5 Complexity of the KLT algorithm

KLT algorithm	FPA	FVA	BP	Foveas	Read/Write	Algorithm Step	STLM	LTLM
a) Feature/template selection								
ReadCamera: input frame from sensor	X				W-1As			1As (Frame)
Extract: Harris corner extraction	X				W-11 × p	ser*: 89As + 20As arr*: 12As + 16As	3As	
b) Feature/template matching								
CalcDisplacement: k KLT steps for s scale		X		s × 12	R-2Aw W-5S	$18Aw + 27At + q \times \begin{pmatrix} 18Aw+ \\ 27At+ \\ 32At+ \\ 25 \end{pmatrix}$	9Aw	
SelectB: keep good pairs			X		R-12 × 5S W-13 × 2p	412		

q is typically 5, s is the 2-based-logarithm of r (the maximal displacement of the video flow). The global transformation based detection is used in the same way as in the previous algorithms (steps c–e also described in Table 9.2)

* Value is depending on the architecture, ser* for serial and arr* for full-grain array implementation

Table 9.6 Complexity of the SIFT algorithm

SIFT algorithm	FPA	FVA	BP	Foveas	Read/Write	Algorithm Step	STLM	LTLM
a) Feature/template selection								
ReadCamera: input frame from sensor	X				W-1As			1As (Frame)
Extract: Gauss scale space, differences, 3D local maxima	X				W-11 × p W-O × ns × As	ser*: $\sum_{os=1}^{O} As^{os} \begin{bmatrix} (ns+2)30+ \\ (ns+2)3+ \\ ns \times 2 \times 27 \end{bmatrix}$ arr*: $O \times As \begin{bmatrix} (s+2)2+ \\ (s+2)1+ \\ s \times 2 \times 27 \end{bmatrix}$	5As	1As (Frame)
Descriptor1: create edge histograms, find peaks						$10At + 36 \times 5$		
Descriptor2: rotate		X		11	R-At		3At	
Descriptor3: create descriptors					W-128 × S	$20At + 10At$	128 × S	
b) Feature/template matching								
Match: matching descriptors			X		R-2×11 × 128S W-13 × 2p	$l2^2 \times 128^2$		

O is the number of octaves used, ns is the number of subscales in each. The global transformation-based detection is used in the same way as in the previous algorithms (steps c–e also described in Table 9.2)

* Value is depending on the architecture, ser* for serial and arr* for full-grain array implementation

Table 9.7 Complexity of the elastic grid-based multi-fovea algorithm

ELG Algorithm	FPA	FVA	BP	Iterations, foveas	Read/Write	Algorithm Step	STLM	LTLM
a) Feature/template selection								
ReadCamera: input frame from sensor	X				W-1As			1As (Frame)
Prefilter: autocorrelation filtering in l1 windows on a fixed grid (l1 = Mc × Nc)		X		l1	R-1At W-2S	[2*4 + 3*4]At + 20	3At	
SelectA: select good locations (l2 from l1)			X		R-l1 × p W-l2 × p	4l1		
b) Feature/template matching								
CalcDisplacement: AbsDiff for templates with joined search, q steps with topological interaction		X		q × l2	R-2Aw R-Aw R-4 × 3S W-3S	8At + 40	4At	(Aw)
"Select" l3 := l2			X		R-l2 × 5S W-l3 × 2p	4l2		
c) Global tr. model est.								
d) Alignment								
Transform: transform previous frame			X		R-1As R-l3×3S W-1As	5As		
e) Detection						ser*: 5As + 10As		
Detect: AbsDiff + threshold + morphology	X				R-1As W-l5 × p W-1As		1As	1As (Frame)
						arr*: 3As		

q is typically three times r (the maximal displacement of the video flow). Steps c and d are different from the global registration-based detection
* Value is depending on the architecture, ser* for serial and arr* for full-grain array implementation

Appendix B

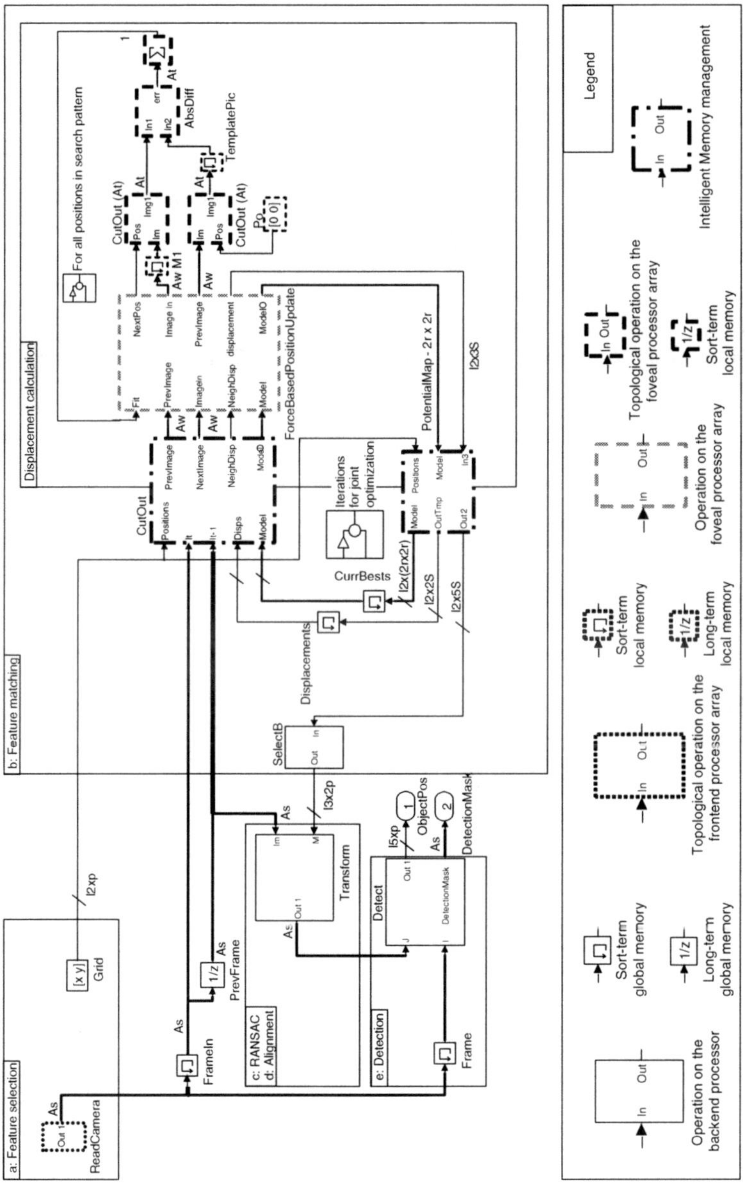

Flowchart of the Elastic Grid algorithm

References

Adiv, Gilad. 1985. Determining three-dimensional motion and structure from optical flow generated by several moving objects. *Pattern Analysis and Machine Intelligence, IEEE Transactions*, PAMI-7, no. 4: 384–401. doi:10.1109/TPAMI.1985.4767678.

Ali, Saad, and Mubarak Shah. 2006. COCOA: tracking in aerial imagery. In *Airborne Intelligence, Surveillance, Reconnaissance (ISR) Systems and Applications III*, 6209:62090D-6. Orlando (Kissimmee), FL, USA: SPIE, May 5. http://link.aip.org/link/?PSI/6209/62090D/1

Argyros, A.A., M.I.A. Lourakis, P.E. Trahanias, and S.C. Orphanoudakis. 1996. Qualitative detection of 3D motion discontinuities. vol. 3, no. 3:1630–1637. doi:10.1109/IROS.1996.569030.

Barron, J.L., D.J. Fleet, S.S. Beauchemin, and T.A. Burkitt. 1992. Performance of optical flow techniques. In *Computer Vision and Pattern Recognition, 1992. Proceedings CVPR '92., 1992 IEEE Computer Society Conference*, 236–242. doi:10.1109/CVPR.1992.223269.

Black, R.J., and A. Jepson. 1994. Estimating multiple independent motions in segmented images using parametric models with local deformations. 220–227. doi:10.1109/MNRAO.1994.346232.

Brown, M., and D.G. Lowe. 2003. Recognising Panoramas. *In Proceedings of the Ninth IEEE International Conference on Computer Vision*, IEEE Computer Society. vol. 2, 1218. http://portal.acm.org/citation.cfm?id = 946247.946772.

Fejes, Sandor, and Larry S. Davis. 1999. Detection of independent motion using directional motion estimation. *Computer Vision and Image Understanding*, 74, no. 2 (May 1): 101–120. doi:10.1006/cviu.1999.0751.

Fischler, Martin A., and Robert C. Bolles. 1981. Random sample consensus: a paradigm for model fitting with applications to image analysis and automated cartography. *Commun. ACM*, 24, no. 6: 381–395. doi:10.1145/358669.358692.

Harris, C., and Stephens, M. 1988. A combined corner and edge detector. In *Proceedings Fourth Alvey Vision Conference*, 147–151. Manchester, UK.

Hartley, Richard, and Andrew Zisserman. 2000. *Multiple view geometry in computer vision*. Cambridge University Press. http://books.google.com/books?hl=en&lr=&id=si3R3Pfa98QC&oi=fnd&pg=PR11&dq=multi+view+geometry&ots=aPo2ktefaM&sig=bIygqBVoMrHq6SnZtgWjrrmnwZ0

Hsieh, J.W. 2004. Fast stitching algorithm for moving object detection and mosaic construction. *Image and Vision Computing*, 22, no. 4: 291–306.

Irani, M., and P. Anandan. 1998. A unified approach to moving object detection in 2D and 3D scenes. *Pattern Analysis and Machine Intelligence, IEEE Transactions*, 20, no. 6: 577–589. doi:10.1109/34.683770.

Jianbo Shi, and C. Tomasi. 1994. Good features to track. *In Computer Vision and Pattern Recognition, 1994. Proceedings CVPR '94., 1994 IEEE Computer Society Conference*, 593–600. doi:10.1109/CVPR.1994.323794.

Kaaniche, K., B. Champion, C. Pegard, and P. Vasseur. 2005. A Vision Algorithm for Dynamic Detection of Moving Vehicles with a UAV. 1878–1883.

Kolmogorov, Vladimir, and Ramin Zabih. 2002. Multi-camera Scene Reconstruction via Graph Cuts. In *Proceedings of the 7th European Conference on Computer Vision-Part III*, 82–96. Springer-Verlag. http://portal.acm.org/citation.cfm?id=756415

Kovesi, P.D. MATLAB *and Octave Functions for Computer Vision and Image Processing*. School of Computer Science and Software Engineering, The University of Western Australia. http://www.csse.uwa.edu.au/~pk/research/matlabfns/

Kumar, R., H. Sawhney, S. Samarasekera, S. Hsu, Hai Tao, Yanlin Guo, K. Hanna, et al. 2001. Aerial video surveillance and exploitation. *Proceedings of the IEEE*, 89, no. 10: 1518–1539.

Lourakis, Manolis I. A., Antonis A. Argyros, and Stelios C. Orphanoudakis. 1998. Independent 3D Motion Detection Using Residual Parallax Normal Flow Fields. 1012–1017. http://citeseer.ist.psu.edu/102877.html

Lowe, David G. 2004. Distinctive image features from scale-invariant keypoints. *International Journal of Computer Vision, 60*, no. 2 (November 1): 91–110. doi:10.1023/B:VISI.0000029664.99615.94.

Lucas, B.D, and T. Kanade. 1981. An Iterative image registration technique with an application to stereo vision. In *International Joint Conference on Artificial Intelligence*, 674–679. Vancouver. http://citeseer.ist.psu.edu/lucas81iterative.html

Mikolajczyk, Krystian, and Cordelia Schmid. 2005. A performance evaluation of local descriptors. *IEEE Trans. Pattern Anal. Mach. Intell. 27*, no. 10: 1615–1630.

Molinier, Matthieu, Tuomas Häme, and Heikki Ahola. 2005. 3D-connected components analysis for traffic monitoring in image sequences acquired from a helicopter. In Image Analysis, 141–150. http://dx.doi.org/10.1007/11499145_16

Morse, B.S., D. Gerhardt, C. Engh, M.A. Goodrich, N. Rasmussen, D. Thornton, and D. Eggett. 2008. Application and evaluation of spatiotemporal enhancement of live aerial video using temporally local mosaics. In *Computer Vision and Pattern Recognition, 2008. CVPR 2008. IEEE Conference on*, 1–8.

Pless, R., T. Brodsky, and Y. Aloimonos. 2000. Detecting independent motion: the statistics of temporal continuity. *Pattern Analysis and Machine Intelligence, IEEE Transactions*, 22, no. 8: 768–773.

Rekeczky, C., I. Szatmari, D. Balya, G. Timar, and A. Zarandy. 2004. Cellular multiadaptive analogic architecture: a computational framework for UAV applications. *Circuits and Systems I: Regular Papers, IEEE Transactions on [Circuits and Systems I: Fundamental Theory and Applications, IEEE Transactions]* 51, no. 5: 864–884. doi:10.1109/TCSI.2004.827629.

Sawhney, H.S., Y. Guo, and R. Kumar. 2000. Independent motion detection in 3D scenes. *Pattern Analysis and Machine Intelligence, IEEE Transactions, 22*, no. 10: 1191–1199.

Sawhney, H.S., and R. Kumar. 1999. True multi-image alignment and its application to mosaicing and lens distortion correction. *Pattern Analysis and Machine Intelligence, IEEE Transactions*, 21, no. 3: 235–243. doi:10.1109/34.754589.

Shan Zhu, and Kai-Kuang Ma. 2000. A new diamond search algorithm for fast block-matching motion estimation. *Image Processing, IEEE Transactions*, 9, no. 2: 287–290. doi:10.1109/83.821744.

Soos, B.G., and C. Rekeczky. 2007. Elastic grid based analysis of motion field for object-motion detection in airborne video flows. In *Circuits and Systems, ISCAS 2007. IEEE International Symposium*, 617–620. doi:10.1109/ISCAS.2007.378813.

Soos, B.G., V. Szabo, and C. Rekeczky. 2009. *Multi-Fovea Architecture and Algorithms for Real-Time Object-Motion Detection in Airborne Surveillance: Comparative Analysis (Technical Report)*. Budapest, Hungary: Pazmány Peter Catholic University.

Szeliski, Richard. 2006. Image alignment and stitching: a tutorial. *Found. Trends. Comput. Graph. Vis.*, 2, no. 1: 1–104.

Torr, PHS. 2002. *A Structure and Motion Toolkit in Matlab: Interactive adventures in S and M*. Microsoft Research.

Weiming Hu, Tieniu Tan, Liang Wang, and S. Maybank. 2004. A survey on visual surveillance of object motion and behaviors. *Systems, Man, and Cybernetics, Part C: Applications and Reviews, IEEE Transactions*, 34, no. 3: 334–352.

Yilmaz, Alper, Omar Javed, and Mubarak Shah. 2006. Object tracking: A survey. *ACM Comput. Surv.*, 38, no. 4: 13. doi:10.1145/1177352.1177355.

Zhang, Zhengyou. 1998. Determining the epipolar geometry and its uncertainty: A review. *Int. J. Comput. Vision*, 27, no. 2: 161–195.

Zhigang Zhu, Hao Tang, G. Wolberg, and J.R. Layne. 2005. Content-based 3D mosaic representation for video of dynamic 3D scenes. In *Applied Imagery and Pattern Recognition Workshop, 2005. Proceedings. 34th*, 6 pp. 203. doi:10.1109/AIPR.2005.25.

Zitova, Barbara, and Jan Flusser. 2003. Image registration methods: a survey. *Image and Vision Computing*, 21, no. 11 (October): 977–1000. doi:10.1016/S0262–8856(03)00137–9.

Chapter 10
Low-Power Processor Array Design Strategy for Solving Computationally Intensive 2D Topographic Problems

Ákos Zarándy and Csaba Rekeczky

Abstract 2D wave type topographic operators are distributed into six classes, based on their implementation methods on different low-power many-core architectures. The following architectures are considered: (1) pipe-line architecture, (2) coarse-grain cellular parallel architecture, (3) fine-grain fully parallel cellular architecture with discrete time processing, (4) fine-grain fully parallel cellular architecture with continuous time processing, and (5) DSP-memory architecture as a reference. Efficient implementation methods of the classes are shown on each architecture. The processor utilization efficiencies, as well as the execution times, and the major constrains are calculated. On the basis of the calculated parameters, an optimal architecture can be selected for a given algorithm.

10.1 Introduction

Cellular neural/nonlinear networks (CNN) were invented in 1988 (Chua and Yang 1988). This new field attracted well beyond hundred researchers in the next two decades, called nowadays the CNN community. They focused on three main areas: the theory, the implementation issues, and the application possibilities. In the implementation area, the first 10 years yielded more than a dozen CNN chips made by only a few designers. Some of them followed the original CNN architecture (Cruz et al. 1994), others made slight modifications, such as the full signal range model (Espejo et al. 1996; Liñan-Cembrano et al. 2003), or discrete time CNN (DTCNN) (Harrer et al. 1994), or skipped the dynamics, and made dense threshold logic in the black-and-white domain only (Paasio et al. 1997). All of these chips had

Á. Zarándy (✉)
Computer and Automation Research Institute of the Hungarian Academy of Sciences
13-17 Kende Street, Budapest, H-1111, Hungary
e-mail: zarandy@sztaki.hu

Cs. Rekeczky
Eutecus Inc., Berkeley, California

C. Baatar et al. (eds.), *Cellular Nanoscale Sensory Wave Computing*, 215
DOI 10.1007/978-1-4419-1011-0_10,

cellular architecture, and implemented the programmable **A** and/or the **B** template matrices of the CNN Universal Machine (Roska and Chua 1993; Chua et al. 1996).

In the second decade, this community slightly shifted the focus of chip implementation. Rather than implementing classic CNN chips with **A** and **B** template matrices, the new target became the efficient implementation of neighborhood processing. Some of these architectures were topographic with different pixel/processor ratios, others were nontopographic. (The notion topographic describes the processor arrangement with respect to the sensor pixel.) Some implementations used analog processors and memories, others digital ones. Certainly, the different architectures had different advantages and drawbacks. One of the goals is to compare these architectures and the actual chip implementations themselves. This attempt is not trivial because their parameter and operation gamut is rather different. To solve this problem, we have categorized the most important 2D wave type operations and examined their implementation methods and efficiency on these architectures.

This study compares the following five architectures, of which the first one is used as the reference of comparison.

1. DSP-memory architecture [in particular DaVinci processors from TI (www.ti.com)];
2. Pipe-line architecture [CASTLE (Keresztes et al. 1999), Falcon (Nagy and Szolgay 2003)];
3. Coarse-grain cellular parallel architecture [Xenon (Foldesy and Zarándy 2008)];
4. Fine-grain fully parallel cellular architecture with discrete time processing [SCAMP (Dudek et al. 2006), Q-Eye (www.anafocus.com)];
5. Fine-grain fully parallel cellular architecture with continuous time processing [ACE-16k (Liñan-Cembrano et al. 2003), ACLA (Dudek 2006)].

On the basis of the result of this analysis, the major implementation parameters (which appeared to be the constrained) of the different architectures for each operation class were identified. These parameters are the resolution, frame-rate, latency, pixel clock, computational demand, flowchart types, power consumption, volume, and design economy. Having these constraints, an optimal architecture can be selected to a given algorithm. The algorithm selection method is described.

The chapter starts with the brief description of the different architectures, which is followed by the categorization of the 2D operators and their implementation methods on them. Then the major parameters of the implementations are compared. Finally, the optimal architecture selection method is introduced.

10.2 Architecture Descriptions

This section describes the architectures examined using the basic spatial gray scale and binary functions (convolution, erosion) of nonpropagating type.

10.2.1 Classic DSP-Memory Architecture

Here we assume a 32-bit DSP architecture with cache memory large enough to store the required number of images and the program internally. In this way, we have to practically estimate/measure the required DSP operations. Most of the modern DSPs have numerous MACs and ALUs. To avoid comparing these DSP architectures, which would lead too far from our original topic, we use the DaVinci video processing DSP by Texas Instrument, as a reference.

We use 3×3 convolution as a measure of gray scale performance. The data requirements of the calculation are 19 bytes (9 pixels, 9 kernel values, result); however, many of these data can be stored in registers, hence, only as an average of a four-data access (3 inputs, because the 6 other ones had already been accessed in the previous pixel position, and one output) is needed for each convolution. From computational point of view, it needs nine multiple-add (MAC) operations. It is very typical that the 32 bit MACs in a DSP can be split into four 8 bit MACs, and other auxiliary ALUs help loading the data to the registers in time. Measurement shows that, for example, the Texas DaVinci family with the TMS320C64x core needs only about 1.5 clock cycles to complete a 3×3 convolution.

The operands of the binary operations are stored in 1 bit/pixel format, which means that each 32 bit word represents a 32×1 segment of an image. Since the DSP's ALU is a 32 bit long unit, it can handle 32 binary pixels in a single clock cycle. As an example, we examine how a 3×3 square-shaped erosion operation is executed. In this case, erosion is a nine input OR operation where the inputs are the binary pixels values within the 3×3 neighborhood. Since the ALU of the DSP does not contain 9 input OR gate, it is executed sequentially on an entire 32×1 segment of the image. The algorithm is simple: the DSP has to prepare the nine different operands, and apply bit-wise OR operations on them.

Figure 10.1 shows the generation method of the first three operands. In the figure, a 32×3 segment of a binary image is shown (9 times), as it is represented in the DSP memory. Some fractions of horizontal neighboring segments are also shown. The first operand can be calculated by shifting the upper line with one bit position to the left and filling in the empty MSB with the LSB of the word from its right neighbor. The second operand is the unshifted upper line. The position and the preparation of the remaining operands are also shown in Fig. 10.1a.

This means that we had to apply 10 memory accesses, 6 shifts, 6 replacements, and 8 OR operations to execute a binary morphological operation for 32 pixels. Because of the multiple cores and the internal parallelism, the Texas DaVinci spends 0.5 clock cycles with the calculation of one pixel.

In the low-power low-cost embedded DSP technology, the trend is to further increase the clock frequency, but most probably, not higher than 1 GHz, otherwise, the power budget cannot be kept. Moreover, the drawback of these DSPs is that their cache memory is too small, which cannot be significantly increased without significant cost rise. The only way to significantly increase the speed is to implement a

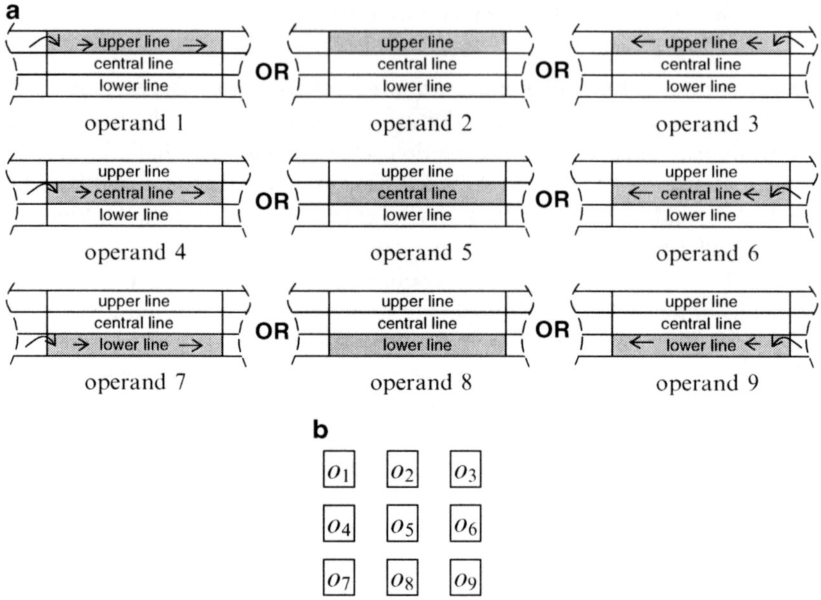

c

$$e_1 = o_1 \text{ OR} o_2 \text{ OR} o_3 \text{ OR} o_4 \text{ OR} o_5 \text{ OR} o_6 \text{ OR} o_7 \text{ OR} o_8 \text{ OR} o_9$$

Fig. 10.1 Illustration of the binary erosion operation on a DSP. (**a**) shows the nine pieces of 32 × 1 segments of the image (operands), as the DSP uses them. The operands are the shaded segments. The arrows indicate shifting of the segments. To make it clearer, consider a 3 × 3 neighborhood as it is shown in (**b**). For one pixel, the form of the erosion calculation is shown in (**c**). $o_1, o_2, \ldots o_9$ are the operands. The DSP does the same, but on 32 pixels parallel

larger number of processors, however, that requires a new way of algorithmic thinking, and software tools.

The DSP-memory architecture is the most versatile from the point of views of both in functionality and programmability. It is easy to program, and there is no limit on the size of the processed images, though it is important to mention that in case of an operation is executed on an image stored in the external memory, its execution time is increasing roughly with an order of magnitude. Although the DSP-memory architecture is considered to be very slow, as it is shown later, it outperforms even the processor arrays in some operations. In QVGA frame size, it can solve quite complex tasks, such as video analytics in security applications on video rate (www.objectvideo.com). Its power consumption is in the 1–3 W range. Relatively small systems can be built by using this architecture. The typical chip count is around 16 (DSP, memory, flash, clock, glue logic, sensor, 3 near sensor components, 3 communication components, 4 power components), whereas this can be reduced to the half in a very basic system configuration.

10.2.2 Pipe-Line Architectures

Here we consider a general digital pipe-line architecture with one processor core per image line arrangement. The basic idea of this pipe-line architecture is to process the images line-by-line, and to minimize both the internal memory capacity and the external IO requirements. Most of the early image processing operations are based on 3×3 neighborhood processing; hence, nine image data are needed to calculate each new pixel value. However, these nine image data would require very high data throughput from the device. As we will see, this requirement can be significantly reduced by applying a smart feeder arrangement.

Figure 10.2 shows the basic building blocks of the pipe-line architecture. It contains two parts, the memory (feeder) and the neighborhood processor. Both the feeder and the neighborhood processor can be configured 8 or 1 bit/pixel wide, depending on whether the unit is used for gray scale or binary image processing. The feeder contains, typically, two consecutive whole rows and a row fraction of the image. Moreover, it optionally contains two more rows of the mask image, depending on the input requirements of the implemented neighborhood operator. In each pixel clock period, the feeder provides 9 pixel values for the neighborhood processor and the mask value optionally if the operation requires it. The neighborhood processor can perform convolution, rank order filtering, or other linear or nonlinear spatial filtering on the image segment in each pixel clock period. Some of these operators (e.g., *hole finder*, or a *CNN* emulation with A and B templates) require two input images. The second input image is stored in the mask. The outputs of the unit are the resulting and, optionally, the input and the mask images. Note that the unit receives and releases synchronized pixels flows sequentially. This enables to cascade

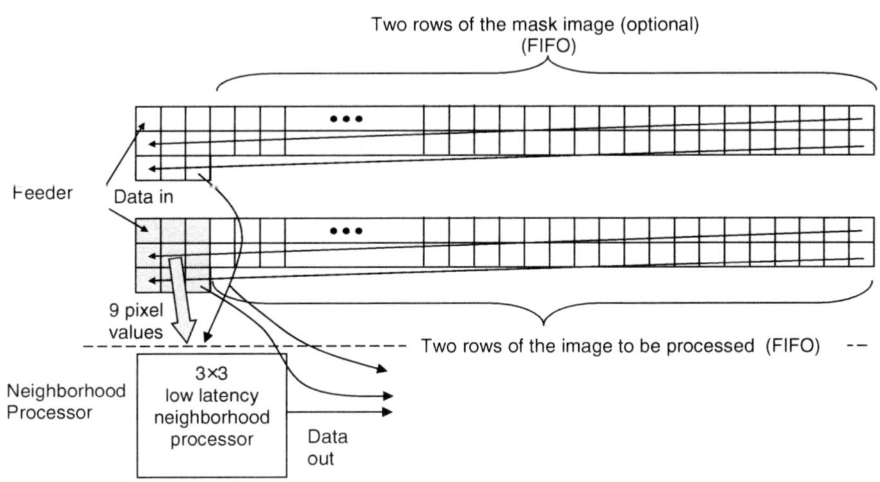

Fig. 10.2 One processor and its memory arrangement in the pipe-line architecture

multiple pieces of the described units. The cascaded units form a chain. In such a chain, only the first and the last units require external data communications, the rest of them receives data from the previous member of the chain and releases the output toward the next one.

An advantageous implementation of the row storage is the application of FIFO memories, where the first three positions are tapped to be able to provide input data for the neighborhood processor. The last positions of rows are connected to the first position of the next row (Fig. 10.2). In this way, pixels in the upper rows are automatically marching down to the lower rows.

The neighborhood processor is of special purpose, which can implement one or a few different kinds of operators with various attributes and parameter. They can implement convolution, rank-order filters, gray scale or binary morphological operations, or any local image processing functions (e.g., Harris corner detection, Laplace operator, gradient calculation). In architectures CASTLE (Keresztes et al. 1999) and Falcon (Nagy and Szolgay 2003), for example, the processors are dedicated to convolution processing where the template values are the attributes. The pixel clock is matched with that of the applied sensor. In case of a 1 megapixel frame at video rate (30 FPS), the pixel clock is about 30 MHz (depending on the readout protocol). This means that all parts of the unit should be able to operate minimum on this clock frequency. In some cases, the neighborhood processor operates on an integer multiplication of this frequency because it might need multiple clock cycles to complete a complex calculation, such as a 3×3 convolution. Considering ASIC or FPGA implementations, clock frequency between 100–300 MHz is a feasible target for the neighborhood processors within tolerable power budget.

The multicore pipe-line architecture is built up from a sequence of such processors. The processor arrangement follows the flowchart of the algorithm. In case of multiple iterations of the same operation, we need to apply as many processor kernels as many iterations we need. This easily ends up in using a few dozens of kernels. Fortunately, these kernels, especially in the black-and-white domain, are relatively inexpensive, either on silicon or in FPGA.

Depending on the application, the data-flow may contain either sequential segments or parallel branches. It is important to emphasize, however, that the frame scanning direction cannot be changed, unless the whole frame is buffered, which can be done in external memory only. Moreover, the frame buffering introduces relatively long (dozens of millisecond) additional latency.

For capability analysis, here we use the Spartan 3ADSP FPGA (XC3SD3400A) from Xilinx (www.xilinx.com) as a reference because this low-cost, medium-performance FPGA was designed especially for embedded image processing. It is possible to implement roughly 120 gray scale processors within this chip, as long as the image row length is below 512, or 60 processors, when the row length is between 512 and 1024.

10.2.3 Coarse-Grain Cellular Parallel Architectures

The coarse-grain architecture is a truly locally interconnected 2D cellular processor arrangement, as opposed to the pipe-line one. A specific feature of the coarse-grain parallel architectures is that each processor cell is topographically assigned to a number of pixels (e.g., an 8×8 segment of the image), rather than to a single pixel only. Each cell contains a processor and some memory, which is large enough to store few bytes for each pixel of the allocated image segment. Exploiting the advantage of the topographic arrangement, the cells can be equipped with photo sensors enabling to implement a single chip sensor-processor device. However, to make this sensor sensitive enough, which is the key in high frame-rate applications, and to keep the pixel density of the array high, at the same time, certain vertical integration techniques are needed for photosensor integration.

In the coarse-grain architectures, each processor serves a larger number of pixels; hence, we have to use more powerful processors than in the one pixel per processor architectures. Moreover, the processors have to switch between serving pixels frequently; hence, more flexibility is needed that an analog processor can provide. Therefore, it is more advantageous to implement 8 bit digital processors, whereas the analog approach is more natural in the one pixel per processor (fine-grain) architectures (See Sect. 10.2.4).

As it can be seen in Fig. 10.3, Xenon chip is constructed in an 8×8, locally interconnected cell arrangement. Each cell contains a subarray of 8×8 photosensors, an analog multiplexer, an 8 bit AD converter, an 8 bit processor with 512

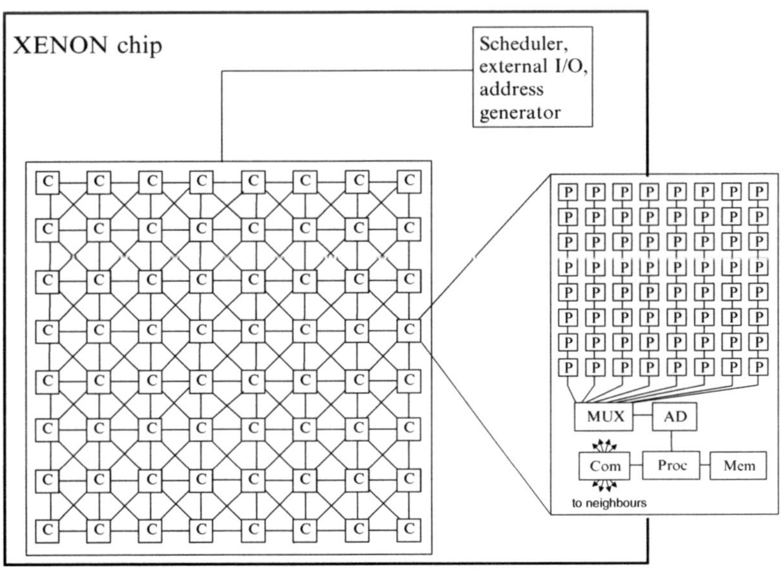

Fig. 10.3 Xenon is a 64 core coarse-grain cellular parallel architecture (C stands for processor cores, whereas P represents pixels)

bytes of memory, and a communication unit of local and global connections. The processor can handle images in 1, 8, and 16 bit/pixel representations; however, it is optimized for 1 and 8 bit/pixel operations. Each processor can execute addition, subtraction, multiplication, multiply–add operations, comparison, in a single clock cycle on 8 bit/pixel data. It can also perform 8 logic operations on 1 bit/pixel data in packed-operation mode in a single cycle. Therefore, in binary mode, one line of the 8×8 subarray is processed jointly, similarly to the way we have seen in the DSP. However, the Xenon chip supports the data shifting and swapping from hardware, which means that the operation sequence, what we have seen in Fig. 10.1, takes 9 clock cycles only. (The swapping and the accessing the memory of the neighbors do not need extra clock cycles.) Besides, the local processor core functions, Xenon can also perform a global OR function. The processors in the array are driven in a single instruction multiple data (SIMD) mode.

Xenon is implemented on a 5×5 mm silicon die with 0.18 μm technology. The clock cycle can go up to 100 MHz. The layout is synthesized; hence the resulting 75 μm equivalent pitch is far from being optimal. It is estimated that through aggressive optimization, it could be reduced to 40 μm (assuming a bump bonded sensor layer), which would make almost double the resolution achievable on the same silicon area. The power consumption of the existing implementation is under 20 mW.

10.2.4 Fine-Grain Fully Parallel Cellular Architectures with Discrete Time Processing

The fine-grain, fully parallel architectures are based on rectangular processor grid arrangements where the 2D data (images) are topographically assigned to the processors. The key feature here is that there is a one-to-one correspondence between the pixels and the processors. This certainly means that at the same time the composing processors can be simpler and less powerful, than in the previous, coarse-grain case. Therefore, fully parallel architectures are typically implemented in analog domain, though bit-sliced digital approach is also feasible.

In the discussed cases, the discrete time-processing type fully parallel architectures are equipped with a general purpose, analog processor, and an optical sensor in each cell. These sensor-processors can handle two types of data (image) representations: gray scale and binary. The instruction set of these processors include addition, subtraction, scaling (with a few discrete factors only), comparison, thresholding, and logic operations. Since it is a discrete time architecture, the processing is clocked. Each operation takes 1–4 clock cycles. The individual cells can be masked. Basic spatial operations, such as convolution, median filtering, or erosion, can be put together as sequences of these elementary processor operations. In this way, the clock cycle counts of a convolution, a rank order filtering, or a morphological filter are between 20 and 40 depending on the number of weighting coefficients.

It is important to note that in case of the discrete time architectures (both coarse- and fine-grain), the operation set is more elementary (lower level) than on the continuous time cores (see Sect. 10.2.5). While in the continuous time case (CNN like processors), the elementary operations are templates (convolution, or feedback convolution) (Roska and Chua 1993); in the discrete time case, the processing elements can be viewed as RISC (reduced instruction set) processor cores with addition, subtraction, scaling, shift, comparison, and logic operations. When a full convolution is to be executed, the continuous time architectures are more efficient. While in the case of operations when both architectures apply a sequence of elementary instructions in an iterative manner (e.g., rank order filters), the RISC is the superior because its elementary operators are more versatile, more accurate, and faster.

The internal analog data representation has both architectural and functional advantages. From architectural point of view, the most important feature is that no AD converter is needed on the cell level, because the sensed optical image can be directly saved in the analog memories, leading to significant silicon space savings. Moreover, the analog memories require smaller silicon area than the equivalent digital counterparts. From the functional point of view, the topographic analog and logic data representations make the implementation of efficient diffusion, averaging, and global OR networks possible.

The drawback of the internal analog data representation and processing is the signal degradation during operation or over time. According to experience, accuracy degradation was more significant in the old ACE16k design (Liñan-Cembrano et al. 2003) than in the recent Q-Eye chip (www.anafocus.com) or SCAMP (Dudek et al. 2006) ones. While in the former case 3–5 gray scale operations led to significant degradations, in the latter ones even 10–20 gray scale operations can conserve the original image features. This makes it possible to implement complex nonlinear image processing functions (e.g., rank order filter) on discrete time architectures, whereas it is practically impossible on the continuous ones (ACE16k).

The two representatives of discrete time solutions, SCAMP and Q-Eye, are slightly similar in design. The SCAMP chip was fabricated by using 0.35 μm technology. The cell array size is 128×128. The cell size is $50 \times 50\,\mu$m, and the maximum power consumption is about 200 mW at 1.25 MHz clock rate. The array of Q-Eye chip has 144×176 cells. It was fabricated on 0.18 μm technology. The cell size is about $30 \times 30\,\mu$m. Its speed and power consumption range is similar to that of the SCAMP chip. Both SCAMP and Q-Eye chips are equipped with single-step mean, diffusion, and global OR calculator circuits. Q-Eye chip also provides hardware support to single-step binary 3×3 morphological operations.

10.2.5 Fine-Grain Fully Parallel Cellular Architecture with Continuous Time Processing

Fully parallel cellular continuous time architectures are based on arrays of spatially interconnected dynamic asynchronous processor cells. Naturally, these architectures exhibit fine-grain parallelism, to be able to perform continuous time spatial waves physically in the continuous value electronic domain. Since these are very carefully optimized, special purpose circuits, they are superefficient in computations they were designed to. We have to emphasize, however, that they are not general purpose image processing devices. Here we mainly focus on two designs. Both of them can generate continuous time spatial-temporal propagating waves in a programmable way. While the output of the first one [ACE-16k (Liñan-Cembrano et al. 2003)] can be in the gray scale domain, the output of the second one [ACLA (Dudek 2006; Lopich and Dudek 2007)] is always in the binary domain.

The ACE-16k (Liñan-Cembrano et al. 2003) is a classical CNN Universal Machine type architecture equipped with feedback and feed-forward template matrices (Roska and Chua 1993), sigmoid type output characteristics, dynamically changing state, optical input, local (cell level) analog and logic memories, local logic, diffusion and averaging network. It can perform full-signal range type CNN operations (Espejo et al. 1996). Therefore, it can be used in retina simulations or other spatial-temporal dynamical system emulations, as well. Its typical feed-forward convolution execution time is in the 5–8 μs range, whereas the wave propagation speed from cell-to-cell is up to 1 μs. Although its internal memories, easily reprogrammable convolution matrices, logic operations, and conditional execution options make it attractive to use as a general purpose high-performance sensor-processor chip for the first sight, its limited accuracy, large silicon area occupation ($\sim$80 $\times$ 80 μm/cell on 0.35 μm 1P5M STM technology), and high-power consumption (4–5 W) prevent the immediate usage in various vision application areas.

The other architecture in this category is the Asynchronous Cellular Logic Array (ACLA) (Dudek 2006; Lopich and Dudek 2007). This architecture is based on spatially interconnected logic gates with some cell level asynchronous controlling mechanisms that allow ultra high-speed spatial binary wave propagation only. Typical binary functionalities implemented on this network are *trigger wave*, *reconstruction*, *hole finder*, *shadow*, etc. Assuming more sophisticated control mechanism on the cell level, it can even perform *skeletonization* or *centroid* calculations. Their implementation is based on a few minimal size logic transistors, which makes them hyperfast, extremely small, and power-efficient. They can reach 500 ps/cell wave propagation speed, with 0.2 mW power consumption for a 128 $\times$ 128 sized array. Their very small area requirement (16 $\times$ 8 μm/cell on 0.35 μm 3M1P AMS technology) makes them a good choice to be implemented as a coprocessor in any fine-grain array processor architecture.

10.3 Implementation and Efficiency Analysis of Various Operators

On the basis of the implementation methods, in this section, we introduce a new 2D operator categorization. Then, the implementation methods on different architectures are described and analyzed from the efficiency aspect.

Here we examine only the 2D single-step neighborhood operators, and the 2D, neighborhood-based wave-type operators. The more complex, but still local operators (such as Canny edge detector) can be built up by using these primitives, whereas other operators (such as Hough or Fourier transform) require global processing, which is not supported by these architectures.

10.3.1 Categorization of 2D Operators

The calculation methods of different 2D operators, due to their different spatial-temporal dynamics, require different computational approaches. The categorization (Fig. 10.4) was done according to their implementation methods on different architectures. It is important to emphasize that we categorize operators (functionalities) here, rather than wave types, because the wave types are not necessarily inherited in the operator itself, but in its implementation method on a particular architecture. As we will see, the same operator is implemented with different spatial wave dynamic patterns on different architectures. The most important 2D operators including all the CNN operators (Zarándy 1999) are considered here.

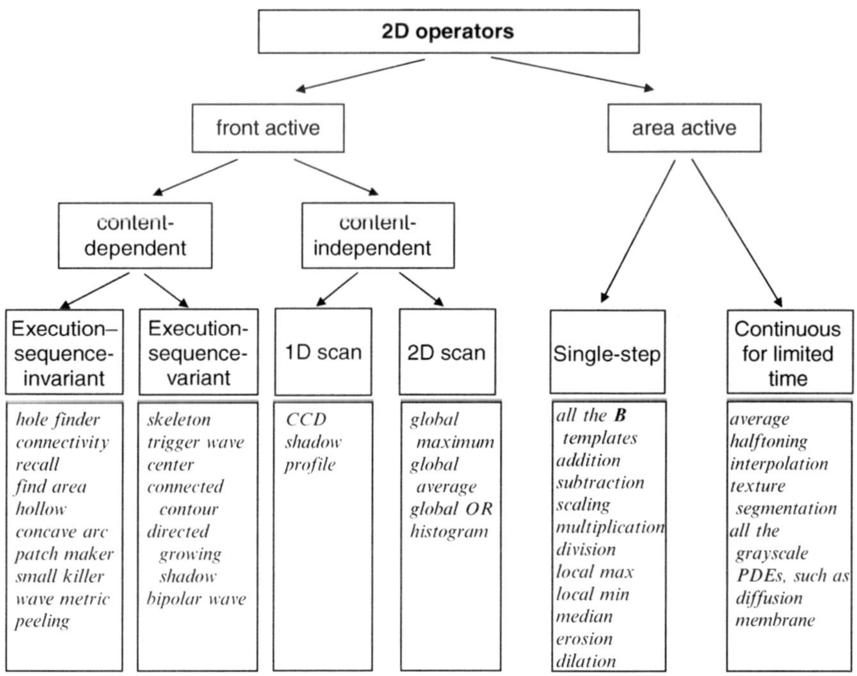

Fig. 10.4 2D local operator categorization

The first distinguishing feature is the location of active pixels (Zarándy 1999). If the active pixels are located along one or few one-dimensional stationary or propagating curves at a time, we call the operator front-active. If the active pixels are everywhere in the array, we call it area-active.

The common property of the front-active propagations is that the active pixels are located only at the propagating wave fronts (Rekeczky and Chua 1999). This means that at the beginning of the wave dynamics (transient), some pixels become active, others remain passive. The initially active pixels may initialize wave fronts that start propagating. A propagating wave front can activate some further passive pixels. This is the mechanism how the wave proceeds. However, pixels apart from a waveform cannot become active (Zarándy 1999). This theoretically enables us to compute only the pixels that are along the front lines, and do not waste efforts to the unchanging others. The question is which architectures can take advantage of such a spatially selective computation.

The front active operators such as *reconstruction*, *hole finder*, or *shadow* are typically binary waves. In CNN terms, they have binary inputs and outputs, positive self-feedback, and space invariant template values. Figure 10.4 contains three exemptions: *global max*, *global average*, and *global OR*. These functions are not wave type operators by nature; however, we will associate a wave with them, that solves them efficiently.

The front active propagations can be content-dependent or content-independent. The content-dependent operator class contains most of the operators where the direction of the propagation depends on the local morphological properties of the objects (e.g., shape, number, distance, size, connectivity) in the image (e.g., *reconstruct*). An operator of this class can be further distinguished as execution-sequence-variant (*skeleton*, etc) or execution-sequence-invariant (*hole finder, recall, connectivity*, etc). In the first case, the final result may depend on the spatial-temporal dynamics of the wave, whereas in the latter, it does not. Since the content-dependent operator class contains the most interesting operators with the most exciting dynamics, they are further investigated in Sect. 10.3.1.1.

We call the operators content-independent when the direction of the propagation and the execution time do not depend on the shape of the objects (e.g., *shadow*). According to propagation, these operators can be either one- [e.g., *CCD*, *shadow*, *profile* (Roska et al. 1998)] or two-dimensional (*global maximum*, *global OR*, *global average*, *histogram*). Content-independent operators are also called single-scan, for their execution requires a single scanning of the entire image. Their common feature is that they reduce the dimension of the input 2D matrices to vectors (*CCD*, *shadow*, *profile*, *histogram*) or scalars (*global maximum*, *global average*, *global OR*). It is worth to mention that on the coarse- and fine-grain topographic array processors, the *shadow*, *profile*, and *CCD* are content-dependent operators, and the number of the iterations (or analog transient time) depends on the image content only. The operation is completed, when the output is ceased to change. Generally, however, it is less efficient to include a test to detect a stabilized output, than to let the operator run in as many cycles as it would run in the worst case.

The area active operator category contains the operators where all the pixels are to be updated continuously (or in each iteration). A typical example is *heat diffusion*. Some of these operators can be solved in a single update of all the pixels [e.g., all the CNN **B** templates (Roska et al. 1998)], whereas others need a limited number of updates (*halftoning, constrained heat diffusion*, etc.).

The fine-grain architectures do update in every pixel location in fully parallel in each time instance. Therefore, the area active operators are naturally the best fit for these computing architectures.

10.3.1.1 Execution-Sequence-Variant Versus Execution-Sequence-Invariant Operators

The crucial difference in fine-grain and pipe-line architectures is in their state overwriting methods. In the fine-grain architecture, the new states of all the pixels are calculated in parallel, and then the previous one is overwritten again in parallel, before the next update cycle is commenced. In the pipe-line architecture, however, the new state is calculated pixel-wise, and it is selectable whether to overwrite a pixel state before the next pixel is calculated (pixel overwriting), or to wait until the new state value is calculated for all the pixels in the frame (frame overwriting). In this context, update means the calculation of the new state for an entire frame. Figures. 10.5 and 10.6 illustrate the difference between the two overwriting

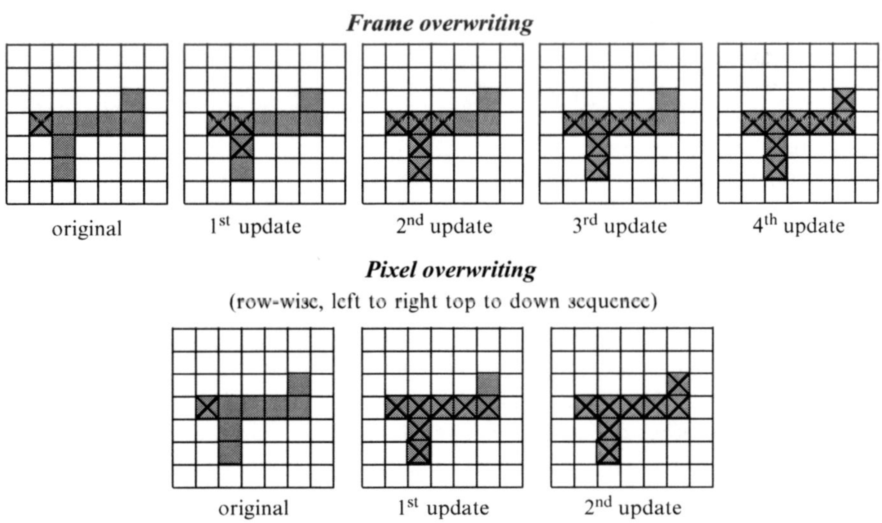

Fig. 10.5 Execution-invariant sequence in different overwriting schemes. Given an image with gray objects against white background. The propagation rule is that the propagation starts from the marked pixel (denoted by X), and it can go on within the gray domain, proceeding one pixel in each update. In the figure, we can see the results of each update. Update means calculating the new states of all the pixels in the frame

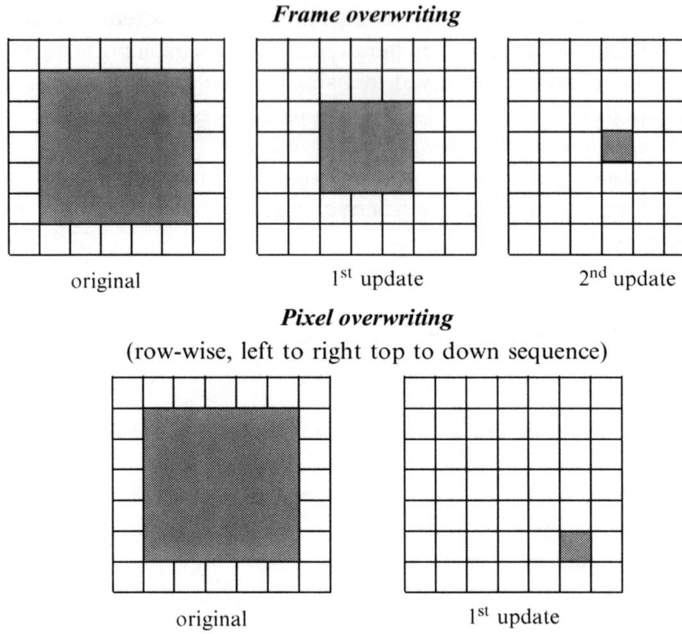

Fig. 10.6 Execution-variant sequence in different overwriting schemes. Given an image with gray objects against white background. The propagation rule is that those pixels of the object, which has both object and background neighbor should became background. In this case, the subsequent peeling leads to find the centroid in the frame overwriting method, while it extracts one pixel of the object in the pixel overwriting mode

schemes. In case of an execution-sequence-variant operation, the result depends on the frame overwriting schemes.

Here the calculation is done pixel-wise, left to right and row-wise, top to down. As we can see, overwriting each pixel before the next pixel's state is calculated (pixel overwriting) speeds up the propagation in the directions of the proceeding of calculation.

On the basis of the above, it is easy to draw the conclusion that the two updating schemes lead to two completely different propagation dynamics and final results in execution-variant cases. One is slower, but controlled; the other one is faster, but uncontrolled. The first can be used in cases when speed maximization is the only criterion, whereas the second is needed when the shape and the dynamics of the propagating wave front count. We called the former case execution-sequence-invariant operators, the latter one execution-sequence-variant operators (Fig. 10.4).

In the fine-grain architecture, we can use frame overwriting scheme only. In the coarse-grain architecture, both pixel overwriting and frame overwriting methods can be selected within the individual subarrays. In this architecture, we may determine even the calculation sequence, which enables speedups in different directions in different updates. Later, we will see an example to illustrate how the *hole finder* operation propagates in this architecture. In the pipe-line architecture, we may decide which one to use; however, we cannot change the direction of the propagation of the calculation, unless paying significant penalty for it in memory size and latency time.

10.3.2 Processor Utilization Efficiency of the Various Operation Classes

In this section, we will analyze the implementation efficiency of various 2D operators from different aspects. We will study both the execution methods and the efficiency from the processor utilization aspect. Efficiency is a key question because in many cases one or a few wave fronts sweep through the image, and one can find active pixels only in the wave fronts, which is less than 1% of the pixels; hence, there is nothing to calculate in the rest of image. We define a measure of efficiency of processor utilization with the following form:

$$\eta = \frac{O_\mathrm{r}}{O_\mathrm{t}} \qquad (10.1)$$

where O_r is the minimum number of required elementary steps to complete an operation, assuming that the inactive pixel locations are not updated, and O_t is the total number of elementary steps performed during the calculation by all the processors in the particular processor architecture.

The efficiency of processor utilization figure will be calculated in the following where it applies, because this is a good parameter (among others) to compare the different architectures.

10.3.2.1 Execution-Sequence-Invariant Content-Dependent Front-Active Operators

A special feature of content-dependent operators is that the path and length of the path of the propagating wave front drastically depend on the image contents itself. For example, the range of the necessary frame overwritings with a *hole finder* operation is from zero overwriting to $n/2$ in a fine-grain architecture, assuming $n \times n$ pixel array size. Hence, neither the propagation time, nor the efficiency can be calculated without knowing the actual image.

Since the gap between the worst and best case is extremely high, it is not meaningful to provide these limits. Rather, it makes more sense to provide approximations for certain image types. But before that, we examine how to implement these operators on the studied architectures. For this purpose, we will use the *hole finder* operator, as an example. Here we will clearly see how the wave propagation follows different paths, as a consequence of varying propagation speed corresponding to different directions. Since this is an execution-sequence-invariant operation, it is certain that wave fronts with different trajectories lead to the same good result.

The *hole finder* operation, which we will study here, is a "grass fire" operation, in which the fire starts from all the boundaries at the beginning of the calculation, and the boundaries of the objects behave like firewalls. In this way, at the end of the operation, only the holes inside objects remain unfilled.

The *hole finder* operation may propagate to any direction. On a *fine-grain architecture*, the wave fronts propagate one pixel steps in each update. Since the wave fronts start from all the edges, they meet in the middle of the image in typically $n/2$ updates, unless there are large structured objects with long bays which may fold the grass fire into long paths. In case of a text, for example, where there are relatively small nonoverlapping objects (with diameter k) with large but not spiral like holes, the wave stops after $n/2 + k$ operations. In case of an arbitrary camera image with an outdoor scene, in most cases 3^*n updates are enough to complete the operation because the image may easily contain large objects blocking the straight paths of the wave front.

On a *pipe-line architecture*, thanks to the pixel overwrite scheme, the first update fills up most of the background (Fig. 10.7). Filling in the remaining background requires typically k updates, assuming the largest concavity size with k pixels. This means that on a pipe-line architecture, roughly $k + 1$ steps are enough, considering small, nonoverlapping objects with size k.

In the *coarse-grain architecture*, we can also apply the pixel overwriting scheme within the $N \times N$ subarrays (Fig. 10.8). Therefore, within the subarray, the wave front can propagate in the same way, as in the pipe-line architecture. However, it cannot propagate beyond the boundary of the subarray, in a single update. In this way, the wave front can propagate N positions in the direction that correspond to the calculation directions, and one pixel in the other directions, in each update. In this way, in n/N updates, the wave-front can propagate n positions in the supported directions. However, the k-sized concavities in other directions would require k

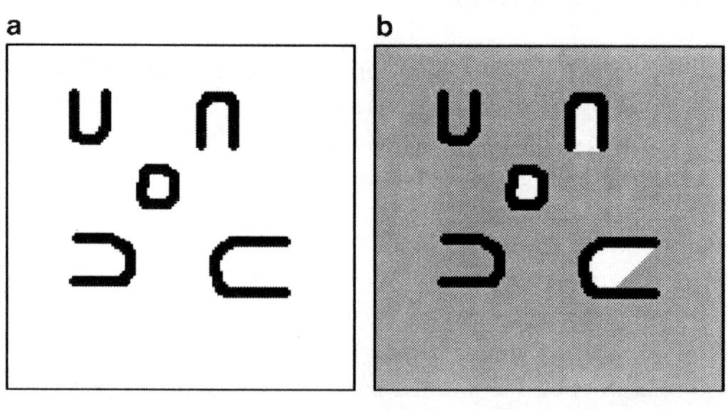

Fig. 10.7 Hole finder operation calculated with a pipe-line architecture. (**a**) Original image. (**b**) result of the first update. (The freshly filled up areas are indicated with gray, just to make it more comprehensible. However, they are black on the black-and-white image, same as the objects.)

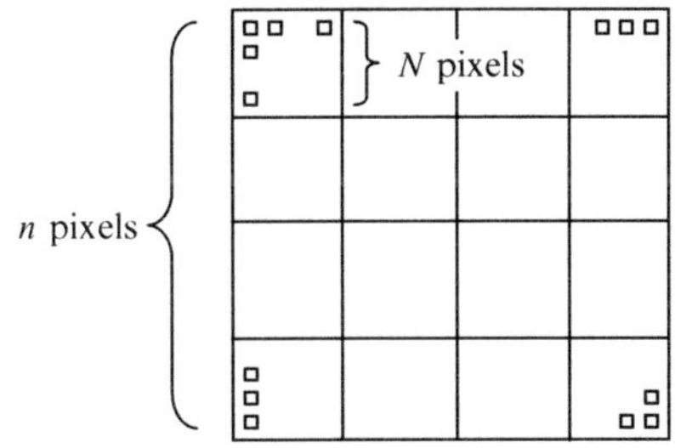

Fig. 10.8 Coarse-grain architecture with $n \times n$ pixels. Each cell is to process an $N \times N$ pixel subarray

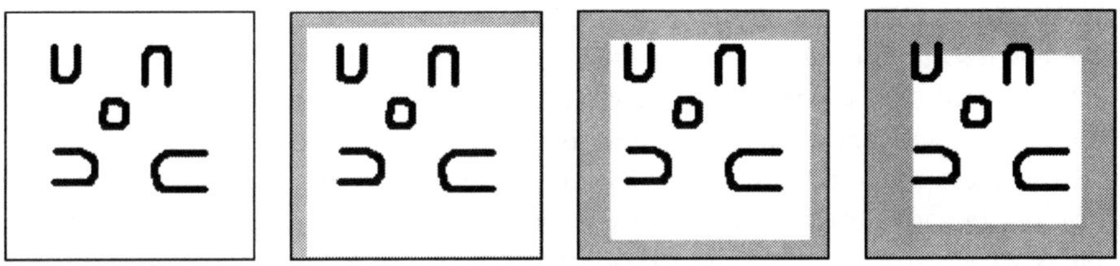

Fig. 10.9 Hole finder operation calculated in a coarse-grain architecture. The first picture shows the original image. The rest shows the sequence of updates, one after the other. The freshly filled-up areas are indicated with gray (instead of black) to make it easier to follow the dynamics of calculation

more steps. To avoid these extra steps, without compromising the speed of the wave-front, we can switch between the top-down and the bottom-up calculation directions after each update. The resulting wave-front dynamics is shown in Fig. 10.9. This means that for an image, containing only few, nonoverlapping small objects with concavities, we need about $n/N + k$ steps to complete the operation.

The *DSP-memory architecture* offers several choices depending on the internal structure of image. The simplest is to apply pixel overwriting scheme, and switch the direction of the calculation. In case of binary image representation, only the vertical directions (up or down) can be efficiently selected, due to the packed 32 pixel line segment storage and handling. In this way, the clean vertical segments (columns of background with maximum one object) are filled up after the second update, and filling up the horizontal concavities would require k steps.

10.3.2.2 Execution-Sequence-Variant Content-Dependent Front Active Operators

The calculation method of the execution-sequence-variant content-dependent front active operators is very similar to that of their execution-sequence-invariant counterparts. The only difference is that in each of the architectures the frame overwriting scheme should be used. This does not make any difference in fine-grain architectures; however, it slows down all the other architectures significantly. In the DSP-memory architectures, it might even make sense to switch to one byte/pixel mode, and calculate updates in the wave fronts only.

10.3.2.3 1D Content-Independent Front Active Operators (1D Scan)

In the 1D content-independent front active category, we use the vertical shadow (north to south) operation as an example. In this category, varying the orientation of propagation may cause drastic efficiency differences on the nontopographic architectures.

On a *fine-grain discrete time architecture* the operator is implemented in a way that in each time instance, each processor should check the value of its upper neighbor. If it is +1 (black), it should change its state to +1 (black), otherwise the state should not change. This can be implemented in one single step in a way, that each cell executes an *OR* operation with its upper neighbor, and overwrites its state with the result. This means that in each time instance the processor array executed n^2 operations, assuming $n \times n$ pixel array size.

In discrete time architectures, each time instance can be considered as a single iteration. In each iteration, the shadow wave front moves by one pixel to the south, that is we need n steps for the wave front to propagate from the top row to the bottom (assuming boundary condition above the top row). In this way, the total number of operations, executed during the calculation is n^3. However, the strictly required number of operations is n^2 because it is enough to do these calculations at the wave front, only ones in each row, starting from the top row, and going down row by row, rolling over the results from the front line to the next one. In this way, the efficiency of the processor utilization in *vertical shadow* calculation in the case of fine-grain discrete time architectures is

$$\eta = \frac{1}{n} \tag{10.2}$$

Considering computational efficiency, the situation is the same in *fine-grain continuous architectures*. However, from the point of power efficiency, the Asynchronous Cellular Logic Network (Lopich and Dudek 2007) is very advantageous because only the active cells in the wave front consume switching power. Moreover, the extraordinary propagation speed (500 ps/cell) compensates for the low processor utilization efficiency.

If we consider a *coarse-grain architecture* (Fig. 10.8), the *vertical shadow* operation is executed in a way that each cell executes the above *OR* operation from its top row, and goes on from the top downwards in each column. This means that $N \times N$ operations are required for a cell to process its subarray. It does not mean, however, that in the first $N \times N$ steps the whole array is processed correctly, because only the first cell row has all the information for locally finalizing the process. For the rest of the rows, their upper boundary condition has not "arrived"; hence at these locations, correct operations cannot be performed. Thus, in the first $N \times N$ steps, the first N rows were completed only. However, the total number of operation executed by the array during this time is

$$O_{NxN} = N^* N^* \frac{n}{N} * \frac{n}{N} = n^* n, \tag{10.3}$$

because there are $n/N{}^*n/N$ processors in the array, and each processor is running all the time. To process also the rest of the lines, we need to perform

$$O_\mathrm{t} = O_{N\times N}{}^* \frac{n}{N} = \frac{n^3}{N}. \tag{10.4}$$

The resulting efficiency is:

$$\eta = \frac{N}{n} \tag{10.5}$$

It is worth to stop at this result for a while. If we consider a fine-grain architecture ($N = 1$), the result is the same as we obtained in (10.2). Its optimum is $N = n$ (one processor per column) when the efficiency is 100%. It turns out that in case of *vertical shadow* processing, the efficiency increases by increasing the number of the processor columns, because in that case, one processor has to deal with less columns. However, the efficiency does not increase when the number of the processor rows is increased. (Indeed, one processor/column is the optimal, as it was shown.) Although the unused processor cells can be switched off with minor extra effort to increase power efficiency, but it would certainly not increase processor utilization.

Pipe-line architecture as well as *DSP-memory architecture* can execute *vertical shadow* operation with 100% processor utilization because there are no multiple processors in a column working parallel.

We have to note, however, that shadows to other three directions are not as simple as the one to downwards. In *DSP architectures*, *horizontal shadows* cause difficulties because the operation is executed parallel on a *32* × *1* line segment; hence, only one of the positions (where the actual wave front is located) performs effectual calculation. If we consider a left to right shadow, this means that once in each line (at left-most black pixel), the shadow propagation should be calculated precisely for each of the 32 positions. Once the "shadow head" (the 32 bit word, which contains the left-most black pixel) is found, and the shadow is calculated within this word, the task is easier, because all the rest of the words in the line should be filled with black pixels, independently of their original content. Thus, the overall resulting cost of a *horizontal shadow* calculation on a *DSP-memory architecture* can be even *20* times higher than that of a *vertical shadow* for a *128* × *128* sized image. Similar situation might happen in *coarse-grain architectures*, if they handle $n \times 1$ binary segments.

While *pipe-line architectures* can execute the *left to right and top to bottom shadows* in a single update at each pixel location, the other directions would require n updates, unless the direction of the pixel flow is changed. The reason of such a high inefficiency is that in each update, the wave front can propagate only one step in the opposite direction.

10.3.2.4 2D Content-Independent Front Active Operators (2D Scan)

The operators belonging to the 2D content-independent front active category require simple scanning of the frame. In *global max* operation, for example, the actual maximum value should be passed from one pixel to another one. After we scanned all the pixels, the last pixel carries the global maximum pixel value.

In *fine-grain architectures*, this can be done in two phases. First, in n comparison steps, each pixel takes over the value of its upper neighbor, if it is larger than its own value. After n steps, each pixel in the bottom row contains the largest value of its column. Then, in the second phase after the next n horizontal comparison steps, the global maximum appears at the end of the bottom row. Thus, to obtain the final result requires $2n$ steps. However, as a fine-grain architecture executes $n \times n$ operations in each step, the total number of the executed operations are $2n^3$. However, the minimum number of requested operations to find the largest value is n^2 only. Therefore, the efficiency in this case is:

$$\eta = \frac{1}{2n} \tag{10.6}$$

The most frequently used operation in this category is *global OR*. To speed up this operation in the fine-grain arrays, a *global OR* net is implemented usually (Liñan-Cembrano et al. 2003). This $n \times n$ input OR gate requires minimal silicon space, and enables to calculate *global OR* in a single step (few microseconds).

However, in that case, when a fine-grain architecture is equipped with *global OR*, the global maximum can be calculated as a sequence of iterated *threshold* and *global OR* operations with interval halving (successive approximation) method applied parallel to the whole array. This means that a global threshold is applied first for the whole image at level $1/2$, and if there are pixels, which are larger than this, we will do the next global thresholding at $3/4$, and so on. Assuming 8 bit accuracy, this means that in 8 iterations (16 operations), the global maximum can be found. The efficiency is much better in this case:

$$\eta = \frac{1}{16}$$

In *coarse-grain architectures*, each cell calculates the *global maximum* in its subarray in $N \times N$ steps. Then n/N vertical steps come, and finally, n/N horizontal steps to find the largest values in the entire array. The total number of steps in this case is $N^2 + 2n/N$, and in each step, $(n/N)^2$ operations are executed. The efficiency is:

$$\eta = \frac{n^2}{(N^2 + 2n/N)^*(n/N)^2} = \frac{1}{(1 + 2n/N^3)} \tag{10.7}$$

Since the sequence of the execution does not matter in this category, it can be solved with 100% efficiency in *pipe-line* and the *DSP-memory architectures*.

10.3.2.5 Area Active Operators

The area active operators require some computation in each pixel in each update; hence, all the architectures work with 100% efficiency. Since the computational load is very high here, it is the most advantageous for the many-core architectures, because the speed advantage of the many processors can be efficiently utilized.

10.3.3 Multiscale Processing

Generally, multiscale processing technique is applied in those situations, when the calculation of an operator on a downscaled image leads to acceptable result from accuracy point of view. Since the calculation of the operator requires significantly smaller computational effort in a lower resolution, in many cases the downscaling, the upscaling (if needed), and the calculation on the downscaled domain requires less computational effort than the calculation of the operator in the original scale. Diffusion is a typical example for this.

Here we discuss how the approximation of the *diffusion* operator leads to a multiscale representation, and analyze its implementation on the discussed architectures. However, with a similar approach, other binary or gray scale front- and area-active operators can be scaled down and executed, as well.

Two ways are used generally to compute the *diffusion* operator on topographic array computers. The first is the iterative way. The second way is to implement it on a hardwired resistive grid, as we have seen in analog fine-grain topographic architectures. Here we deal with the first option.

The problem with the iterative implementation of *diffusion* equation is that after a few iterations the differences of the neighboring pixels become very small, and the propagation slows down. Moreover, if there are some computational errors, due to the limited precision of the processors, calculation of the *diffusion* equation will be useless and irrelevant, after a while. To obtain accurate solution would require floating point number representation and a large number of iterations. However, one can approximate it by using multiscale approach, as it is shown in Fig. 10.10. As we can see, ten iterations on a full scale image result in small blurring only, whereas the same ten iterations on a downscaled image lead to large scale *diffusion*. The downscaling and the upscaling with linear interpolation need less computational effort, than a single iteration of the diffusion. Moreover, the calculation of an iteration on the downscaled image requires only $1/s^2$ (s is the downscaling factor) of computational power. Naturally, it should be kept in mind that this method can be used in that cases only when the accuracy of the approximated diffusion operator is good satisfactory in a particular application.

The multiscale iterative *diffusion* can be implemented on classic DSP-memory architectures, multicore pipe-line architectures (Fig. 10.11), and on coarse-grain architectures as well. In fine-grain architectures, the multiscale approach cannot be efficiently implemented.

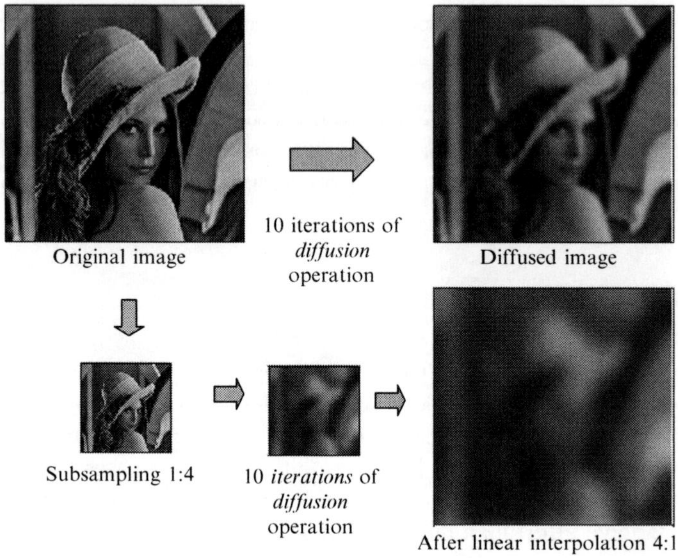

Fig. 10.10 Iterative approximation of the diffusion operator combining different spatial resolutions

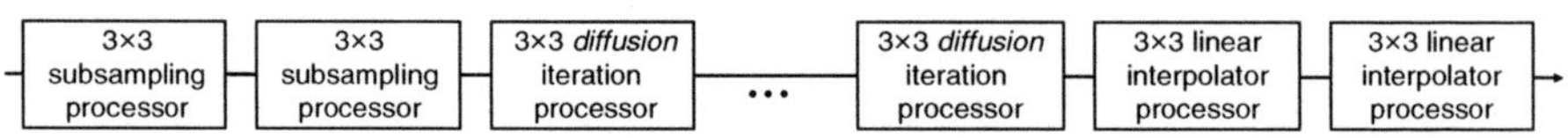

Fig. 10.11 Implementation of multiscale diffusion calculation approach on a pipe-line architecture. In this example, it starts with two subsampling steps. The pixel clock drops into 1/16th. Then the computationally hard diffusion calculations can be applied much easier since more time is available for each pixel. The processing is completed with the two interpolation steps

10.4 Comparison of the Architectures

As we have stated in the previous section, front active wave operators run well under 100% efficiency on topographic architectures, since only the wave fronts need calculation, and the processors of the array in nonwave front positions do dummy cycles only or may be switched off. On the contrary, the computational capability (GOPS) and the power efficiency (GOPS/W) of multicore arrays are significantly higher than those of DSP-memory architectures. This section shows the efficiency figures of these architectures in different categories. To make fair comparison with relevant industrial devices we have selected two market-leading, video processing units, a DaVinci video processing DSP from Texas Instruments (TMS320DM6443) (www.ti.com), and a Spartan 3 DSP FPGA from Xilinx (XC3SD3400A) (www.xilinx.com). Both of these products' functionalities, capabilities and prices were optimized to efficiently perform embedded video analytics.

Table 10.1 Computational parameters of the different architectures for arithmetic (3 × 3 convolution) and logic (3 × 3 binary erosion) operations

	DSP (DaVinci[a])	Pipe-line (FPGA[b])	Coarse-grain (Xenon)	Fine-grain (SCAMP/Q-Eye)
Silicon technology (nm)	90	65	180	350/180
Silicon area (mm^2)			100	100/50
Power consumption (W)	1.25	2–3	0.08	0.20
Arithmetic proc. clock speed (MHz)	600	250	100	1,2/2.5
Number of arithmetic proc.	8	120	256	16384
Efficiency of arithmetic calc. (%)	75[c]	100	80[e]	50[d]
Arit. computational speed	3.6 GMAC	30 GMAC	20 GMAC	~20GOPS[f]
3 × 3 convolution time (μs)	42.3[g]	4.9	12.1	22[f]
Arithmetic speedup	1	8.6	3.5	1.9
Morphological proceeding clock speed (MHz)	600	83	100	1,2/5
Number of morphological proceeding	64	864	2048	147456
Morphological processor kernel type (bits)	2 × 32	96 × 9	256 × 8	16384 × 9
Efficiency of morphological calculation (%)	28[c]	100	90[e]	100
Morphological computational power (GOPS)	10	71	184	737
3 × 3 morphological operation time (μs)	13.6[g]	2.05	1.1	0.2
Morphological speedup	1	6.6	12.4	68.0

[a] Texas Instrument DaVinci video processor (TMS320DM64x)
[b] Xilinx Spartan 3ADSP FPGA (XC3SD3400A)
[c] Processors are faster than cache access
[d] Data access from neighboring cell is an additional clock cycle
[e] Due to pipe-line stages in the processor kernel, no effective calculation in each clock cycle
[f] No multiplication, scaling with few discrete values
[g] These data-intensive operators slow down to one-third or even one-fifth when the image does not fit to the internal memory (typically above 128 × 128 with a DaVinci, which has 64 kb internal memory)

Table 10.1 summarizes the basic parameters of the different architectures, and indicates the processing time of a 3 × 3 convolution, and a 3 × 3 erosion. To make the comparison easier, values are calculated for images of 128 × 128 resolution. For this purpose, we considered 128 × 128 Xenon and Q-Eye chips. Some of these data are from catalogues, other ones are from measurements, or estimation. As fine-grain architecture examples, we included both the SCAMP and Q-Eye architectures.

We can see from Table 10.1, the DSP was implemented on 90 nm, whereas the FPGA on 65 nm technologies. In contrast, Xenon, Q-Eye, and SCAMP were implemented on more conservative technologies and their power budget is an order of magnitude smaller. When we compare the computational power figures, we also have to take these parameters into consideration.

Table 10.1 shows the speed advantages of the different architectures, compared to DSP-memory architecture both in 3×3 neighborhood arithmetic (8 bit/pixel) and morphological (1 bit/pixel) cases. This indicates the speed advantage of the area active single step, and the front active content-dependent execution-sequence-variant operators. In Table 10.2, we summarize the speed relations of the rest of the wave type operations. The table indicates the computed values, using the formulas that we have derived in the previous section. In some cases, however, the coarse- and especially the fine-grain arrays contain some special accelerator circuits, which takes the advantage of the topographic arrangement and the data representation (e.g., global OR network, mean network, diffusion network). These are marked by notes, and the real speedup with the special hardware is shown in parenthesis.

In our comparison tables, we have represented a typical FPGA as a vehicle to implement the pipe-line architectures. The only reason is that all the currently available pipe-line architectures are implemented in FPGAs is mainly attributed to much lower costs and quicker time-to-market development cycles. However, they could also be certainly implemented in ASIC, which would significantly reduce their power consumption, and decrease their large-volume prices making it possible to process even multimega pixel images at a video rate.

Table 10.3 shows the computational power, the consumed power and the power efficiency of the selected architectures. As we can see, the three topographic arrays have over hundred times power efficiency advantage compared to DSP-memory architectures. This can be explained with their local data access, and relatively low clock frequency. In case of ASIC implementation, the power efficiency of the pipe-line architecture would also be increased with a similar factor.

Figure 10.12 shows the relation between the frame-rate and the resolution in a video analysis task. Each of the processors had to calculate 20 *convolutions*, 2 *diffusions*, 3 *means*, 40 *morphologies*, and 10 *global OR*s. Only the DSP-memory and pipe-line architectures support trading between resolution and frame-rate. The characteristics of these architectures form lines. The chart shows the performance of the three discussed chips too. The chips are represented here with their real sizes.

As it can be seen in Fig. 10.12, both SCAMP and Xenon have the same speed as the DSP. In the case of Xenon, this is so, because its array size is 64×64 only. In the case of SCAMP, the processor was designed for very accurate low-power calculation by using a conservative technology.

10.5 Optimal Architecture Selection

So far, we have studied how to implement the different wave type operators on different architectures, identified constrains and bottlenecks, and analyzed the efficiency of these implementations. After having these results in our hand, we can define rules for optimal image processing architecture selection for topographic problems.

Table 10.2 Speed relations in the different function groups calculated for 128 × 128 sized images

	DSP (DaVinci[a])	Pipe-line (FPGA[b])	Coarse-grain (Xenon)	Fine-grain discrete time (SCAMP/ Q-eye)	Fine-grain continuous time (ACLA)
1D content-independent front active operators					
Processor utility efficiency	100%	100%	N/n: 6.25%	$1/n$: 0.8%	$1/n$: 0.8%
Speedup in advantageous direction (vertical)	1	6.6	0.77	0.53	188
Speedup in disadvantageous direction (horizontal)	1	1	2	10.6	3750
2D content-independent front active operators					
Processor utility efficiency	100%	100%	$1/(1+2n/N^3)$: *66%*	$1/2n$: 0.4%	n/a
Speedup (*global OR*)	1	6.6	8.2 (13[c])	0.27 (20[c])	n/a
Speedup (*global max*)	1	8.6	2.3	n/a	n/a
Speedup (*average*)	1	8.6	2.3	n/a (2.5)[d]	n/a
Execution-sequence-invariant content-dependent front active operators					
Hole finder with $k = 10$ sized small objects	4 updates	$k+1$ updates (11)	$n/N+k$ (26)	$n/2+k$ updates (74)	$n/2+k$ updates (74)
Speedup	1	2.4	1.9	3.7	1500
Area active					
Processor utility efficiency (%)	100	100	100	100	
Speedup	1	8.6	3.5	1.9 (210[e])	n/a
Multiscale					
1:4 Scaling	1	8.6	3.5	0.1	n/a

The notes indicate the functionalities by which the topographic arrays are speeded up with special purpose devices.

[a] Texas Instrument DaVinci video processor (TMS320DM64x)

[b] Xilinx Spartan 3ADSP FPGA (XC3SD3400A)

[c] Hard wired global OR device speeds up this function (<1 μs concerning the whole array)

[d] Hard wired mean calculator device makes this function available (~2 μs concerning the whole array)

[e] Diffusion calculated on resistive network (<2 μs concerning the whole array)

Table 10.3 Computational power, the consumed electronic power, and their proportion in different architectures for convolution operations. As a comparison, the cell multiprocessor developed by IBM-Sony-Toshiba (Kahle et al. 2005) is also given

	GOPS	W	GOPS/W
DaVinci	3.6	1.25	2.88
Pipe-line (FPGA)	30	3	10
Xenon (64 × 64)	10	0.02	500
SCAMP (128 × 128)	20	0.2	100
Q-Eye	25	0.2	125
Cell multiprocessor	225	85	2.6

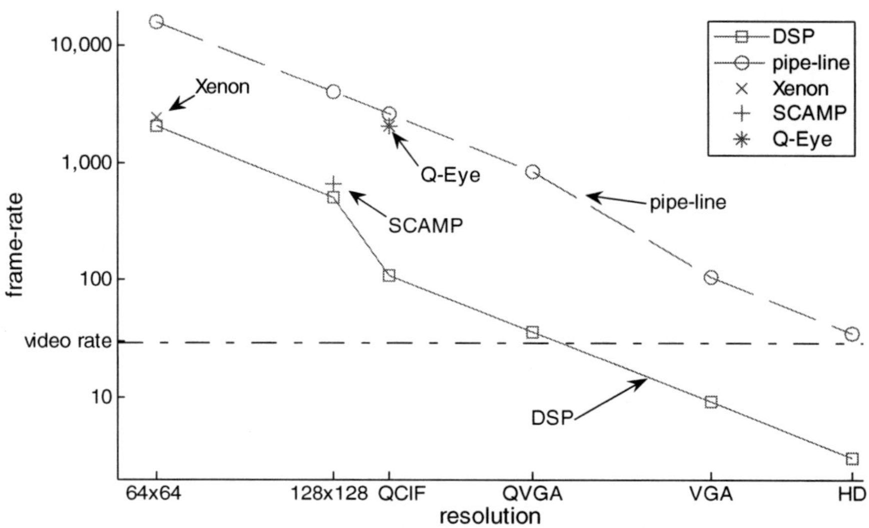

Fig. 10.12 Frame-rate versus resolution in a typical image analysis task. Both of the axes are in logarithmic scale

Image processing devices are usually special purpose architectures, optimized for solving specific problems or a family of similar algorithms. Figure 10.13 shows a method of special purpose processor architecture selection. It always starts with the understanding of the problem in all aspect. Then, different algorithms suitable for solving the problem are derived. The algorithms are described with flowchart, with the list of the used operations, and with the specification of the most important parameters. In this way, a set of formal data describes the algorithms, which are as follows: *resolution, frame-rate, pixel clock, latency, computational demand* (type and number of operators), and *flowchart*. Other application-specific (secondary) parameters are also given: maximal *power consumption*, maximal *volume*, *economy* etc. The algorithm derivation is a human activity supported by various simulators for evaluation and verification purposes.

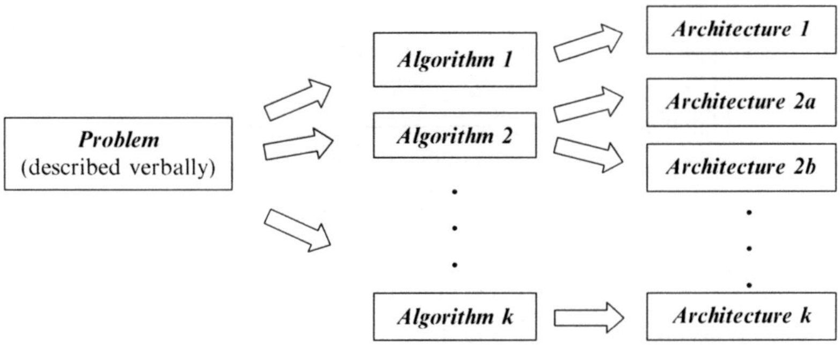

Fig. 10.13 Methodology of special purpose processor architecture selection

The next step is the architecture selection. By using the previously compiled data, we can define a methodology for the architecture selection step. As we will see, based on the formal specifications, we can derive the possible architectures. There might not be any, or there might be several, according to the demands of the specification.

The first step of the methodology is the comprehensive analysis of the parameter set. Fortunately, in many cases, it immediately leads to a single possible architecture. If it does not lead to any architecture, in a second step, we have to seek for options, how to fulfill tough specification demands. If it leads to multiple architectures, a ranking is needed based on secondary parameters.

The three most important parameters are the *frame-rate*, the *resolution*, and their product, the minimal value of the *pixel clock*.[1] In many cases, especially in challenging applications, these parameters determine the available solutions. Figure 10.14 shows frame-rate – resolution matrix. The matrix is divided into 16 segments, and each segment indicates the potential architectures that can operate in that particular parameter environment. The matrix shows the minimal pixel clock figures (red) in the grid points also.

In Fig. 10.14, the pipe-line and the DSP can be positioned freely between frame-rate and resolution without constrains. Thus, they appear everywhere, under a certain pixel clock rate. The digital coarse-grain sensor-processor arrays appear in the low resolution applications (left column), whereas the analog (mixed-signal) fine-grain sensor-processor arrays appear in both the low and medium resolution columns.

[1] The minimal value of the pixels clock is equivalent to the product of the frame-rate and the number of pixels (resolution). If the image source is a sensor, the pixel clock is defined by the sensor readout speed. In balanced sensor applications (high-speed applications are usually balanced), the integration time and the readout time are roughly the same. Since there are short blank periods in the sensor readout protocol for synchronization purposes, the pixel clock is slightly higher then the minimal pixel clock in balanced application. However, in low light applications, the sensor integration time takes much longer than the readout time. In these cases, the sensor pixel clock can be orders of magnitude higher than the minimal pixel clock.

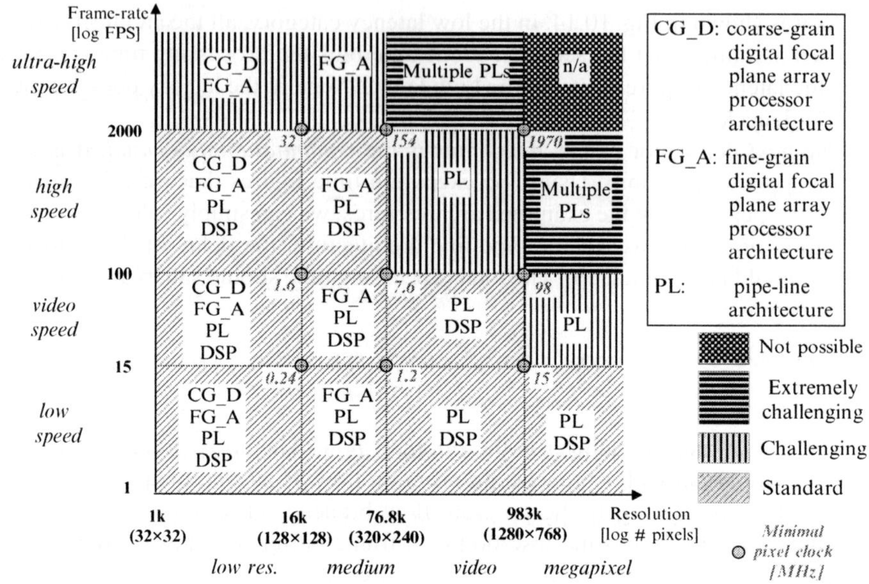

Fig. 10.14 Feasible architectures in the frame-rate-resolution matrix

The next important parameter is the *latency*. Latency is critical when the vision device is in a control loop, because large delays might make the control loops instable. It is worth to distinguish three latency requirement regions:

- very low latency (*latency* <2 ms, e.g., missile, UAV, high-speed robot controlling);
- low latency (2 ms < *latency* < 50 ms, e.g., robotic, automotive);
- high latency (50 ms < *latency*, e.g., security, industrial quality check).

Latency has two components. The first is the readout time of the sensor, and the second is the completion of the processing on the entire frame. The readout time is negligible in the fine-grain mixed-signal architectures, since the analog sensor readout should be transferred to an analog memory through a fully parallel bus. The readout time is also very small (∼100 μs) in the coarse-grain digital processor array, because there is an embedded AD converter array to do conversion in parallel. The DSPs and the pipe-line processor arrays use external image sensors, in which the readout time usually is in the millisecond range. Therefore, in case of very low latency requirements, the mixed-signal and the digital focal plane arrays can be used. (There are some ultrahigh frame-rate sensors with high-speed readout, which can be combined with pipe-line processors. However, these can be applied in very special applications only due to their high complexities.)

As it is shown in Fig. 10.14, in the low latency category, all the architectures can be used assuming that the sensor readout time plus the processing time is smaller than the latency requirements. In the high latency region, the latency does not mean any bottleneck.

The next descriptor of the algorithms is the *computational demand*. It is a list of the applied operations. Using the execution time figures that we calculated for different operations on the examined architectures, we can simply calculate the total execution time. (In case of the pipe-line architecture, the delay of the individual stages should be summed up). The total processing time should satisfy the following two relations:

$$t_{total\_processing} < t_{latency} - t_{readout}$$
$$t_{total\_processing} < 1/frame\_rate$$

The last primary parameter is the *program flow*. The array processors and the DSP are not sensitive for branches in program flow. However, the pipe-line architectures are challenged by the program flow branches. The implementation of a conditional branch needs the insertion of a frame buffer that drastically increases the latency and makes the design more complex.

There are three secondary design parameters. The first is the *power consumption*. Generally, the ASIC solutions need much less power than the FPGA or DSP solutions. The second is the *cubature* of the circuit. Smaller *cubature* can be achieved with sensor-processor arrays because the combination of these two functionalities reduces the chip count. The third parameter is the *economy*. In case of low volume, the DSP is the cheapest because the invested engineering cost is the smaller there. In case of medium volume, the FPGA is the most economical, whereas in case of high volume, the ASIC solutions are the cheapest.

10.6 Conclusions

We have categorized the 2D operators into six sets, based on their implementation methods on different image processing architectures. By using this categorization, the efficiency figures of 2D operators were calculated considering different architectures. This enabled us to compare the architectures, and provide a guide for selecting the optimal architecture for a given algorithm. Moreover, we have measured, collected, or calculated some key parameters of existing implementations. Comparing the different architectures, we can draw the following conclusions:

- The computational speed on digital coarse-grain architectures is roughly the same as on fine-grain architectures (Fig. 10.12). The accuracy of the digital one is better; however, the required silicon area is also larger (Table 10.1).
- The analog/mixed signal fine-grain architecture can take advantage of utilizing various specific processing networks, like mean grid, diffusion grid, global OR grid, etc (Table 10.2).

- In focal-plane sensor-processor application where the specification requires lower precision, the analog fine-grain implementations are more advantageous.
- In applications where high-precision calculation is required, the coarse-grain architecture is more advantageous.
- It is important to note that in the case of array processors, the speed up rate changes with the processor array size. In some cases, the speed advantage is proportional to the number of the processors in the array (area active single step, and the front active content-dependent execution-sequence-variant operators), whereas in the rest of the cases, it is proportional with the number of the processors located in one row/column (Table 10.2).

As it is shown in Table 10.3, the GOPS/W figure of the studied topographic many-core architectures are orders of magnitude better than the single or many core high-end processors used nowadays in PCs and servers, This makes any of those much more suitable for embedded mobile applications, compared to a DSP or a RISC processor.

References

Chua L O and Yang L, "Cellular neural networks: theory and applications", *IEEE Transactions on Circuits and Systems*, vol. 35, no. 10, October 1988, pp. 1257–1290.

Chua L O, Roska T, Kozek T, Zarándy Á, "CNN universal chips crank up the computing power", *IEEE Circuits and Devices*, July 1996, pp. 18–28.

Cruz J M, Chua L O, Roska T, "A fast, complex and efficient test implementation of the CNN universal machine", Proc. of the third IEEE Int. Workshop on Cellular Neural Networks and their Application (CNNA-94), pp. 61-66, Rome Dec. 1994.

Dudek P, and Carey S J, "A general-purpose 128 × 128 SIMD processor array with integrated image sensor", Electronics Letters, vol. 42, no. 12, June 2006, pp. 678–679.

Dudek P, "An asynchronous cellular logic network for trigger-wave image processing on fine-grain massively parallel arrays", IEEE Transactions on Circuits and Systems II: Analog and Digital Signal Processing, vol. 53 no. 5, 2006 pp. 354–358.

Espejo S, Carmona R, Domínguez-Castro R, Rodríguez-Vázquez, A, "A VLSI-oriented continuous-time CNN model", International Journal of Circuit Theory and Applications, vol. 24, May–June 1996, pp. 341–356.

Espejo S, Carmona R, Domingúez-Castro R, Rodrigúez-Vázquez A, "CNN universal chip in CMOS technology", Int. J. of Circuit Theory & Appl., vol. 24, 1996, pp. 93–111.

Foldesy P, Zarándy Á, Rekeczky Cs, Roska T, "Configurable 3D integrated focal-plane sensor-processor array architecture", Int. J. Circuit Theory and Applications (CTA), 2008, pp. 573–588.

Harrer H, Nossek J A, Roska T, Chua L O, "A Current-mode DTCNN Universal Chip", Proc. of IEEE Intl. Symposium on Circuits and Systems, 1994, pp. 135–138.

Kahle J A, Day M N, Hofstee H P, Johns C R, Maeurer T R, Shippy D, "Introduction to the Cell multiprocessor", IBM J. Res. & Dev., vol. 49, no. 4/5, July/September 2005.

Keresztes P, Zarándy Á, Roska T, Szolgay P, Bezák T, Hídvégi T, Jónás P, Katona A, "An emulated digital CNN implementation", Journal of VLSI Signal Processing Special Issue: Spatiotemporal Signal Processing with Analogic CNN Visual Microprocessors, (JVSP Special Issue), Kluwer, 1999 November–December.

Liñan-Cembrano G, Rodríguez-Vázquez A, Espejo-Meana S, Domínguez-Castro R, "ACE16k: A 128 × 128 focal plane analog processor with digital I/O", International Journal of Neural System, vol. 13, no. 6, 2003, pp. 427–434.

Lopich, Dudek P, "Implementation of an asynchronous cellular logic network as a co-processor for a general-purpose massively parallel array", ECCTD 2007, Seville, Spain.

Nagy Z, and Szolgay P, "Configurable Multi-Layer CNN-UM Emulator on FPGA", IEEE Transactions on Circuits and Systems I: Fundamental Theory and Applications, vol. 50, 2003, pp. 774–778.

Paasio A, Dawindzuk, Halonen K, Porra V, "Minimum size 0.5 micron CMOS programmable 48 × 48 CNN test chip", European Conference on Circuit Theory and Design, Budapest, 1997, pp. 154–115.

Rekeczky C S, and Chua L O, "Computing with front propagation: active contour and skeleton models in continuous-time CNN", Journal of VLSI Signal Processing Systems, Vol. 23, No. 2/3, November-December 1999, pp. 373–402.

Roska T and Chua L O, "The CNN universal machine: an analogic array computer", *IEEE Transactions on Circuits and Systems* - II, vol. 40, March 1993, pp. 163–173.

Roska T, Kék L, Nemes L, Zarándy Á, Brendel M, Szolgay P, "CNN Software Library (Templates and Algorithms) Version 7.2", (DNS-1-1998), Budapest, MTA SZTAKI, 1998, http://cnn-technology.itk.ppke.hu/Library_v2.1b.pdf

Zarándy Á, "The art of CNN template design", Int. J. Circuit Theory and Applications - Special Issue: Theory, Design and Applications of Cellular Neural Networks: Part II: Design and Applications, (CTA Special Issue - II), Vol. 17, No. 1, 1999, pp. 5–24.

Zarándy Á, Keresztes P, Roska T, Szolgay P, "CASTLE: An emulated digital architecture; design issues, new results", Proceedings of 5th IEEE International Conference on Electronics, Circuits and Systems, (ICECS'98), Vol. 1, 1998, pp. 199–202, Lisboa.

Index

Lightning Source UK Ltd.
Milton Keynes UK
14 March 2010

151311UK00007BB/144/P

9 781441 910103